Ben Goldacre

BAD PHARMA

Ben Goldacre is a doctor and a writer. His first book, *Bad Science*, was an international bestseller and has been translated into twenty-five languages. He is thirty-eight and lives in London.

ALSO BY BEN GOLDACRE

Bad Science

BAD PHARMA

BAD PHARMA

How Drug Companies Mislead

Doctors and Harm Patients

BEN GOLDACRE

ff

Faber and Faber, Inc.

An affiliate of Farrar, Straus and Giroux

New York

Faber and Faber, Inc.
An affiliate of Farrar, Straus and Giroux
18 West 18th Street, New York 10011

Originally published in 2012 by Fourth Estate, Great Britain
Published in the United States in 2012 by Faber and Faber, Inc.
Revised edition originally published in 2013 by Fourth Estate, Great Britain
Revised edition published in the United States by Faber and Faber, Inc.
First American paperback edition, 2014

The Library of Congress has cataloged the hardcover edition as follows:
Goldacre, Ben.
Bad pharma : how drug companies mislead doctors and harm
patients / Ben Goldacre. — First American edition.
    pages cm
Includes bibliographical references and index.
ISBN 978-0-86547-800-8 (alk. paper)
1. Drugs—Testing. 2. Drugs—Testing—Moral and ethical aspects.
3. Clinical trials—Moral and ethical aspects. 4. Pharmaceutical
industry—Moral and ethical aspects. 5. Drugs—Quality control.
I. Title.

RM301.27 .G65 2013
615.1072'4—dc23

                                                    2012038902

Paperback ISBN: 978-0-86547-806-0

www.fsgbooks.com
www.twitter.com/fsgbooks • www.facebook.com/fsgbooks

1   3   5   7   9   10   8   6   4   2

To whom it may concern

CONTENTS

INTRODUCTION

Medicine is broken: the plane flies, but it crashes much more often than it needs to. And I genuinely believe that if patients and the public ever fully understand what has been done to them – what doctors, academics and regulators have permitted – they will be angry. On this, only you can judge.

We like to imagine that medicine is based on evidence and the results of fair tests. In reality, those tests are often profoundly flawed. We like to imagine that doctors are familiar with the research literature, when in reality much of it is hidden from them by drug companies. We like to imagine that doctors are well-educated, when in reality much of their education is funded by industry. We like to imagine that regulators let only effective drugs onto the market, when in reality they approve hopeless drugs, with data on side effects casually withheld from doctors and patients.

I'm going to tell you how medicine works, just over the page, in one paragraph that will seem so absurd, so ludicrously appalling, that when you read it, you'll probably assume I'm exaggerating. We're going to see that the whole edifice of medicine is broken, because the evidence we use to make decisions is hopelessly and systematically distorted; and this is no small thing. Because in medicine, doctors and patients use abstract data to make decisions in the very real world of flesh and blood. If those decisions are misguided, they can result in death, and suffering, and pain.

This isn't a simple story of cartoonish evil, and there will be no conspiracy theories. Drug companies are not withholding the secret to curing cancer, nor are they killing us all with vaccines. Those kinds of stories have, at best, a poetic truth: we all know, intuitively, from the fragments we've picked up, that something is wrong in medicine. But most of us, doctors included, don't know exactly what.

These problems have been protected from public scrutiny because they're too complex to capture in a soundbite, or even 3,000 words. This is why they've gone unfixed by politicians, at least to some extent; but it's also why you're holding a book of over four hundred pages. The people you should have been able to trust to fix these problems have failed you, and because you have to understand a problem properly in order to fix it yourself, this book contains all that you need to know.

So, to be clear, this whole book is about meticulously defending every assertion in the paragraph that follows.

Drugs are tested by the people who manufacture them, in poorly designed trials, on hopelessly small numbers of weird, unrepresentative patients, and analysed using techniques which are flawed by design, in such a way that they exaggerate the benefits of treatments. Unsurprisingly, these trials tend to produce results that favour the manufacturer. When trials throw up results that companies don't like, they are perfectly entitled to hide them from doctors and patients, so we only ever see a distorted picture of any drug's true effects. Regulators see most of the trial data, but only from early on in a drug's life, and even then they don't give this data to doctors or patients, or even to other parts of government. This distorted evidence is then communicated and applied in a distorted fashion. In their forty years of practice after leaving medical school, doctors hear about what works through ad hoc oral traditions, from sales reps, colleagues or journals. But those colleagues can be in the pay of drug companies – often undisclosed – and the journals are too. And so are the patient groups. And finally, academic papers, which everyone thinks of as objective, are often covertly planned and written by people who work directly for the companies, without disclosure. Sometimes whole academic journals are even owned outright by one drug company. Aside from all this, for several of the most important and enduring problems in medicine, we have no idea what the best treatment is, because it's not in anyone's financial interest to conduct any trials at all. These are ongoing problems, and although people have claimed to fix many of them, for the most part they have failed; so all these problems persist, but worse than ever, because now people can pretend that everything is fine after all.

That's a lot to stand up, and the details are much more horrific than that paragraph makes it sound. There are some individual stories that will make you seriously question the integrity of the individuals involved; some that will make you angry; and some, I suspect, that might make you very sad. But I hope you will come to see that this is not just a book about bad people. In fact, it's possible for good people, in perversely designed systems, to casually perpetrate acts of great harm on strangers, sometimes without ever realising it. The current regulations – for companies, doctors and researchers – create perverse incentives; and we'll have better luck fixing those broken systems than we will ever have trying to rid the world of avarice.

Some people will say that this book is an attack on the pharmaceutical industry, and of course it is. But it's not only that. First, as you will see, the problems are diffuse, and doctors, regulators, academic journals, pharmacists, patient groups, and many more all play their part. Second, it's not unbounded. I suspect that most of the people who work in this industry are fundamentally good-hearted, and there is no medicine without medicines. Drug companies around the world have produced some of the most amazing innovations of the past fifty years, saving lives on an epic scale. But that does not allow them to hide data, mislead doctors and harm patients.

Today, when an academic or doctor tells you that they are working for the pharmaceutical industry, they often do so with a look of quiet embarrassment. I want to work towards a world where doctors and academics can feel actively optimistic about collaborating with industry, to make better treatments and better patients. This will require big changes, and some of them have been a very long time coming.

To that end, because the stories I am telling you are so worrying, I've tried to go beyond simply documenting the problems. Where there are obvious fixes, I've set out what they are. But I've also included, at the end of each chapter, some suggestions on what you can do to improve things. These are tailored to whoever you might be: a doctor, a patient, a politician, a researcher, a regulator or a drug company.

More than anything, though, I don't want you to lose sight of one thing: this is a pop science book. The tricks and distortions documented in these pages are beautiful, and intricate, and fascinating in their details. The true scale of this murderous disaster only fully reveals itself when the details are untangled. Good science has been perverted on an industrial scale, but this has happened slowly, and evolved naturally, over time. This has all been perpetrated by ordinary people, but many of them may not even know what they've done.

I want you to find them, and tell them.

What's coming

The book follows a simple trajectory.

We start by defending our central claim: industry-sponsored

studies are more likely to produce results that flatter the sponsor's drug, which has now been demonstrated, beyond any doubt, by current research. In this section we also encounter the idea of a 'systematic review' for the first time. A systematic review is an unbiased survey of all the evidence on a given question. It is the best-quality evidence that can be used, and where they exist, systematic reviews are used for evidence throughout this book, with individual studies described only to give you a flavour of how the research has been done, or how mischief has been made.

Then we look at how the pharmaceutical industry manages to create all these positive trials for its drugs. Our first stop is to review the evidence showing that unflattering trial data can simply be withheld from doctors and patients. Companies are perfectly entitled to conduct seven studies, but only publish the two positive ones, and this behaviour is commonplace. What's more, it happens in every field of science and medicine: from basic laboratory research, where selective publication fills the literature with false positive findings, wasting everyone's time; through early research trials, where evidence that drugs might be dangerous is hidden from view; and on to major trials used to inform everyday clinical practice. Because so much trial data is withheld from doctors and patients, we can have no clear idea of the true effects of the treatments that we use every day in medicine. The stories in this section go from antidepressants, through statins, cancer drugs, diet pills, and right up to Tamiflu. Governments around the world have spent billions of dollars to stockpile this flu drug, in fear of a pandemic, and yet the evidence on whether it reduces the rate of pneumonia and death is being withheld right now, to this day.

Next, we take a step back, to look at where drugs come from. We cover the drug development process, from the moment someone dreams up a new molecule, through tests in labs, on animals, the first tests in humans, and then the early trials necessary to show that a drug works on patients. Here you will find, I suspect, some surprises. Risky 'first-in-man' drug tests are conducted on homeless people; but more than that, full clinical trials are being globalised, a new development that has arisen very suddenly, in only the last couple of years. This raises serious ethical problems, because trial participants in developing countries are often unlikely to benefit from expensive new drugs; but it also raises interesting new problems for trusting the data.

Then we look at regulation, and the hoops you must go through to get a drug onto the market. We will see that the bar is very low: that drugs must only prove that they are better than nothing, even when there are highly effective treatments on the market already. This means that real patients are given dummy placebo pills for no good reason, but also that drugs appear on the market which are worse than the treatments we already have. We will see that companies break their promises over follow-up studies, and that regulators let them do this. We will also see how data on side effects and effectiveness can be with-held from regulators, and that regulators, in turn, are obsessively secretive, withholding the data they do have from doctors and patients. Last, we will see the harm done by this secrecy: 'many eyes' are often very powerful, to spot problems with medicines, and some of the most frightening harms have been missed by regulators, and only identified by academics who were forced to fight hard for access to data.

Then we take a tour through 'bad trials'. It is tempting to believe that a simple clinical trial is always a fair test of a treatment: and if done properly, it is. But several tricks have been introduced, over the course of many years, which allow researchers to overstate and exaggerate the benefits of the treatments they are testing. When you get here, you might think that some of these are innocent mistakes; in all seriousness, while I doubt that, I'm more interested in how clever they are. More important, we will see how obvious these tricks are, and how people who should know better at every step of the chain, from ethics committees through to academic journals, have allowed companies and researchers to engage in these shameful outright distortions.

After a brief detour to discuss how some of the problems around bad evidence, and missing evidence, could be addressed, we move on to marketing, which is where most previous books on drug companies have focused their attention.

Here we will see that pharmaceutical companies spend tens of billions of pounds every year trying to change the treatment decisions of doctors: in fact, they spend twice as much on marketing and advertising as they do on the research and development of new drugs. Since we all want doctors to prescribe medicine based on evidence, and evidence is universal, there is only one possible reason for such huge spends: to distort evidence-based practice. All of this money comes directly from patients and governments, so we ourselves are paying for this privilege. Doctors spend forty years practising medicine, with very little formal education after their initial training. Medicine changes completely in four decades, and as they try to keep up, doctors are bombarded with information: from ads that misrepresent the benefits and risks of new medicines; from sales reps who spy on patients' confidential prescribing records; from colleagues who are quietly paid by drug companies; from 'teaching' that is sponsored by industry; from independent 'academic' journal articles that are quietly written by drug company employees; and worse.

Finally, we will see what can be done. While the deceit of a marketing drive can be ignored by an ethical doctor, the problems caused by distorted evidence affect everybody, without exception. The most expensive doctors in the world can only make decisions about your care on the basis of the evidence publicly available to them, and nobody has a special inside track. If this evidence is distorted, then we are all exposed to avoidable suffering, pain and death. The whole system needs to be fixed, and until it is, we are all, very truly, in this together.

How to read this book

I have set out to explain the flaws in our system for assessing the benefits of treatments, and for disseminating the evidence we collect. Often, one specific drug is used as an example to explain a wider systemic flaw, but this is not an almanac of good and bad drugs, and you certainly shouldn't change your medication on the basis of what you read here. If you are concerned by what you read, then there is advice on what you can do to improve things for yourself, and for others, at the end of the book. Most of the harms discussed arise from our failure to do the best we can, not from handing out treatments that are worse than nothing. If that disappoints you, if you want more melodrama, then I suggest you go and read a book by a quack.

I deliberately haven't gone overboard to explain every medical term, to save space and avoid distractions: this doesn't mean that you miss out. If a symptom, for example, isn't explained or defined, that means you genuinely don't need this detail to understand the story; but I've left the long word in to help medics or academics find their feet, and to anchor the general principle in a specific corner of medicine for them. Acronyms and abbreviations are defined as we go, and used in a haphazard way after that, because this is how people talk in the real world. There's a glossary at the back for some common ideas, really just in case you read sections out of order, but there's nothing in there that doesn't come up in the main text.

Similarly, I haven't given the full names of most clinical trials, because they are conventionally known by their acronyms, and most medical textbooks wouldn't bother either: the 'ISIS trial', the 'CAST trial', in the minds of most doctors and academics, are the real names. If you're very interested, you can search for them online or in the endnotes, but they're not relevant to your enjoyment or understanding of the arguments in this book. Drugs present a different problem, because they have two names: the generic name, which is the correct scientific name for the molecule; and then the brand name used by the company manufacturing it in their packaging and advertising, which is usually a bit catchier. In general, doctors and academics think you should always use the scientific name, because it tells you a little about the class of the molecule, and is less ambiguous; while journalists and patients will more often use brand names. But everybody is inconsistent about which name they use, and in this book, so am I. Again, this simply reflects how people talk about medicines in the real world.

All the specific studies discussed are referenced at the back of the book. Where there was a choice, I've tried to select papers in open-access journals, so that they can be read for free by all. I've also tried to reference papers that give a good overview of a field, or good books on a subject, so that you can read more on whole areas if you want to.

Last: to an extent, this is a field where you need to know everything, to understand how it impacts on everything else. I've bent over backwards to introduce the ideas in the best order, but if all this material is completely new to you, then yo u might spot some extra connections – or feel greater outrage in your belly – reading it a second time. I have not assumed any prior knowledge. I have, however, assumed that you might be willing to deploy a little intellectual horsepower here and there. Some of this stuff is hard. That's precisely why these problems have been ignored, and that's why I've had to explain it to you here, in this book. If you want to catch people with their trousers down, you have to go into their home.

Enjoy.

Ben Goldacre
August 2012

For the paperback edition of the book, I've made occasional small changes to the text, many of them in light of helpful comments from readers: this was to disambiguate, improve, toughen and occasionally correct small errors (such as $n/3$ instead of $3/n$ on page 151, which broke my heart). Over the past year, even more evidence has been published about the ongoing problems described in the book. I've resisted the urge to expand every section, but some of this new evidence is covered in the new Afterword, 'What Happened Next'. For the key problem of withheld trial data, there has been an international campaign over the course of 2013, with some small forward movement; this is also described at the end, and I hope it will rouse you to action.

Ben Goldacre
August 2013

BAD PHARMA

1

Missing Data

Sponsors get the answer they want

Before we get going, we need to establish one thing beyond any doubt: industry-funded trials are more likely to produce a positive, flattering result than independently funded trials. This is our core premise, and you're about to read a very short chapter, because this is one of the most well-documented phenomena in the growing field of 'research about research'. It has also become much easier to study in recent years, because the rules on declaring industry funding have become a little clearer.

We can begin with some recent work: in 2010, three researchers from Harvard and Toronto found all the trials looking at five major classes of drug – antidepressants, ulcer drugs and so on – then measured two key features: were they positive, and were they funded by industry?[1] They found over five hundred trials in total: 85 per cent of the industry-funded studies were positive, but only 50 per cent of the government-funded trials were. That's a very significant difference.

In 2007, researchers looked at every published trial that set out to explore the benefit of a statin.[2] These are cholesterol-lowering drugs which reduce your risk of having a heart attack, they are prescribed in very large quantities, and they will loom large in this book. This study found 192 trials in total, either comparing one statin against another, or comparing a statin against a different kind of treatment. Once the researchers controlled for other factors (we'll delve into what this means later), they found that industry-funded trials were twenty times more likely to give results favouring the test drug. Again, that's a very big difference.

We'll do one more. In 2006, researchers looked into every trial of psychiatric drugs in four academic journals over a ten-year period, finding 542 trial outcomes in total. Industry sponsors got favourable outcomes for their own drug 78 per cent of the time, while independently funded trials only gave a positive result in 48 per cent of cases. If you were a competing drug put up against the sponsor's drug in a trial, you were in for a pretty rough ride: you would only win a measly 28 per cent of the time.[3]

These are dismal, frightening results, but they come from individual studies. When there has been lots of research in a field, it's always possible that someone – like me, for example – could cherry-pick the results, and give a partial view. I could, in essence, be doing exactly what I accuse the pharmaceutical industry of doing, and only telling you about the studies that support my case, while hiding the reassuring ones from you.

To guard against this risk, researchers invented the systematic review. We'll explore this in more detail soon (p. 14), since it's at the core of modern medicine, but in essence a systematic review is simple: instead of just mooching through the research literature, consciously or unconsciously picking out papers here and there that support your pre-existing beliefs, you take a scientific, systematic approach to the very process of looking for scientific evidence, ensuring that your evidence is as complete and representative as possible of all the research that has ever been done.

Systematic reviews are very, very onerous. In 2003, by

coincidence, two were published, both looking specifically at the question we're interested in. They took all the studies ever published that looked at whether industry funding is associated with pro-industry results. Each took a slightly different approach to finding research papers, and both found that industry-funded trials were, overall, about four times more likely to report positive results.[4] A further review in 2007 looked at the new studies that had been published in the four years after these two earlier reviews: it found twenty more pieces of work, and all but two showed that industry-sponsored trials were more likely to report flattering results.[5]

I am setting out this evidence at length because I want to be absolutely clear that there is no doubt on the issue. Industry-sponsored trials give favourable results, and that is not my opinion, or a hunch from the occasional passing study. This is a very well-documented problem, and it has been researched extensively, without anybody stepping out to take effective action, as we shall see.

There is one last study I'd like to tell you about. It turns out that this pattern of industry-funded trials being vastly more likely to give positive results persists even when you move away from published academic papers, and look instead at trial reports from academic conferences, where data often appears for the first time (in fact, as we shall see, sometimes trial results only appear at an academic conference, with very little information on how the study was conducted).

Fries and Krishnan studied all the research abstracts presented at the 2001 American College of Rheumatology meetings which reported any kind of trial, and acknowledged industry sponsorship, in order to find out what proportion had results that favoured the sponsor's drug. There is a small punchline coming, and to understand it we need to cover a little of what an academic paper looks like. In general, the results section is extensive: the raw numbers are given for each outcome, and for each possible causal factor, but not just as raw figures. The 'ranges' are given, subgroups are perhaps explored, statistical tests are conducted, and each detail of the result is described in table form, and in shorter narrative form in the text, explaining the most important results. This lengthy process is usually spread over several pages.

In Fries and Krishnan [2004] this level of detail was unnecessary. The results section is a single, simple, and – I like to imagine – fairly passive-aggressive sentence:

> The results from every RCT (45 out of 45) favored the drug of the sponsor.

This extreme finding has a very interesting side effect, for those interested in time-saving shortcuts. Since every industry-sponsored trial had a positive result, that's all you'd need to know about a piece of work to predict its outcome: if it was funded by industry, you could know with absolute certainty that the trial found the drug was great.

How does this happen? How do industry-sponsored trials almost always manage to get a positive result? It is, as far as anyone can be certain, a combination of factors. It may be that companies are more likely to run trials when they're most confident their treatment is going to 'win': this sounds reasonable, although even it conflicts with the ethical principle that you should only do a trial when there's genuine uncertainty about which treatment is best (otherwise you're exposing half of your participants to a treatment you already know to be inferior). Sometimes the chances of one treatment winning can be increased by outright design flaws. You can compare your new drug with something you know to be rubbish – an existing drug at an inadequate dose, perhaps, or a placebo sugar pill that does almost nothing. You can choose your patients very carefully, so they are more likely to get better on your treatment. You can peek at the results halfway through, and stop your trial early if they look good (which is – for interesting reasons we shall discuss – statistical poison). And so on.

But before we get to these fascinating methodological twists and quirks, these nudges and bumps that stop a trial from being a fair test of whether a treatment works or not, there is something very much simpler at hand.

Sometimes drug companies conduct lots of trials, and when they see that the results are unflattering, they simply fail to publish them. This is not a new problem, and it's not limited to medicine. In fact, this issue of negative results that go missing in action cuts into almost every corner of science. It distorts findings in fields as diverse as brain imaging and economics, it makes a mockery of all our efforts to exclude bias from our studies, and despite everything that regulators, drug companies and even some academics will tell you, it is a problem that has been left unfixed for decades.

In fact, it is so deep-rooted that even if we fixed it today – right now, for good, forever, without any flaws or loopholes in our legislation – that still wouldn't help, because we would still be practising medicine, cheerfully making decisions about which treatment is best, on the basis of decades of medical evidence which is – as you've now seen – fundamentally distorted.

But there is a way ahead.

Why missing data matters

Reboxetine is a drug I myself have prescribed. Other drugs had done nothing for this particular patient, so we wanted to try something new. I'd read the trial data before I wrote the prescription, and found only well-designed, fair tests, with overwhelmingly positive results. Reboxetine was better than placebo, and as good as any other antidepressant in head-to-head comparisons.

It's approved for use by the Medicines and Healthcare products Regulatory Agency (the MHRA) in the UK, but wisely, the US FDA chose not to approve it. (This is no proof of the FDA being any smarter; there are plenty of drugs available in the US that the UK never approved.) Reboxetine was clearly a safe and and effective treatment. The patient and I discussed the evidence briefly, and agreed it was the right treatment to try next. I signed a prescription saying I wanted my patient to have this drug.

But we had both been misled. In October 2010 a group of researchers were finally able to bring together all the trials that had ever been conducted on reboxetine.[6] Through a long process of investigation – searching in academic journals, but also arduously requesting data from the manufacturers and gathering documents from regulators – they were able to assemble all the data, both from trials that were published, and from those that had never appeared in academic papers.

When all this trial data was put together it produced a shocking picture. Seven trials had been conducted comparing reboxetine against placebo. Only one, conducted in 254 patients, had a neat, positive result, and that one was published in an academic journal, for doctors and researchers to read. But six more trials were conducted, in almost ten times as many patients. All of them showed that reboxetine was no better than a dummy sugar pill. None of these trials was published. I had no idea they existed.

It got worse. The trials comparing reboxetine against other drugs showed exactly the same picture: three small studies, 507 patients in total, showed that reboxetine was just as good as any other drug. They were all published. But 1,657 patients' worth of data was left unpublished, and this unpublished data showed that patients on reboxetine did worse than those on other drugs. If all this wasn't bad enough, there was also the side-effects data. The drug looked fine in the trials which appeared in the academic literature: but when we saw the unpublished studies, it turned out that patients were more likely to have side effects, more likely to drop out of taking the drug, and more likely to withdraw from the trial because of side effects, if they were taking reboxetine rather than one of its competitors.

If you're ever in any doubt about whether the stories in this book make me angry – and I promise you, whatever happens, I will keep to the data, and strive to give a fair picture of everything we know – you need only look at this story. I did everything a doctor is supposed to do. I read all the papers, I critically appraised them, I understood them, I discussed them with the patient, and we made a decision together, based on the evidence. In the published data, reboxetine was a safe and effective drug. In reality, it was no better than a sugar pill, and worse, it does more harm than good. As a doctor I did something which, on the balance of all the evidence, harmed my patient, simply because unflattering data was left unpublished.

If you find that amazing, or outrageous, your journey is just beginning. Because nobody broke any law in that situation, reboxetine is still on the market, and the system that allowed all this to happen is still in play, for all drugs, in all countries in the world. Negative data goes missing, for all treatments, in all areas of science. The regulators and professional bodies we would reasonably expect to stamp out such practices have failed us.

In a few pages, we will walk through the literature that demonstrates all of this beyond any doubt, showing that 'publication bias' – the process whereby negative results go unpublished – is endemic throughout the whole of medicine and academia; and that regulators have failed to do anything about it, despite decades of data showing the size of the problem. But before we get to that research, I need you to feel its implications, so we need to think about why missing data matters.

Evidence is the only way we can possibly know if something works – or doesn't work – in medicine. We proceed by testing things, as cautiously as we can, in head-to-head trials, and gathering together *all* of the evidence. This last step is crucial: if I withhold half the data from you, it's very easy for me to convince you of something that isn't true. If I toss a coin a hundred times, for example, but only tell you about the results when it lands heads-up, I can convince you that this is a two-headed coin. But that doesn't mean I really do have a two-headed coin: it means I'm misleading you, and you're a fool for letting me get away with it. This is exactly the situation we tolerate in medicine, and always have. Researchers are free to do as many trials as they wish, and then choose which ones to publish.

The repercussions of this go way beyond simply misleading doctors about the benefits and harms of interventions for patients, and way beyond trials. Medical research isn't an abstract academic pursuit: it's about people, so every time we fail to publish a piece of research we expose real, living people to unnecessary, avoidable suffering.

TGN1412

In March 2006, six volunteers arrived at a London hospital to take part in a trial. It was the first time a new drug called TGN1412 had ever been given to humans, and they were paid £2,000 each.[7] Within an hour these six men developed headaches, muscle aches, and a feeling of unease. Then things got worse: high temperatures, restlessness, periods of forgetting who and where they were. Soon they were shivering, flushed, their pulses racing, their blood pressure falling. Then, a cliff: one went into respiratory failure, the oxygen levels in his blood falling rapidly as his lungs filled with fluid. Nobody knew why. Another dropped his blood pressure to just 65/40, stopped breathing properly, and was rushed to an intensive care unit, knocked out, intubated, mechanically ventilated. Within a day all six were disastrously unwell: fluid on their lungs, struggling to breathe, their kidneys failing, their blood clotting uncontrollably throughout their bodies, and their white blood cells disappearing. Doctors threw everything they could at them: steroids, anti-histamines, immune-system receptor blockers. All six were ventilated on intensive care. They stopped producing urine; they were all put on dialysis; their blood was replaced, first slowly, then rapidly; they needed plasma, red cells, platelets. The fevers continued. One developed pneumonia. And then the blood stopped getting to their peripheries. Their fingers and toes went flushed, then brown, then black, and then began to rot and die. With heroic effort, all escaped, at least, with their lives.

The Department of Health convened an Expert Scientific Group to try to understand what had happened, and from this two concerns were raised.[8] First: can we stop things like this from happening again? It's plainly foolish, for example, to give a new experimental treatment to all six participants in a 'first-in-man' trial at the same time, if that treatment is a completely unknown quantity. New drugs should be given to participants in a staggered process, slowly, over a day. This idea received considerable attention from regulators and the media.

Less noted was a second concern: could we have foreseen this disaster? TGN1412 is a molecule that attaches to a receptor called CD28 on the white blood cells of the immune system. It was a new and experimental treatment, and it interfered with the immune system in ways that are poorly understood, and hard to model in animals (unlike, say, blood pressure, because immune systems are very variable between different species). But as the final report found, there was experience with a similar intervention: it had simply not been published. One researcher presented the inquiry with unpublished data on a study he had conducted in a single human subject a full ten years earlier, using an antibody that attached to the CD3, CD2 and CD28 receptors. The effects of this antibody had parallels with those of TGN1412, and the subject on whom it was tested had become unwell. But nobody could possibly have known that, because these results were never shared with the scientific community. They sat unpublished, unknown, when they could have helped save six men from a terrifying, destructive, avoidable ordeal.

That original researcher could not foresee the specific harm he contributed to, and it's hard to blame him as an individual, because he operated in an academic culture where leaving data unpublished was regarded as completely normal. The same culture exists today. The final report on TGN1412 concluded that sharing the results of all first-in-man studies was essential: they should be published, every last one, as a matter of routine. But phase 1 trial results weren't published then, and they're still not published now. In 2009, for the first time, a study was published looking specifically at how many of these first-in-man trials get published, and how many remain hidden.[9] They took all such trials approved by one ethics committee over a year. After four years, nine out of ten remained unpublished; after eight years, four out of five were still unpublished.

In medicine, as we shall see time and again, research is not abstract: it relates directly to life, death, suffering and pain. With every one of these unpublished studies we are potentially exposed, quite unnecessarily, to another TGN1412. Even a huge international news story, with horrific images of young men brandishing blackened feet and hands from hospital beds, wasn't enough to get movement, because the issue of missing data is too complicated to fit in one sentence.

When we don't share the results of basic research, such as a small first-in-man study, we expose people to unnecessary risks in the future. Was this an extreme case? Is the problem limited to early, experimental, new drugs, in small groups of trial participants? No.

In the 1980s, US doctors began giving anti-arrhythmic drugs to all patients who'd had a heart attack. This practice made perfect sense on paper: we knew that anti-arrhythmic drugs helped prevent abnormal heart rhythms; we also knew that people who've had a heart attack are quite likely to have abnormal heart rhythms; we also knew that often these went unnoticed, undiagnosed and untreated. Giving anti-arrhythmic drugs to everyone who'd had a heart attack was a simple, sensible preventive measure.

Unfortunately, it turned out that we were wrong. This prescribing practice, with the best of intentions, on the best of principles, actually killed people. And because heart attacks are very common, it killed them in very large numbers: well over 100,000 people died unnecessarily before it was realised that the fine balance between benefit and risk was completely different for patients without a proven abnormal heart rhythm.

Could anyone have predicted this? Sadly, yes, they could have. A trial in 1980 tested a new anti-arrhythmic drug, lorcainide, in a small number of men who'd had a heart attack – less than a hundred – to see if it was any use. Nine out of forty-eight men on lorcainide died, compared with one out of forty-seven on placebo. The drug was early in its development cycle, and not long after this study it was dropped for commercial reasons. Because it wasn't on the market, nobody even thought to publish the trial. The researchers assumed it was an idiosyncrasy of their molecule, and gave it no further thought. If they had published, we would have been much more cautious about trying other anti-arrhythmic drugs on people with heart attacks, and the phenomenal death toll – over 100,000 people in their graves prematurely – might have been stopped sooner. More than a decade later, the researchers finally did publish their results, with a *mea culpa*, recognising the harm they had done by not sharing them earlier:

> When we carried out our study in 1980, we thought that the increased death rate that occurred in the lorcainide group was an effect of chance. The development of lorcainide was abandoned for commercial reasons, and this study was therefore

11

never published; it is now a good example of 'publication bias'. The results described here might have provided an early warning of trouble ahead.[10]

As we shall shortly see, this problem of unpublished data is widespread throughout medicine, and indeed the whole of academia, even though the scale of the problem, and the harm it causes, have been documented beyond any doubt. We will see stories on basic cancer research, Tamiflu, cholesterol blockbusters, obesity drugs, antidepressants and more, with evidence that goes from the dawn of medicine to the present day, and data that is still being withheld, right now, as I write, on widely used drugs which many of you reading this book will have taken this morning. We will also see how regulators and academic bodies have repeatedly failed to address the problem.

Because researchers are free to bury any result they please, patients are exposed to harm on a staggering scale throughout the whole of medicine, from research to practice. Doctors can have no idea about the true effects of the treatments they give. Does this drug really work best, or have I simply been deprived of half the data? Nobody can tell. Is this expensive drug worth the money, or have the data simply been massaged? No one can tell. Will this drug kill patients? Is there any evidence that it's dangerous? No one can tell.

This is a bizarre situation to arise in medicine, a discipline where everything is supposed to be based on evidence, and where everyday practice is bound up in medico-legal anxiety. In one of the most regulated corners of human conduct we've taken our eyes off the ball, and allowed the evidence driving practice to be polluted and distorted. It seems unimaginable. We will now see how deep this problem goes.

Why we summarise data

Missing data has been studied extensively in medicine. But before I lay out that evidence, we need to understand exactly why it matters, from a scientific perspective. And for that we need to understand systematic reviews and 'meta-analysis'. Between them, these are two of the most powerful ideas in modern medicine. They are incredibly simple, but they were invented shockingly late.

When we want to find out if something works or not, we do a trial. This is a very simple process, and the first recorded attempt at some kind of trial was in the Bible (Daniel 1:12, if you're interested). First, you need an unanswered question: for example, 'Does giving steroids to a woman delivering a premature baby increase the chances of that baby surviving?' Then you find some relevant participants, in this case, mothers about to deliver a premature baby. You'll need a reasonable number of them, let's say two hundred for this trial. Then you divide them into two groups at random, give the mothers in one group the current best treatment (whatever that is in your town), while the mothers in the other group get current best treatment plus some steroids. Finally, when all two hundred women have gone through your trial, you count up how many babies survived in each group.

This is a real-world question, and lots of trials were done on this topic, starting from 1972 onwards: two trials showed that steroids saved lives, but five showed no significant benefit. Now, you will often hear that doctors disagree when the evidence is mixed, and this is exactly that kind of situation. A doctor with a strong pre-existing belief that steroids work – perhaps preoccupied with some theoretical molecular mechanism, by which the drug might do something useful in the body – could come along and say: 'Look at these two positive trials! Of course we must give steroids!' A doctor with a strong prior intuition that steroids were rubbish might point at the five negative trials and say: 'Overall the evidence shows no benefit. Why take a risk?'

Up until very recently, this was basically how medicine progressed. People would write long, languorous review articles – essays surveying the literature – in which they would cite the trial data they'd come across in a completely unsystematic fashion, often reflecting their own prejudices and values. Then, in the 1980s, people began to do something called a 'systematic review'. This is a clear, systematic survey of the literature, with the intention of getting all the trial data you can possibly find on one topic, without being biased towards any particular set of findings. In a systematic review, you describe exactly how you looked for data: which databases you searched, which search engines and indexes you used, even what words you searched for. You pre-specify the kinds of studies that can be included in your review, and then you present everything you've found, including the papers you rejected, with an explanation of why. By doing this, you ensure that your methods are fully transparent, replicable and open to criticism, providing the reader with a clear and complete picture of the evidence. It may sound like a simple idea, but systematic reviews are extremely rare outside clinical medicine, and are quietly one of the most important and transgressive ideas of the past forty years.

When you've got all the trial data in one place, you can conduct something called a meta-analysis, where you bring all the results together in one giant spreadsheet, pool all the data and get one single, summary figure, the most accurate summary of all the data on one clinical question. The output of this is called a 'blobbogram', and you can see one on the following page, in the logo of the Cochrane Collaboration, a global, non-profit academic organisation that has been producing gold-standard reviews of evidence on important questions in medicine since the 1980s.

This blobbogram shows the results of all the trials done on

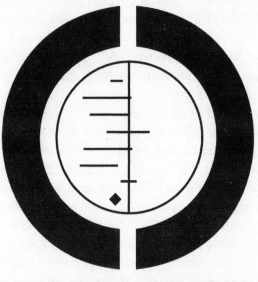

THE COCHRANE
COLLABORATION®

giving steroids to help premature babies survive. Each horizontal line is a trial: if that line is further to the left, then the trial showed steroids were beneficial and saved lives. The central, vertical line is the 'line of no effect': and if the horizontal line of the trial touches the line of no effect, then that trial showed no statistically significant benefit. Some trials are represented by longer horizontal lines: these were smaller trials, with fewer participants, which means they are prone to more error, so the estimate of the benefit has more uncertainty, and therefore the horizontal line is longer. Finally, the diamond at the bottom shows the 'summary effect': this is the overall benefit of the intervention, pooling together the results of all the individual trials. These are much narrower than the lines for individual trials, because the estimate is much more accurate: it is summarising the effect of the drug in many more patients. On this blobbogram you can see – because the diamond is a long way from the line of no effect – that giving steroids is hugely beneficial. In fact, it reduces the chances of a premature baby dying by almost half.

The amazing thing about this blobbogram is that it had to be invented, and this happened very late in medicine's history. For many years we had all the information we needed to know that steroids saved lives, but nobody knew they were effective, because nobody did a systematic review until 1989. As a result, the treatment wasn't given widely, and huge numbers of babies died unnecessarily; not because we didn't have the information, but simply because we didn't synthesise it together properly.

In case you think this is an isolated case, it's worth examining exactly how broken medicine was until frighteningly recent times. The diagram on the following page contains two blobbograms, or 'forest plots', showing all the trials ever conducted to see whether giving streptokinase, a clot-busting drug, improves survival in patients who have had a heart attack.[11]

Look first only at the forest plot on the left. This is a conventional forest plot, from an academic journal, so it's a little busier than the stylised one in the Cochrane logo. The principles, however, are exactly the same. Each horizontal line is a trial, and you can see that there is a hodgepodge of results, with some trials showing a benefit (they don't touch the vertical line of no effect, headed '1') and some showing no benefit (they do cross that line). At the bottom, however, you can see the summary effect – a dot on this old-fashioned blobbogram, rather than a diamond. And you can see very clearly that overall, streptokinase saves lives.

So what's that on the right? It's something called a cumulative meta-analysis. If you look at the list of studies on the left of the diagram, you can see that they are arranged in order of date. The cumulative meta-analysis on the right adds in each new trial's results, as they arrived over history, to the previous trials' results. This gives the best possible running estimate, each year,

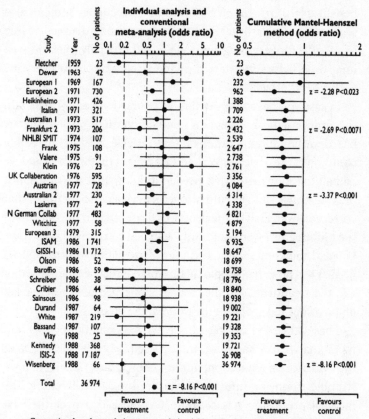

—*Conventional and cumulative meta-analysis of 33 trials of intravenous streptokinase for acute myocardial infarction. Odds ratios and 95% confidence intervals for effect of treatment on mortality are shown on a logarithmic scale*

of how the evidence would have looked at that time, if anyone had bothered to do a meta-analysis on all the data available to them. From this cumulative blobbogram you can see that the horizontal lines, the 'summary effects', narrow over time as more and more data is collected, and the estimate of the overall benefit of this treatment becomes more accurate. You can also see that these horizontal lines stopped touching the vertical line of no effect a very long time ago – and crucially, they do so a long time before we started giving streptokinase to everyone with a heart attack.

In case you haven't spotted it for yourself already – to be fair, the entire medical profession was slow to catch on – this chart has devastating implications. Heart attacks are an incredibly common cause of death. We had a treatment that worked, and we had all the information we needed to know that it worked, but once again we didn't bring it together systematically to get that correct answer. Half of the people in those trials at the bottom of the blobbogram were randomly assigned to receive no streptokinase, I think unethically, because we had all the information we needed to know that streptokinase worked: they were deprived of effective treatments. But they weren't alone, because so were most of the rest of the people in the world at the time.

These stories illustrate, I hope, why systematic reviews and meta-analyses are so important: we need to bring together *all* of the evidence on a question, not just cherry-pick the bits that we stumble upon, or intuitively like the look of. Mercifully the medical profession has come to recognise this over the past couple of decades, and systematic reviews with meta-analyses are now used almost universally, to ensure that we have the most accurate possible summary of all the trials that have been done on a particular medical question.

But these stories also demonstrate why missing trial results are so dangerous. If one researcher or doctor 'cherry-picks', when summarising the existing evidence, and looks only at the trials that support their hunch, then they can produce a misleading picture of the research. That is a problem for that one individual (and for anyone who is unwise or unlucky enough to be influenced by them). But if we are *all* missing the negative trials, the entire medical and academic community, around the world, then when we pool the evidence to get the best possible view of what works – as we must do – we are all completely misled. We get a misleading impression of the treatment's effectiveness: we incorrectly exaggerate its benefits; or perhaps even find incorrectly that an intervention was beneficial, when in reality it did harm.

Now that you understand the importance of systematic reviews, you can see why missing data matters. But you can also appreciate that when I explain *how much* trial data is missing, I am giving you a clean overview of the literature, because I will be explaining that evidence using systematic reviews.

How much data is missing?

If you want to prove that trials have been left unpublished, you have an interesting problem: you need to prove the existence of studies you don't have access to. To work around this, people have developed a simple approach: you identify a group of trials you know have been conducted and completed, then check to see if they have been published. Finding a list of completed trials is the tricky part of this job, and to achieve it people have used various strategies: trawling the lists of trials that have been approved by ethics committees (or 'institutional review boards' in the USA), for example; or chasing up the trials discussed by researchers at conferences.

In 2008 a group of researchers decided to check for publication of every trial that had ever been reported to the US Food and Drug Administration for all the antidepressants that came onto the market between 1987 and 2004.[12] This was no small task. The FDA archives contain a reasonable amount of information on all the trials that were submitted to the regulator in order to get a licence for a new drug. But that's not all the trials, by any means, because those conducted after the drug has come onto the market will not appear there; and the information that is provided by the FDA is hard to search, and often scanty. But it is an important subset of the trials, and more than enough for us to begin exploring how often trials go missing, and why. It's also a representative slice of trials from all the major drug companies.

The researchers found seventy-four studies in total, representing 12,500 patients' worth of data. Thirty-eight of these trials had positive results, and found that the new drug worked; thirty-six were negative. The results were therefore an even split between success and failure for the drugs, in reality. Then the researchers set about looking for these trials in the published academic literature, the material available to doctors and patients. This provided a very different picture. Thirty-seven of the positive trials – all but one – were published in full, often with much fanfare. But the trials with negative results had a very different fate: only three were published. Twenty-two were simply lost to history, never appearing anywhere other than in those dusty, disorganised, thin FDA files. The remaining eleven which had negative results in the FDA summaries did appear in the academic literature, but were written up as if the drug was a success. If you think this sounds absurd, I agree: we will see in Chapter 4, on 'bad trials', how a study's results can be reworked and polished to distort and exaggerate its findings.

This was a remarkable piece of work, spread over twelve drugs from all the major manufacturers, with no stand-out bad guy. It very clearly exposed a broken system: in reality we have thirty-eight positive trials and thirty-six negative ones; in the academic literature we have forty-eight positive trials and three negative ones. Take a moment to flip back and forth between those in your mind: 'thirty-eight positive trials, thirty-six negative'; or 'forty-eight positive trials and only three negative'.

If we were talking about one single study, from one single group of researchers, who decided to delete half their results because they didn't give the overall picture they wanted, then we would quite correctly call that act 'research misconduct'. Yet somehow when exactly the same phenomenon occurs, but with whole studies going missing, by the hands of hundreds and thousands of individuals, spread around the world, in both the public and private sector, we accept it as a normal part of life.[13] It passes by, under the watchful eyes of regulators and professional bodies who do nothing, as routine, despite the undeniable impact it has on patients.

Even more strange is this: we've known about the problem of negative studies going missing for almost as long as people have been doing serious science.

This was first formally documented by an American psychologist called Theodore Sterling in 1959.[14] He went through every paper published in the four big psychology journals of the time, and found that 286 out of 294 reported a statistically significant result. This, he explained, was plainly fishy: it couldn't possibly be a fair representation of every study that had been conducted, because if we believed that, we'd have to believe that almost every theory ever tested by a psychologist in an experiment had turned out to be correct. If psychologists really were so great at predicting results, there'd hardly be any point in bothering to run experiments at all. In 1995, at the end of his career, the same researcher came back to the same question, half a lifetime later, and found that almost nothing had changed.[15]

Sterling was the first to put these ideas into a formal academic context, but the basic truth had been recognised for many centuries. Francis Bacon explained in 1620 that we often mislead ourselves by only remembering the times something worked, and forgetting those when it didn't.[16] Dr Thomas Fowler in 1786 listed the cases he'd seen treated with arsenic, and pointed out that he could have glossed over the failures, as others might be tempted to do, but had included them.[17] To do otherwise, he explained, would have been misleading.

Yet it was only three decades ago that people started to realise that missing trials posed a serious problem for medicine. In 1980 Elina Hemminki found that almost half the trials conducted in the mid-1970s in Finland and Sweden had been left unpublished.[18] Then, in 1986, an American researcher called Robert Simes decided to investigate the trials on a new treatment for ovarian cancer. This was an important study, because it looked at a life-or-death question. Combination chemotherapy for this kind of cancer has very tough side effects, and knowing this, many researchers had hoped it might be better to give a single 'alkylating agent' drug first, before moving on to full chemotherapy. Simes looked at all the trials published on this question in the academic literature, read by doctors and academics. From this, giving a single drug first looked like a great idea: women with advanced ovarian cancer (which is not a good diagnosis to have) who were on the alkylating agent alone were significantly more likely to survive longer.

Then Simes had a smart idea. He knew that sometimes trials can go unpublished, and he had heard that papers with less 'exciting' results are the most likely to go missing. To prove that this has happened, though, is a tricky business: you need to find a fair, representative sample of all the trials that have been conducted, and then compare their results with the smaller pool of trials that have been published, to see if there are any embarrassing differences. There was no easy way to get this information from the medicines regulator (we will discuss this problem in some detail later), so instead he went to the International Cancer Research Data Bank. This contained a register of interesting trials that were happening in the USA, including most of the ones funded by the government, and many others from around the world. It was by no means a complete list, but it did have one crucial feature: the trials were registered before their results came in, so any list compiled from this source would be, if not complete, at least a representative sample of all the research that had ever been done, and not biased by whether their results were positive or negative.

When Simes compared the results of the published trials against the pre-registered trials, the results were disturbing. Looking at the academic literature – the studies that researchers and journal editors chose to publish – alkylating agents alone looked like a great idea, reducing the rate of death from advanced ovarian cancer significantly. But when you looked only at the pre-registered trials – the unbiased, fair sample of all the trials ever conducted – the new treatment was no better than old-fashioned chemotherapy.

Simes immediately recognised – as I hope you will too – that the question of whether one form of cancer treatment is better than another was small fry compared to the depth charge he was about to set off in the medical literature. Everything we thought we knew about whether treatments worked or not was probably distorted, to an extent that might be hard to measure, but that would certainly have a major impact on patient care. We were seeing the positive results, and missing the negative ones. There was one clear thing we should do about this: start a registry of all clinical trials, demand that people register their study before they start, and insist that they publish the results at the end.

That was 1986. Since then, a generation later, we have done very badly. In this book, I promise I won't overwhelm you with data. But at the same time, I don't want any drug company, or government regulator, or professional body, or anyone who doubts this whole story, to have any room to wriggle. So I'll now go through all the evidence on missing trials, as briefly as possible, showing the main approaches that have been used. All of what you are about to read comes from the most current systematic reviews on the subject, so you can be sure that it is a fair and unbiased summary of the results.

One research approach is to get all the trials that a medicines regulator has record of, from the very early ones done for the purposes of getting a licence for a new drug, and then check to see if they all appear in the academic literature. That's the method we saw used in the paper mentioned above, where researchers sought out every paper on twelve antidepressants, and found that a 50/50 split of positive and negative results turned into forty-eight positive papers and just three negative ones. This method has been used extensively in several different areas of medicine:

- Lee and colleagues, for example, looked for all of the 909 trials submitted alongside marketing applications for all ninety new drugs that came onto the market from 2001 to 2002: they found that 66 per cent of the trials with significant results were published, compared with only 36 per cent of the rest.[19]
- Melander, in 2003, looked for all forty-two trials on five antidepressants that were submitted to the Swedish drug regulator in the process of getting a marketing authorisation: all twenty-one studies with significant results were published; only 81 per cent of those finding no benefit were published.[20]
- Rising et al., in 2008, found more of those distorted write-ups that we'll be dissecting later: they looked for all trials on two years' worth of approved drugs. In the FDA's summary of the results, once those could be found, there were 164 trials. Those with favourable outcomes were a full four times more likely to be published in academic papers than those with negative outcomes. On top of that, four of the trials with negative outcomes changed, once they appeared in the academic literature, to favour the drug.[21]

If you prefer, you can look at conference presentations: a huge amount of research gets presented at conferences, but our current best estimate is that only about half of it ever appears in the academic literature.[22] Studies presented only at conferences are almost impossible to find, or cite, and are especially hard to assess, because so little information is available on the specific methods used in the research (often as little as a paragraph). And as you will see shortly, not every trial is a fair test of a treatment. Some can be biased by design, so these details matter.

The most recent systematic review of studies looking at what happens to conference papers was done in 2010, and it found thirty separate studies looking at whether negative conference presentations – in fields as diverse as anaesthetics, cystic fibrosis, oncology, and A&E – disappear before becoming fully fledged academic papers.[23] Overwhelmingly, unflattering results are much more likely to go missing.

If you're very lucky, you can track down a list of trials whose existence was publicly recorded before they were started, perhaps on a register that was set up to explore that very question. From the pharmaceutical industry, up until very recently, you'd be very lucky to find such a list in the public domain. For publicly funded research the story is a little different, and here we start to learn a new lesson: although the vast majority of trials are conducted by the industry, with the result that they set the tone for the community, this phenomenon is not limited to the commercial sector.

- By 1997 there were already four studies in a systematic review on this approach. They found that studies with significant results were two and a half times more likely to get published than those without.[24]
- A paper from 1998 looked at all trials from two groups of trialists sponsored by the US National Institutes of Health over the preceding ten years, and found, again,

25

that studies with significant results were more likely to be published.[25]

- Another looked at drug trials notified to the Finnish National Agency, and found that 47 per cent of the positive results were published, but only 11 per cent of the negative ones.[26]
- Another looked at all the trials that had passed through the pharmacy department of an eye hospital since 1963: 93 per cent of the significant results were published, but only 70 per cent of the negative ones.[27]

The point being made in this blizzard of data is simple: this is not an under-researched area; the evidence has been with us for a long time, and it is neither contradictory nor ambiguous.

Two French studies in 2005 and 2006 took a new approach: they went to ethics committees, and got lists of all the studies they had approved, and then found out from the investigators whether the trials had produced positive or negative results, before finally tracking down the published academic papers.[28] The first study found that significant results were twice as likely to be published; the second that they were four times as likely. In Britain, two researchers sent a questionnaire to all the lead investigators on 101 projects paid for by NHS R&D: it's not industry research, but it's worth noting anyway. This produced an unusual result: there was no statistically significant difference in the publication rates of positive and negative papers.[29]

But it's not enough simply to list studies. Systematically taking all the evidence that we have so far, what do we see overall?

It's not ideal to lump every study of this type together in one giant spreadsheet, to produce a summary figure on publication bias, because they are all very different, in different fields, with different methods. This is a concern in many meta-analyses (though it shouldn't be overstated: if there are lots of trials comparing one treatment against placebo, say, and they're all using the same outcome measurement, then you might be fine just lumping them all in together).

But you can reasonably put some of these studies together in groups. The most current systematic review on publication bias, from 2010, from which the examples above are taken, draws together the evidence from various fields.[30] Twelve comparable studies follow up conference presentations, and taken together they find that a study with a significant finding is 1.62 times more likely to be published. For the four studies taking lists of trials from before they started, overall, significant results were 2.4 times more likely to be published. Those are our best estimates of the scale of the problem. They are current, and they are damning.

All of this missing data is not simply an abstract academic matter: in the real world of medicine, published evidence is used to make treatment decisions. This problem goes to the core of everything that doctors do, so it's worth considering in some detail what impact it has on medical practice. First, as we saw in the case of reboxetine, doctors and patients are misled about the effects of the medicines they use, and can end up making decisions that cause avoidable suffering, or even death. We might also choose unnecessarily expensive treatments, having been misled into thinking they are more effective than cheaper older drugs. This wastes money, ultimately depriving patients of other treatments, since funding for health care is never infinite.

It's also worth being clear that this data is withheld from everyone in medicine, from top to bottom. Most countries have organisations to create careful, unbiased summaries of all the evidence on new treatments to determine whether they are cost effective. In the UK the organisation is called NICE (the National Institute for Health and Clinical Excellence); in Germany it is called IQWiG, while in the US insurers may make their own assessments. But these organisations are unable either to identify or to access data that has been withheld by researchers or companies on a drug's effectiveness; they have no more legal right to that data than you or I do. In fact, as we shall see, some regulators, despite having access to this information, have refused to share it with the public or doctors. Others have hidden the information they hold behind walls of chaos. This is an extraordinary and perverse situation.

So, while doctors are kept in the dark, patients are exposed to inferior treatments, ineffective treatments, unnecessary treatments, and unnecessarily expensive treatments that are no better than cheap ones; governments pay for unnecessarily expensive treatments, and mop up the cost of harms created by inadequate or harmful treatment; and individual participants in trials, such as those in the TGN1412 study, are exposed to terrifying, life-threatening ordeals, resulting in lifelong scars, again quite unnecessarily.

At the same time, the whole of the research project in medicine is retarded, as vital negative results are held back from those who could use them. This affects everyone, but it is especially egregious in the world of 'orphan diseases', medical problems that affect only small numbers of patients, because these corners of medicine are already short of resources, and are neglected by the research departments of most drug companies, since the opportunities for revenue are thinner. People working on orphan diseases will often research existing drugs that have been tried and failed in other conditions, but that have theoretical potential for the orphan disease. If the data from earlier work on these drugs in other diseases is missing, then the job of researching them for the orphan disease is both harder and more dangerous: perhaps they have already been shown to have benefits or effects that would help accelerate research; perhaps they have already been shown to be actively harmful when used on other diseases, and there are important safety signals that would help protect future research participants from harm. Nobody can tell you.

Finally, and perhaps most shamefully, when we allow unflattering data to go unpublished, we betray the patients who participated in these studies: the people who have given their bodies, and sometimes their lives, in the implicit belief that they are doing something to create new knowledge, that will benefit others in the same position as them in the future. In fact, their belief is not implicit: often it's exactly what we tell them, as researchers, and it is a lie, because the data might be withheld, and we know it.

So whose fault is this?

Why do negative trials disappear?

In a moment we will see more clear cases of drug companies withholding data – in stories where we can identify individuals – sometimes with the assistance of regulators. When we get to these, I hope your rage might swell. But first, it's worth taking a moment to recognise that publication bias occurs outside commercial drug development, and in completely unrelated fields of academia, where people are motivated only by reputation, and their own personal interests.

In many respects, after all, publication bias is a very human process. If you've done a study and it didn't have an exciting, positive result, then you might wrongly conclude that your experiment isn't very interesting to other researchers. There's also the issue of incentives: academics are often measured, rather unhelpfully, by crude metrics like the numbers of citations for their papers, and the number of 'high-impact' studies they get into glamorous well-read journals. If negative findings are harder to publish in bigger journals, and less likely to be cited by other academics, then the incentives to work at disseminating them are lower. With a positive finding, meanwhile, you get a sense of discovering something new. Everyone around you is excited, because your results are exceptional.

One clear illustration of this problem came in 2010. A mainstream American psychology researcher called Daryl Bem published a competent academic paper, in a well-respected journal, showing evidence of precognition, the ability to see into the future.* These studies were well-designed, and the findings were statistically significant, but many people weren't very convinced, for the same reasons you aren't: if humans really could see into the future, we'd probably know about it already; and extraordinary claims require extraordinary evidence, rather than one-off findings.

But in fact the study has been replicated, though Bem's positive results have not been. At least two groups of academics have rerun several of Bem's experiments, using the exact same methods, and both found no evidence of precognition. One group submitted their negative results to the *Journal of Personality and Social Psychology* – the very same journal that published Bem's paper in 2010 – and that journal rejected their paper out of hand. The editor even came right out and said it: we never publish studies that replicate other work.

Here we see the same problem as in medicine: positive findings are more likely to be published than negative ones. Every now and then, a freak positive result is published showing, for example, that people can see into the future. Who knows how

---

* Instead of designing elaborate new studies to see whether people could consciously see forward in time, Bem simply ran some classic psychology experiments backwards. So, for example, he conducted a well-known experiment on subliminal influence, where you show people two mirror images of the same picture, and then ask them which they prefer; but you flash up an unpleasant subliminal image underneath one or other image for just a few milliseconds before they make their choice. In the normal run of this study, the subliminal image makes people less likely to choose that option. In the Bem study, the unpleasant subliminal images were flashed up just *after* the participants made their choice of favourite image. However unlikely it sounds, Bem found that these subliminal images still had an effect on people's choices.

many psychologists have tried, over the years, to find evidence of psychic powers, running elaborate, time-consuming experiments, on dozens of subjects – maybe hundreds – and then found no evidence that such powers exist? Any scientist trying to publish such a 'So what?' finding would struggle to get a journal to take it seriously, at the best of times. Even with the clear target of Bem's paper on precognition, which was widely covered in serious newspapers across Europe and the USA, the academic journal with a proven recent interest in the question of precognition simply refused to publish a paper with a negative result. Yet replicating these findings was key – Bem himself said so in his paper – so keeping track of the negative replications is vital too.

People working in real labs will tell you that sometimes an experiment can fail to produce a positive result many times before the outcome you're hoping for appears. What does that mean? Sometimes the failures will be the result of legitimate technical problems; but sometimes they will be vitally important statistical context, perhaps even calling the main finding of the research into question. Many research findings, remember, are not absolute black-and-white outcomes, but fragile statistical correlations. Under our current system, most of this contextual information about failure is just brushed under the carpet, and this has huge ramifications for the cost of replicating research, in ways that are not immediately obvious. For example, researchers failing to replicate an initial finding may not know if they've failed because the original result was an overstated fluke, or because they've made some kind of mistake in their methods. In fact, the cost of proving that a finding was wrong is vastly greater than the cost of making it in the first place, because you need to run the experiment many more times to prove the *absence* of a finding, simply because of the way that the statistics of detecting weak effects work; and you also need to be absolutely certain that you've excluded all technical problems, to avoid getting egg on your face if your replication turns out to have been inadequate. These barriers to refutation may partly explain why it's so easy to get away with publishing findings that ultimately turn out to be wrong.[31]

Publication bias is not just a problem in the more abstract corners of psychology research. In 2012 a group of researchers reported in the journal *Nature* how they tried to replicate fifty-three early laboratory studies of promising targets for cancer treatments: forty-seven of the fifty-three could not be replicated.[32] This study has serious implications for the development of new drugs in medicine, because such unreplicable findings are not simply an abstract academic issue: researchers build theories on the back of them, trust that they're valid, and investigate the same idea using other methods. If they are simply being led down the garden path, chasing up fluke errors, then huge amounts of research money and effort are being wasted, and the discovery of new medical treatments is being seriously retarded.

The authors of the study were clear on both the cause of and the solution for this problem. Fluke findings, they explained, are often more likely to be submitted to journals – and more likely to be published – than boring, negative ones. We should give more incentives to academics for publishing negative results; but we should also give them more opportunity.

This means changing the behaviour of academic journals, and here we are faced with a problem. Although they are usually academics themselves, journal editors have their own interests and agendas, and have more in common with everyday journalists and newspaper editors than some of them might wish to admit, as the episode of the precognition experiment above illustrates very clearly. Whether journals like this are a sensible model for communicating research at all is a hotly debated subject in academia, but this is the current situation. Journals are the gatekeepers, they make decisions on what's relevant and interesting for their audience, and they compete for readers.

This can lead them to behave in ways that don't reflect the best interests of science, because an individual journal's desire to provide colourful content might conflict with the collective need to provide a comprehensive picture of the evidence. In newspaper journalism, there is a well-known aphorism: 'When a dog bites a man, that's not news; but when a man bites a dog . . . ' These judgements on newsworthiness in mainstream media have even been demonstrated quantitatively. One study in 2003, for example, looked at the BBC's health news coverage over several months, and calculated how many people had to die from a given cause for one story to appear. 8,571 people died from smoking for each story about smoking; but there were three stories for every death from new variant CJD, or 'mad cow disease'.[33] Another, in 1992, looked at print-media coverage of drug deaths, and found that you needed 265 deaths from paracetamol poisoning for one story about such a death to appear in a paper; but every death from MDMA received, on average, one piece of news coverage.[34]

If similar judgements are influencing the content of academic journals, then we have a problem. But can it really be the case that academic journals are the bottleneck, preventing doctors and academics from having access to unflattering trial results about the safety and effectiveness of the drugs they use? This argument is commonly deployed by industry, and researchers too are often keen to blame journals for rejecting negative findings en masse. Luckily, this has been the subject of some research; and overall, while journals aren't blameless, it's hard to claim that they are the main source of this serious public-health problem. This is especially so since there are whole academic journals dedicated to publishing clinical trials, with a commitment to publishing negative results written into their constitutions.

But to be kind, for the sake of completeness, and because industry and researchers are so keen to pass the blame on to academic journals, we can see if what they claim is true.

One survey simply asked the authors of unpublished work if they had ever submitted it for publication. One hundred and twenty-four unpublished results were identified, by following up on every study approved by a group of US ethics committees, and when the researchers contacted the teams behind the unpublished results, it turned out that only six papers had ever actually been submitted and rejected.[35] Perhaps, you might say, this was a freak finding. Another approach is to follow up all the papers submitted to one journal, and see if those with negative results are rejected more often. Where this has been tried, the journals seem blameless: 745 manuscripts submitted to the *Journal of the American Medical Association (JAMA)* were followed up, and there was no difference in acceptance rate for significant and non-significant findings.[36] The same thing has been tried with papers submitted to the *BMJ*, the *Lancet, Annals of Internal Medicine* and the *Journal of Bone and Joint Surgery.*[37] Again and again, no effect was found. Some have argued that this might still represent evidence of editorial bias if academics know that manuscripts with negative results have to be of higher quality before submission, to get past editors' prejudices. It's also possible that the journals played fair when they knew they were being watched, although turning around an entire publishing operation for one brief performance would be tough.

These studies all involved observing what has happened in normal practice. One last option is to run an experiment, sending identical papers to various journals, but changing the direction of the results at random, to see if that makes any difference to the acceptance rates. This isn't something you'd want to do very often, because it wastes a lot of people's time; but since publication bias matters, it has been regarded as a justifiable intrusion on a few occasions.

In 1990 a researcher called Epstein created a series of fictitious papers, with identical methods and presentation, differing only in whether they reported positive or negative results. He sent them at random to 146 social-work journals: the positive papers were accepted 35 per cent of the time, and the negative ones 26 per cent of the time, a difference that wasn't large enough to be statistically significant.[38]

Other studies have tried something similar on a smaller scale, not submitting a paper to a journal, but rather, with the assistance of the journal, sending spoof academic papers to individual peer reviewers: these people do not make the final decision on publication, but they do give advice to editors, so a window into their behaviour would be useful. These studies have had more mixed results. In one from 1977, sham papers with identical methods but different results were sent to seventy-five reviewers. Some bias was found from reviewers against findings that disagreed with their own views.[39]

Another study, from 1994, looked at reviewers' responses to a paper on TENS machines: these are fairly controversial devices sold for pain relief. Thirty-three reviewers with strong views one way or the other were identified, and again it was found that their judgements on the paper were broadly correlated with their pre-existing views, though the study was small.[40] Another paper did the same thing with papers on quack treatments; it found that the direction of findings had no effect on reviewers from mainstream medical journals deciding whether to accept them.[41]

One final randomised trial from 2010 tried on a grand scale to see if reviewers really do reject ideas based on their pre-existing beliefs (a good indicator of whether journals are biased by results, when they should be focused simply on whether a study is properly designed and conducted). Fabricated papers were sent to over two hundred reviewers, and they were all identical, except for the results they reported: half of the reviewers got results they would like, half got results they wouldn't. Reviewers were more likely to recommend publication if they received the version of the manuscript with results they'd like

(97 per cent vs 80 per cent), more likely to detect errors in a manuscript whose results they didn't like, and rated the methods more highly in papers whose results they liked.[42]

Overall, though, even if there are clearly rough edges in some domains, these results don't suggest that the journals are the main cause of the problem of the disappearance of negative trials. In the experiments isolating the peer reviewers, those individual referees were biased in some studies, but they don't have the last word on publication, and in all the studies which look at what happens to negative papers submitted to journals in the real world, the evidence suggests that they proceed into print without problems. Journals may not be entirely innocent – *The Trouble with Medical Journals* by Richard Smith, previously editor of the *BMJ*, is an excellent overview of their faults;[43] and we will see later how they fail to police important and simple flaws in studies. But it would be wrong to lay the blame for publication bias entirely at their door.

In the light of all this, the data on what researchers say about their own behaviour is very revealing. In various surveys they have said that they thought there was no point in submitting negative results, because they would just be rejected by journals: 20 per cent of medical researchers said so in 1998;[44] 61 per cent of psychology and education researchers said so in 1991;[45] and so on.[46] If asked why they've failed to send in research for publication, the most common reasons researchers give are negative results, a lack of interest, or a lack of time.

This is the more abstract end of academia – largely away from the immediate world of clinical trials – but it seems that academics are mistaken, at best, about the reasons why negative results go missing. Journals may pose some barriers to publishing negative results, but they are hardly absolute, and much of the problem lies in academics' motivations and perceptions.

More than that, in recent years, the era of open-access academic journals has got going in earnest: there are now several, such as *Trials*, which are free to access, and have a core editorial policy that they will accept any trial report, regardless of result, and will actively solicit negative findings. Trial registers like clinicaltrials.gov will also post results. With offers like this on the table, it is very hard to believe that anyone would really struggle to disseminate the findings of a trial with a negative result if they wanted to. And yet, despite this, negative results continue to go missing, with vast multinational companies simply withholding results on their drugs, even though academics and doctors are desperate to see them.

You might reasonably wonder whether there are people who are supposed to prevent this kind of data from being withheld. The universities where research takes place, for example; or the regulators; or the 'ethics committees', which are charged with protecting patients who participate in research. Unfortunately, our story is about to take a turn to the dark side. We will see that many of the very people and organisations we would have expected to protect patients from the harm inflicted by missing data have, instead, shirked their responsibilities; and worse than that, we will see that many of them have actively conspired in helping companies to withhold data from patients. We are about to hit some big problems, some bad people, and some simple solutions.

How ethics committees and universities have failed us

By now, you will, I hope, share my view that withholding results from clinical trials is unethical, for the simple reason that hidden data exposes patients to unnecessary and avoidable harm. But the ethical transgressions here go beyond the simple harm inflicted on future patients.

Patients and the public participate in clinical trials at some considerable cost to themselves: they expose themselves to hassle and intrusion, because clinical trials almost always require that you have more check-ups on your progress, more blood tests, and more examinations; but participants may also expose themselves to more risk, or the chance of receiving an inferior treatment. People do this out of altruism, on the implicit understanding that the results from their experience will contribute to improving our knowledge of what works and what doesn't, and so will help other patients in the future. In fact, this understanding isn't just implicit: in many trials it's explicit, because patients are specifically told when they sign up to participate that the data will be used to inform future decisions. If this isn't true, and the data can be withheld at the whim of a researcher or a company, then the patients have been actively lied to. That is very bad news.

So what are the formal arrangements between patients, researchers and sponsors? In any sensible world, we'd expect universal contracts, making it clear that all researchers are obliged to publish their results, and that industry sponsors – which have a huge interest in positive results – must have no control over the data. But despite everything we know about industry-funded research being systematically biased, this does not happen. In fact, quite the opposite is true: it is entirely normal for researchers and academics conducting industry-funded trials to sign contracts subjecting them to gagging clauses which forbid them to publish, discuss or analyse data from the trials they have conducted, without the permission of the funder. This is such a secretive and shameful situation that even trying to document it in public can be a fraught business, as we shall now see.

In 2006 a paper was published in *JAMA* describing how common it was for researchers doing industry-funded trials to have these kinds of constraints placed on their right to publish the results.[47] The study was conducted by the Nordic Cochrane

Centre, and it looked at all the trials given approval to go ahead in Copenhagen and Frederiksberg. (If you're wondering why these two places were chosen, it was simply a matter of practicality, and the bizarre secrecy that shrouds this world: the researchers applied elsewhere without success, and were specifically refused access to data in the UK, despite being British researchers based in Oxford University.[48]) These trials were overwhelmingly sponsored by the pharmaceutical industry (98 per cent), and the rules governing the management of the results tell a familiar story.

For sixteen of the forty-four trials the sponsoring company got to see the data as it accumulated, and in a further sixteen they had the right to stop the trial at any time, for any reason. This means that a company can see if a trial is going against it, and can interfere as it progresses. As we will see later (early stopping, breaking protocols, pp. 184, 200), this distorts a trial's results with unnecessary and hidden biases. For example, if you stop a trial early because you have been peeking at the preliminary results, then you can either exaggerate a modest benefit, or bury a worsening negative result. Crucially, the fact that the sponsoring company had this opportunity to introduce bias wasn't mentioned in any of the published academic papers reporting the results of these trials, so nobody reading the literature could possibly know that these studies were subject – by design – to such an important flaw.

Even if the study was allowed to finish, the data could still be suppressed. There were constraints on publication rights in forty of the forty-four trials, and in half of them the contracts specifically stated that the sponsor either owned the data outright (what about the patients, you might say?), or needed to approve the final publication, or both. None of these restrictions was mentioned in any of the published papers, and in fact, none of the protocols or papers said that the sponsor had full access to all the data from the trial, or the final say on whether to publish.

It's worth taking a moment to think about what this means. The results of all these trials were subject to a bias that will significantly distort the academic literature, because trials that show early signs of producing a negative result (or trials that do produce a negative result) can be deleted from the academic record; but nobody reading these trials could possibly have known that this opportunity for censorship existed.

The paper I've just described was published in *JAMA*, one of the biggest medical journals in the world. Shortly afterwards, a shocking tale of industry interference appeared in the *BMJ*.[49] Lif, the Danish pharmaceutical industry association, responded to the paper by announcing in the *Journal of the Danish Medical Association* that it was 'both shaken and enraged about the criticism, that could not be recognised'. It demanded an investigation of the scientists, though it failed to say by whom, or of what. Then Lif wrote to the Danish Committee on Scientific Dishonesty, accusing the Cochrane researchers of scientific misconduct. We can't see the letter, but the Cochrane researchers say the allegations were extremely serious – they were accused of deliberately distorting the data – but vague, and without documents or evidence to back them up.

Nonetheless, the investigation went on for a year, because in academia people like to do things properly, and assume that all complaints are made in good faith. Peter Gøtzsche, the director of the Cochrane centre, told the *BMJ* that only Lif's third letter, ten months into this process, made specific allegations that could be investigated by the committee. Two months later the charges were dismissed. The Cochrane researchers had done nothing wrong. But before they were cleared, Lif copied the letters alleging scientific dishonesty to the hospital where four of them worked, and to the management organisation running that hospital, and sent similar letters to the Danish Medical Association, the Ministry of Health, the Ministry of Science, and so on. Gøtzsche and his colleagues said that they felt 'intimidated and harassed' by Lif's behaviour. Lif continued to insist that the researchers were guilty of misconduct even after the investigation was completed. So, researching in this area is not easy: it's hard to get funding, and the industry will make your work feel like chewing on a mouthful of wasps.

Even though the problem has been widely recognised, attempts to fix it have failed.[50] The International Committee of Medical Journal Editors, for example, stood up in 2001, insisting that the lead author of any study it published must sign a document stating that the researchers had full access to the data, and full control over the decision to publish. Researchers at Duke University, North Carolina, then surveyed the contracts between medical schools and industry sponsors, and found that this edict was flouted as a matter of routine. They recommended boilerplate contracts for the relationship between industry and academia. Was this imposed? No. Sponsors continue to control the data.

Half a decade later, a major study in the *New England Journal of Medicine* investigated whether anything had changed.[51] Administrators at all 122 accredited medical schools in the US were asked about their contracts (to be clear, this wasn't a study of what they *did*; rather it was a study of what they were willing to say in public). The majority said contract negotiations over the right to publish data were 'difficult'. A worrying 62 per cent said it was OK for the clinical trial agreement between academics and industry sponsor to be confidential. This is a serious problem, as it means that anyone reading a study cannot know how much interference was available to the sponsor, which is important context for the person reading and interpreting the research. Half of the centres allowed the sponsor to draft the research paper, which is another interesting hidden problem in medicine, as biases and emphases can be quietly introduced (as we shall see in more detail in Chapter 6). Half said it was OK for contracts to forbid researchers sharing data after the research was completed and published, once again hindering research. A quarter said it was acceptable for the sponsor to insert its own statistical analyses into the manuscript. Asked about disputes, 17 per cent of administrators had seen an argument about who had control of data in the preceding year.

Sometimes, disputes over access to such data can cause serious problems in academic departments, when there is a divergence of views on what is ethical. Aubrey Blumsohn was a senior lecturer at Sheffield University, working on a project funded by Procter & Gamble to research an osteoporosis drug called risedronate.[52] The aim was to analyse blood and urine samples from an earlier trial, led by Blumsohn's head of department, Professor Richard Eastell. After signing the contracts, P&G sent over some 'abstracts', brief summaries of the findings, with Blumsohn's name as first author, and some summary results tables. That's great, said Blumsohn, but I'm the researcher here: I'd like to see the actual raw data and analyse it myself. The company declined, saying that this was not their policy. Blumsohn stood his ground, and the papers were left unpublished. Then, however, Blumsohn saw that Eastell had published another paper with P&G, stating that all the researchers had 'had full access to the data and analyses'. He complained, knowing this was not true. Blumsohn was suspended by Sheffield University, which offered him a gagging clause and £145,000, and he was eventually forced out of his job. Eastell, meanwhile, was censured by the General Medical Council, but only after a staggering delay of several years, and he remains in post.

So contracts that permit companies and researchers to withhold or control data are common, and they're bad news. But that's not just because they lead to doctors and patients being misled about what works. They also break another vitally important contract: the agreement between researchers and the patients who participate in their trials.

People participate in trials believing that the results of that research will help to improve the treatment of patients like them in the future. This isn't just speculation: one of the few studies to ask patients why they have participated in a trial found that 90 per cent believed they were making a 'major' or 'moderate' contribution to society, and 84 per cent felt proud that they were making this contribution.[53] Patients are not stupid or naïve to believe this, because it is what they are told on the consent forms they sign before participating in trials. But they are mistaken, because the results of trials are frequently left unpublished, and withheld from doctors and patients. These signed consent forms therefore mislead people on two vitally important points. First, they fail to tell the truth: that the person conducting the trial, or the person paying for it, may decide not to publish the results, depending on how they look at the end of the study. And worse than that, they also explicitly state a falsehood: researchers tell patients that they are participating in order to create knowledge that will be used to improve treatment in future, even though the researchers know that in many cases, those results will never be published.

There is only one conclusion that we can draw from this: consent forms routinely lie to patients participating in trials. This is an extraordinary state of affairs, made all the more extraordinary by the huge amounts of red tape that everyone involved in a trial must negotiate, closely monitoring endless arcane pieces of paperwork, and ensuring that patients are fully informed on the minutiae of their treatment. Despite all this regulatory theatre, which hinders good research on routine practice (as we shall see – p. 230), we allow these forms to tell patients outright lies about the control of data, and we fail to police one of the most important ethical problems in the whole of medicine. The deceit of these consent forms is, to me, a good illustration of how broken and outdated the regulatory frameworks of medicine have become. But it also, finally, poses a serious problem for the future of research.

We desperately need patients to continue to believe that they are contributing to society, because trial recruitment is in crisis, and it is increasingly hard to persuade patients to participate at all. In one recent study, a third of all trials failed to reach their original recruitment target, and more than half had to be awarded an extension.[54] If word gets out that trials are often more promotional than genuinely scientific, recruitment will become even more difficult. The answer is not to hide this problem, but rather to fix it.

So universities and ethics committees may have failed us, but there is one group of people we might expect to step up, to try to show some leadership on missing trial data. These are the medical and academic professional bodies, the associations, societies, academies, and royal colleges representing doctors and scientists around the world.

These organisations have the opportunity to set the tone of academic and clinical medicine, in their codes of conduct, their aspirations, and in some cases their rules, since some have the ability to impose sanctions, and all have the ability to exclude those who fail to meet basic ethical standards. We have established, I hope, beyond any doubt, that non-publication of trials in humans is research misconduct, that it misleads doctors and harms patients around the world. Have these organisations used their powers, stood up and announced, prominently and fiercely, that this must stop, and that they will take action?

One has: the Faculty of Pharmaceutical Medicine, a small organisation of doctors – largely working in industry – with 1,400 members in the UK. If you know of any others, in any country on the globe, or if you can lobby just one more to make a positive statement on this widespread problem, I would like to hear about it.

What can be done?

There are several simple solutions to these problems, which fall into three categories. There is no argument against any of the following suggestions that I am aware of. The issue of missing data has been neglected through institutional inertia, and reluctance by senior academics to challenge industry. Their failure to act harms patients every day.

Gagging clauses

If there is nothing to be ashamed of in gagging clauses – if companies, and legislators, and academics, and university contracts departments, all believe that they are acceptable – then everything should be out in the open, prominently flagged up, so that everyone outside those systems can decide if they agree.

1. Until they can be eradicated, where gagging clauses exist, patients should be told, at the time they are invited to participate in a trial, that the company sponsoring it is allowed to hide the results if it doesn't like them. The consent form should also explain clearly that withholding negative results will lead to doctors and patients being misled about the effectiveness of the treatment being trialled, and so exposed to unnecessary harm. Trial participants can then decide for themselves if they think these contracts are acceptable.
2. Wherever the sponsoring company has the contractual right to gag publication, even if it doesn't exercise that right, the fact that a gagging clause existed should be stated clearly: in the academic paper; in the trial protocol; and in the publicly available trial registry entry that goes up before the trial starts. Readers of the trial findings can then decide for themselves if they trust that

sponsor and research group to publish all negative findings, and interpret any reported positive findings in that light.

3. All university contracts should follow the same boilerplate format, and forbid gagging clauses. Failing that, all universities should be forced to prominently declare which contracts with gagging clauses they have permitted, and to publish those clauses online for all to see, so that all can be alerted that the institution is producing systematically biased research, and discount any findings from them accordingly.

4. In legislation, gagging clauses should be made illegal, with no possibility of quibbles. If there is a dispute about analysis or interpretation between the people running the trial and the people paying for it, it should take place in the published academic literature, or some other public forum. Not in secret.

Professional bodies

1. All professional bodies should take a clear stand on unpublished trial data, declare it clearly as research misconduct, and state that it will be handled like any other form of serious misconduct that misleads doctors and harms patients. That they have not done so is a stain on the reputation of these organisations, and on their senior members.

2. All research staff involved in any trial on humans should be regarded as jointly and severally liable for ensuring that it is published in full within one year of completion.

3. All those responsible for withholding trial data should have their names prominently posted in a single database, so that others can be made aware of the risk of working with them, or allowing them access to patients for research, in future.

Ethics committees

1. No person should be allowed to conduct trials in humans if a research project they are responsible for is currently withholding trial data from publication more than one year after completion. Where any researcher on a project has a previous track record of delayed publication of trial data, the ethics committee should be notified, and this should be taken into account, as with any researcher guilty of research misconduct.
2. No trial should be approved without a firm guarantee of publication within one year of completion.

How regulators and journals have failed us

So far we've established that ethics committees, universities and the professional bodies of medical researchers have all failed to protect patients from publication bias, even though the selective non-publication of unflattering data is a serious issue for medicine. We know that it distorts and undermines every decision that researchers, doctors and patients make, and that it exposes patients to avoidable suffering and death. This is not seriously disputed, so you might reasonably imagine that governments, regulators and medical journals must all have tried to address it.

They have tried, and failed. Worse than simply failing, they have repeatedly provided what we might regard as 'fake fixes': we have seen small changes in regulations and practices, announced with great fanfare, but then either ignored or bypassed. This has given false reassurance to doctors, academics and the public that the problem has been fixed. What follows is the story of these fake fixes.

Registers

The earliest and simplest solution proposed was to open registers of trials: if people are compelled to publish their protocol, in full, before they start work, then we at least have the opportunity to go back and check to see if they've published the trials that they've conducted. This is very useful, for a number of reasons. A trial protocol describes in great technical detail everything that researchers will do in a trial: how many patients they'll recruit, where they'll come from, how they'll be divided up, what treatment each group will get, and what outcome will be measured to establish if the treatment was successful. Because of this, it can be used to check whether a trial was published, but also whether its methods were distorted along the way, in a manner that would allow the results to be exaggerated (as described in Chapter 4).

The first major paper to call for a registry of clinical trial protocols was published in 1986,[55] and it was followed by a flood. In 1990 Iain Chalmers (we can call him Sir Iain Chalmers if you like*) published a classic paper called 'Underreporting

---

* Iain Chalmers was knighted for setting up the Cochrane Collaboration. Being very practical people, the researchers at Cochrane wanted to know if there was any real value in this, so they ran a randomised trial. Were recipients of letters from 'Iain Chalmers' more or less likely to respond, they wondered, if he signed his name 'Sir Iain Chalmers'? A simple system was set up, and just before they were posted, outgoing letters were randomly signed 'Sir Iain Chalmers' or just 'Iain Chalmers'. The researchers then compared the number of replies to each signature: and the 'Sir' made no difference at all. This study is published in full – despite reporting a negative result – in the *Journal of the Royal Society of Medicine*, and it is not a flippant subject for research. There are many knights in medicine; there are troubling things you can do to increase your chance of becoming one; and lots of people think, 'If I was a knight, people would take my good ideas much more seriously.' The paper is titled 'Yes Sir, No Sir, Not Much Difference Sir'.[56] After reading it, you can relax your ambitions.

Research Is Scientific Misconduct',[57] and he has traced the chequered history of trials registers in the UK.[58] In 1992, as the Cochrane Collaboration began to gather influence, representatives of the Association of the British Pharmaceutical Industry (ABPI) asked to meet Chalmers.[59] After explaining the work of Cochrane, and the vital importance of summarising all the trial results on a particular drug, he explained very clearly to them how biased under-reporting of results harms patients.

The industry's representatives were moved, and soon they took action. Mike Wallace, the chief executive of Schering and a member of that ABPI delegation, agreed with Chalmers that withholding data was ethically and scientifically indefensible, and said that he planned to do something concrete to prevent it, if only to protect the industry from having the issue forced upon it in less welcome terms. Wallace stepped out of line from his colleagues, and committed to registering every trial conducted by his company with Cochrane. This was not a popular move, and he was reprimanded by colleagues, in particular those from other companies.

But soon GlaxoWellcome followed suit, and in 1998 its chief executive, Richard Sykes, wrote an editorial in the *BMJ* called 'Being a modern pharmaceutical company involves making information available on clinical trial programmes'.[60] 'Programmes' was the crucial word, because as we've seen, and as we shall see in greater detail later, you can only make sense of individual findings if you assess them in the context of all the work that has been done on a drug.

GlaxoWellcome set up a clinical trials registry, and Elizabeth Wager, the head of the company's medical writers group, pulled together a group from across the industry to develop ethical guidelines for presenting research. The ABPI, seeing individual companies take the lead, saw the writing on the wall: it decided to commend GlaxoWellcome's policy to the whole industry, and launched this initiative at a press conference where

Chalmers – a strong critic – sat on the same side of the table as the industry. AstraZeneca, Aventis, MSD, Novartis, Roche, Schering Healthcare and Wyeth began registering some of their trials – only the ones involving UK patients, and retrospectively – but there was movement at last.

At the same time, there was movement in America. The 1997 FDA Modernization Act created clinicaltrials.gov, a register run by the US government National Institutes of Health. This legislation required that trials should be registered, but only if they related to an application to put a new drug on the market, and even then, only if it was for a serious or life-threatening disease. The register opened in 1998, and the website clinicaltrials.gov went online in 2000. The entry criteria were widened in 2004.

But soon it all began to fall apart. GlaxoWellcome merged with SmithKline Beecham to become GlaxoSmithKline (GSK), and initially the new logo appeared on the old trials register. Iain Chalmers wrote to Jean-Paul Garnier, the chief executive of the new company, to thank him for maintaining this valuable transparency: but no reply ever came. The registry website was closed, and the contents were lost (though GSK was later forced to open a new register, as part of a settlement with the US government over the harm caused by its withholding of data on new drug trials just a couple of years later). Elizabeth Wager, the author of the Good Publication Practice guidelines for drug companies, was out of a job, as her writing department at GSK was closed. Her guidelines were ignored.[61]

From the moment that these registries were first suggested, and then opened, it was implicitly assumed that the shame of producing this public record, and then not publishing your study, would be enough to ensure that people would do the right thing. But the first problem for the US register, which could have been used universally, was that people simply chose not to use it. The regulations required only a very narrow range of trials to be posted, and nobody else was in a hurry to post their trials if they didn't have to.

In 2004 the International Committee of Medical Journal Editors (ICMJE) – a collection of editors from the most influential journals in the world – published a policy statement, announcing that none of them would publish any clinical trials after 2005, unless they had been properly registered before they began.[62] They did this, essentially, to force the hand of the industry and researchers: if a trial has a positive result, then people desperately want to publish it in the most prestigious journal they can find. Although they had no legal force, the journal editors did have the thing that companies and researchers wanted most: the chance of a major journal publication. By insisting on pre-registration, they were doing what they could to force researchers and industry sponsors to register all trials. Everyone rejoiced: the problem had been fixed.

If you think it seems odd – and perhaps unrealistic – that fixing this crucial flaw in the information architecture of a $700 billion industry should be left to an informal gathering of a few academic editors, with no legislative power, then you'd be right. Although everybody began to talk as if publication bias was a thing of the past, in reality it was continuing just as before, because the journal editors simply ignored their own threats and promises. Later (pp. 245, 305) we will see the phenomenal financial inducements on offer to editors for publishing positive industry papers, which can extend to millions of dollars in reprint and advertising revenue. But first we should look at what they actually did after their solemn promise in 2005.

In 2008 a group of researchers went through every single trial published in the top ten medical journals, every one of which was a member of the ICMJE, after the deadline for pre-registration. Out of 323 trials published during 2008 in these high-impact academic journals, only half were adequately registered (before the trial, with the main outcome measure properly specified), and trial registration was entirely lacking for over a quarter.[63] The ICMJE editors had simply failed to keep their word.

Meanwhile, in Europe, there were some very bizarre developments. With enormous fanfare, the European Medicines Agency created a registry of trials called EudraCT. EU legislation requires all trials to be posted here if they involve any patients in Europe, and many companies will tell you that they've met their responsibilities for transparency by doing so. But the contents of this EU register have been kept entirely secret. I can tell you that it contains around 30,000 trials, since that figure is in the public domain, but that is literally all I know, and all anyone can know. Despite EU legislation requiring that the public should be given access to the contents of this register, it remains closed. This creates an almost laughable paradox: the EU clinical trials register is a transparency tool, held entirely in secret. Since March 2011, after heavy media criticism (from me at any rate), a subset of trials has slowly been made public through a website called EudraPharm. As of summer 2012, although the agency now claims that its register is accessible to all, at least 10,000 trials are still missing, and the search engine doesn't work properly.[64] It's absolutely one of the strangest things I've ever seen, and nobody other than the EU even regards this peculiar exercise as a trials register: I certainly don't, I doubt you do, and both the ICMJE and the World Health Organization have explicitly stated that EudraCT is not a meaningful register.

But new work was being done in the US, and it seemed sensible. In 2007 the FDA Amendment Act was passed. This is much tighter: it requires registration of all trials of any drug or device, at any stage of development other than 'first-in-man' tests, if they have any site in the US, or involve any kind of application to bring a new drug onto the market. It also imposes a startling new requirement: all results of all trials must be posted to clinicaltrials.gov, in abbreviated summary tables, within one year of completion, for any trial on any *marketed* drug that completes after 2007.

Once again, to great fanfare, everyone believed that the problem had been fixed. But it hasn't been, for two very important reasons.

First, unfortunately, despite the undoubted goodwill, requiring the publication of all trials starting from 'now' does absolutely nothing for medicine today. There is no imaginable clinic, anywhere in the world, at which medicine is practised only on the basis of trials that completed within the past three years, using only drugs that came to market since 2008. In fact, quite the opposite is true: the vast majority of drugs currently in use came to market over the past ten, twenty or thirty years, and one of the great challenges for the pharmaceutical industry today is to create drugs that are anything like as innovative as those that were introduced in what has come to be known as the 'golden era' of pharmaceutical research, when all the most widely used drugs, for all the most common diseases, were developed. Perhaps they were the 'low-lying fruit', plucked from the research tree, but in any case, these are the tablets we use.

And crucially, it is for these drugs – the ones we actually use – that we need the evidence: from trials completed in 2005 or 1995. These are the drugs that we are prescribing completely blind, misled by a biased sample of trials, selectively published, with the unflattering data buried in secure underground data archives somewhere in the hills (I am told) of Cheshire.

But there is a second, more disturbing reason why these regulations should be taken with a pinch of salt: they have been widely ignored. A study published in January 2012 looked at the first slice of trials subject to mandatory reporting, and found that only one in five had met its obligation to post results.[65] Perhaps this is not surprising: the fine for non-compliance is $10,000 a day, which sounds spectacular, until you realise that it's only $3.5 million a year, which is chickenfeed for a drug bringing in $4 billion a year. And what's more, no such fine has ever been levied, throughout the entire history of the legislation.

So that, in total, is why I regard the ICMJE, the FDA and the EU's claims to have addressed this problem as 'fake fixes'. In fact, they have done worse than fail: they have given false reassurance that the problem has been fixed, false reassurance that it has gone away, and they have led us to take our eyes off the ball. For half a decade now, people in medicine and academia have talked about publication bias as if it was yesterday's problem, discovered in the 1990s and early 2000s, and swiftly fixed.

But the problem of missing data has not gone away, and soon we will see exactly how shameless some companies and regulators can be, in the very present day.

What can be done?

The ICMJE should keep its promises, the EU should be less risible, and the FDA Amendment Act 2007 should be muscularly enforced. But there are many more disappointments to come, so I will leave my action plan on missing data for later in this chapter.

Blood from a stone: trying to get data from regulators

So far we've established that doctors and patients have been failed by a range of different people and organisations, all of whom we might have expected to step up and fix the problem of missing data, since it harms patients in large numbers around the world. We have seen that governments take no action against those who fail to publish their results, despite the public pretence to the contrary; and that they take no action against those who fail to register their trials. We have seen that medical journal editors continue to publish unregistered trials, despite the public pretence that they have taken a stand. We have seen that ethics committees fail to insist on universal publication, despite their stated aim of protecting patients. And we have seen that professional bodies fail to take action against what is obviously research misconduct, despite evidence showing that the problem of missing data is of epidemic proportions.[66]

While the published academic record is hopelessly distorted, you might hope that there is one final route which patients and doctors could use to get access to the results of clinical trials: the regulators, which receive large amounts of data from drug companies during the approval process, must surely have obligations to protect patients' safety? But this, sadly, is just one more example of how we are failed by the very bodies that are supposed to be protecting us.

In this section, we will see three key failures. First, the regulators may not have the information in the first place. Second, the way in which they 'share' summary trial information with doctors and patients is broken and shabby. And finally, if you try to get all of the information that a drug company has provided – the long-form documents, where the bodies are often buried – then regulators present bizarre barriers, blocking and obfuscating for several years at a time, even on drugs that turn out to be ineffective and harmful. Nothing of what I am about to tell you is in any sense reassuring.

One: Information is withheld from regulators

Paroxetine is a commonly used antidepressant, from the class of drugs known as 'selective serotonin reuptake inhibitors', or SSRIs. You will hear more about this class of drugs later in this book, but here we will use paroxetine to show how companies have exploited our longstanding permissiveness about missing trials, and found loopholes in our inadequate regulations on trial disclosure. We will see that GSK withheld data about whether paroxetine works as an antidepressant, and even withheld data about its harmful side effects, but most important, we will see that what it did was all entirely legal.

To understand why, we first need to go through a quirk of the licensing process. Drugs do not simply come onto the market for use in all medical conditions: for any specific use of any drug, in any specific disease, you need a separate marketing authorisation. So, a drug might be licensed to treat ovarian cancer, for example, but not breast cancer. That doesn't mean the drug doesn't work in breast cancer. There might well be some evidence that it's great for treating that disease too, but maybe the company hasn't gone to the trouble and expense of getting a formal marketing authorisation for that specific use. Doctors can still go ahead and prescribe it for breast cancer, if they want, because the drug is available for prescription, and there are boxes of it sitting in pharmacies waiting to go out (even though, strictly speaking, it's only got marketing approval for use in ovarian cancer). In this situation the doctor will be prescribing the drug legally, but 'off-label'.

This is fairly common, as getting a marketing authorisation for a specific use can be time-consuming and expensive. If doctors know that there's a drug which has been shown in good-quality trials to help treat a disease, it would be perverse and unhelpful of them not to prescribe it, just because the company hasn't applied for a formal licence to market it for that specific use. I'll discuss the ins and outs of all this in more detail later. But for now, what you need to know is that the use of a drug in children is treated as a separate marketing authorisation from its use in adults.

This makes sense in many cases, because children can respond to drugs in very different ways to adults, so the risks and benefits might be very different, and research needs to be done in children separately. But this licensing quirk also brings some disadvantages. Getting a licence for a specific use is an arduous business, requiring lots of paperwork, and some specific studies. Often this will be so expensive that companies will not bother to get a licence specifically to market a drug for use in children, because that market is usually much smaller.

But once a drug is available in a country for one specific thing, as we have seen, it can then be prescribed for absolutely anything. So it is not unusual for a drug to be licensed for use in adults, but then prescribed for children on the back of a hunch; or a judgement that it should at least do no harm; or studies that suggest benefit in children, but that would probably be insufficient to get through the specific formal process of getting marketing authorisation for use in kids; or even good studies, but in a disease where the market is so small that the company can't be bothered to get a marketing approval for use in children.

Regulators have recognised that there is a serious problem with drugs being used in children 'off-label', without adequate research, so recently they have started to offer incentives for companies to conduct the research, and formally seek these licences. The incentives are patent extensions, and these can be lucrative. All drugs slip into the public domain about a decade after coming onto the market, and become like paracetamol, which anyone can make very cheaply. If a company is given a six-month extension on a drug, for all uses, then it can make a lot more money from that medicine. This seems a good example of regulators being pragmatic, and thinking creatively about what carrots they can offer. Licensed use in children will probably not itself make a company much extra money, since doctors are prescribing the drug for children already, even without a licence or good evidence, simply because there are no other options. Meanwhile, six months of extra patent life for a blockbuster drug will be very lucrative, if its adult market is large enough.

There's a lot of debate about whether the drug companies have played fair with these offers. For example, since the FDA started offering this deal, about a hundred drugs have been given paediatric licences, but many of them were for diseases that aren't very common in children, like stomach ulcers, or arthritis. There have been far fewer applications for less lucrative products that could be used in children, such as more modern medicines called 'large-molecule biologics'. But there it is.

When GSK applied for a marketing authorisation in children for paroxetine, an extraordinary situation came to light, triggering the longest investigation in the history of UK drugs regulation. This investigation was published in 2008, and examined whether GSK should face criminal charges.[67] It turned out that what the company had done – withholding important data about safety and effectiveness that doctors and patients clearly needed to see – was plainly unethical, and put children around the world at risk; but our laws are so weak that GSK could not be charged with any crime.

Between 1994 and 2002 GSK conducted nine trials of paroxetine in children.[68] The first two failed to show any benefit, but the company made no attempt to inform anyone of this by changing the 'drug label' that is sent to all doctors and patients. In fact, after these trials were completed, an internal company management document stated: 'it would be commercially unacceptable to include a statement that efficacy had not been demonstrated, as this would undermine the profile of paroxetine'. In the year after this secret internal memo, 32,000 prescriptions were issued to children for paroxetine in the UK alone: so, while the company knew the drug didn't work in children, it was in no hurry to tell doctors that, despite knowing that large numbers of children were taking it. More trials were conducted over the coming years – nine in total – and none showed that the drug was effective at treating depression in children.

It gets much worse than that. These children weren't simply receiving a drug that the company knew to be ineffective for them: they were also being exposed to side effects. This should be self-evident, since any effective treatment will have some side effects, and doctors factor this in, alongside the benefits (which in this case were non-existent). But nobody knew how bad these side effects were, because the company didn't tell doctors, or patients, or even the regulator about the worrying safety data from its trials. This was because of a loophole (on which more, in two pages' time): you only had to tell the regulator about side effects reported in studies looking at *the specific uses for which the drug has a marketing authorisation*. Because the use of paroxetine in children was 'off-label', GSK had no legal obligation at the time to tell anyone about what it had found.

People had worried for a long time that paroxetine might increase the risk of suicide, though that is quite a difficult side effect to detect in an antidepressant, because people with depression are at a much higher risk of suicide than the general population anyway, as a result of their depression. There are also some grounds to believe that as patients first come out of their depression, and leave behind the sluggish lack of motivation that often accompanies profound misery, there may be a period during which they are more capable of killing themselves, just because of the depression slowly lifting.

Furthermore, suicide is a mercifully rare event, which means you need a lot of people on a drug to detect an increased risk. Also, suicide is not always recorded accurately on death certificates, because coroners and doctors are reluctant to give a verdict that many would regard as shameful, so the signal you are trying to detect in the data – suicide – is going to be corrupted. Suicidal thoughts or behaviours that don't result in death are more common than suicide itself, so they should be easier to detect, but they too are hard to pick up in routinely collected data, because they're often not presented to doctors, and where they are, they can be coded in health records in all sorts of different ways, if they appear at all. Because of all these difficulties, you would want to have every scrap of data you could possibly cobble together on the question of whether these drugs cause suicidal thoughts or behaviour in children; and you would want a lot of experienced people, with a wide range of skills, all looking at the data and discussing it.

In February 2003, GSK spontaneously sent the MHRA a package of information on the risk of suicide on paroxetine, containing some analyses done in 2002 from adverse-event data in trials the company had held, going back a decade. This analysis showed that there was no increased risk of suicide. But it was misleading: although it was unclear at the time, data from trials in children had been mixed in with data from trials in adults, which had vastly greater numbers of participants. As a result, any sign of increased suicide risk among children on paroxetine had been completely diluted away.

Later in 2003, GSK had a meeting with the MHRA to discuss another issue involving paroxetine. At the end of this meeting, the GSK representatives gave out a briefing document, explaining that the company was planning to apply later that year for a specific marketing authorisation to use paroxetine in children. They mentioned, while handing out the document, that the MHRA might wish to bear in mind a safety concern the company had noted: an increased risk of suicide among children with depression who received paroxetine, compared with those on dummy placebo pills.

This was vitally important side-effect data, being presented, after an astonishing delay, casually, through an entirely inappropriate and unofficial channel. GSK knew that the drug was being prescribed in children, and it knew that there were safety concerns in children, but it had chosen not to reveal that information. When it did share the data, it didn't flag it up as a clear danger in the current use of the drug, requiring urgent attention from the relevant department in the regulator; instead it presented it as part of an informal briefing about a future application. Although the data was given to completely the wrong team, the MHRA staff present at this meeting had the wit to spot that this was an important new problem. A flurry of activity followed: analyses were done, and within one month a letter was sent to all doctors advising them not to prescribe paroxetine to patients under the age of eighteen.

How is it possible that our systems for getting data from companies are so poor that they can simply withhold vitally important information showing that a drug is not only ineffective, but actively dangerous? There are two sets of problems here: first, access for regulators; and second, access for doctors.

There is no doubt that the regulations contain ridiculous loopholes, and it's dismal to see how GSK cheerfully exploited them. As I've mentioned, the company had no legal duty to give over the information, because prescription of the drug in children was outside of paroxetine's formally licensed uses – even though GSK knew this was widespread. In fact, of the nine studies the company conducted, only one had its results reported to the MHRA, because that was the only one conducted in the UK.

After this episode, the MHRA and the EU changed some of their regulations, though not adequately, and created an obligation for companies to hand over safety data for uses of a drug outside its marketing authorisation, closing the paroxetine loophole.

This whole episode illustrates a key problem, and it is one that recurs throughout this section of the book: you need *all* of the data in order to see what's happening with a drug's benefits, and risks. Some of the trials that GSK conducted were published in part, but that is obviously not enough: we already know that if we see only a biased sample of the data, we are misled. But we also need all the data for the more simple reason that we need *lots* of data: safety signals are often weak, subtle and difficult to detect. Suicidal thoughts and plans are rare in children – even those with depression, even those on paroxetine – so all the data from a large number of participants needed to be combined before the signal was detectable in the noise. In the case of paroxetine, the dangers only became apparent when the adverse events from all of the trials were pooled and analysed together.

That leads us to the second obvious flaw in the current system: the results of these trials – the safety data and the effectiveness data – are given in secret to the regulator, which then sits and quietly makes a decision. This is a huge problem, because you need many eyes on these difficult problems. I don't think that the people who work in the MHRA are bad, or incompetent: I know a lot of them, and they are smart, good people. But we shouldn't trust them to analyse this data alone, in the same way that we shouldn't trust any single organisation to analyse data alone, with nobody looking over its shoulder, checking the working, providing competition, offering helpful criticism, speeding it up, and so on.

This is even worse than academics failing to share their primary research data, because at least in an academic paper you get a lot of detail about what was done, and how. The output of a regulator is often simply a crude, brief summary: almost a 'yes' or 'no' about side effects. This is the opposite of science, which is only reliable because everyone shows their working, explains *how they know* that something is effective or safe, shares their methods and their results, and allows others to decide if they agree with the way they processed and analysed the data.

Yet for the safety and efficacy of drugs, one of the most important of all analyses done by science, we turn our back on this process completely: we allow it to happen behind closed doors, because drug companies have decided that they want to share their trial results discretely with the regulators. So the most important job in evidence-based medicine, and a perfect example of a problem that benefits from many eyes and minds, is carried out alone and in secret.

This perverse and unhealthy secrecy extends way beyond regulators. NICE, the UK's National Institute for Health and Clinical Excellence, is charged with making recommendations about which treatments are most cost-effective, and which work best. When it does this, it's in the same boat as you or me: it has absolutely no statutory right to see data on the safety or effectiveness of a drug, if a company doesn't want to release it, even though the regulators have all of that data. For 'single technology appraisals', on one treatment, they ask the company to make available to them the information the company thinks is relevant. For 'guidelines' on treatment in a whole area of medicine, they are more vulnerable to what is published in journals. As a result, even NICE can end up working on distorted, edited, biased samples of the data.

Sometimes NICE is able to access some extra unpublished data from the drug companies: this is information that doctors and patients aren't allowed to see, despite the fact that they are the people making decisions about whether to prescribe the drugs, or are actually taking them. But when NICE does get information in this way, it can come with strict conditions on confidentiality, leading to some very bizarre documents being published. On the next page, for example, is the NICE document discussing whether it's a good idea to have Lucentis, an extremely expensive drug, costing well over £1,000 per treatment, that is injected into the eye for a condition called acute macular degeneration.

As you can see, the NICE document on whether this treatment is a good idea is censored. Not only is the data on the effectiveness of the treatment blanked out by thick black rectangles, in case any doctor or patient should see it; absurdly, even the names of some trials are missing, preventing the reader from even knowing of their existence, or cross referencing information about them. Most disturbing of all, as you can see in the last bullet point, the data on adverse events is also censored: I'm

- *Subgroup analysis:* In the MARINA, FOCUS and ANCHOR trials, the difference in the primary outcome between the ranibizumab groups and the comparator group was statistically significant for each lesion subgroup. In the reduced dose frequency PIER study, the difference between 0.5mg ranibizumab ███████████████████ versus sham injection ███████████████ was █████████████████ ████ for the group of patients with ████████

- *Contrast sensitivity:* Contrast sensitivity ████████ in the ranibizumab groups ██████████████████████████████████ ████████████ in the sham injection ██████████████ PDT groups ████████████████████ The reduced dose frequency PIER study found ███████████████████████████████ or ███████████████████ in contrast sensitivity compared with ███████████████████████████

- *Anatomical changes:* The MARINA, █████████ and ANCHOR trials demonstrated statistically significant differences between 0.3 mg or 0.5 mg ranibizumab and the comparator for the area of CNV, area of leakage from CNV plus intense progressive retinal pigment epithelium staining, or area of classic CNV.

- *Visual function questionnaire scores:* ████████████████ results were reported for ████████████ ranibizumab compared with sham injection █████████████████████ NEI VFQ-25 ██████████████████, ██████████████ and ██████████ A████████████████ in ████████ was found with ████████████████████████ when compared with PDT. There were ███████████████████████ between ranibizumab and sham injection in the reduced dose PIER study.

- *Adverse events:* Adverse events were common but most were mild to moderate. Serious ocular events were rare in the MARINA and ANCHOR trials. Incidences of severe ocular inflammation varied between treatment arms, and were highest in the 0.5mg ranibizumab groups. The rate of serious ocular adverse events was ██████████████████████████ in the ranibizumab plus PDT group compared with ████████████████ PDT. Endophthalmitis was reported by very few patients in the active treatment arms of the ranibizumab trials and none in the control arms. The condition occurred in up to 1.4% of 0.5mg dose ranibizumab patients in the ANCHOR trial, and the rate per injection was 0.05% in the MARINA trial. Endophthalmitis occurred in ██████████ of patients across the ██████ and ████ trials ██████████ Very few deaths were reported in the ranibizumab trials, with numbers of deaths being reproducing the whole page here because I worry that it would otherwise be almost too bizarre for you to believe. This level of censorship isn't an everyday phenomenon; but it illustrates the absurdity of what medicine has come to accept, in professional documents that most people wouldn't bother to read.[69]

Why shouldn't we all – doctors, patients and NICE – have access to information on trials that regulators see? This is something I asked both Kent Woods from the MHRA and Hans Georg Eichler, Medical Director of the European Medicines Agency, in 2010. Both, separately, gave me the same answer: people outside the agencies cannot be trusted with this information, because they might misinterpret it, either deliberately or through incompetence. Both, separately – though I guess they must chat at parties – raised the MMR vaccine scare as the classic example of how the media can contrive a national panic on the basis of no good evidence, creating dangerous public-health problems along the way. What if they released raw safety data, and people who don't know how to analyse it properly found imaginary patterns, and created scares that put patients off taking life-saving medication?

I accept that this is a risk, but I also believe their priorities are wrong: I think that the advantages of many eyes working on these vitally important problems are enormous, and the possibility of a few irrational scaremongers is no excuse for hiding data. Drug companies and regulators also both say that you can already get all the information you need from regulators' websites, in summary form.

We shall now see that this is untrue.

Two: Regulators make it hard to access the data they do have

When exposed to criticism, drug companies often become indignant, and declare that they already share enough data for doctors and patients to be informed. 'We give everything to the regulator,' they say, 'and you can get it from them.' Similarly, regulators insist that all you need to do is look on their website, and you will easily find all the data you need. In reality, there is a messy game, in which doctors and academics trying to find all the data on a drug are sent around the houses, scrabbling for information that is both hard to find and fatally flawed.

First, as we've already seen, regulators don't have all the trials, and they don't share all the ones that they do. Summary documents are available on the early trials used to get a drug onto the market in the first place, but only for the specific licensed uses of the drug. Even where the regulator has been given safety data for off-label uses (following the paroxetine case above) the information from these trials still isn't made publicly available through the regulator: it simply sits quietly in the regulator's files.

For example: duloxetine is another drug in fairly widespread use, which is usually given as an antidepressant. During a trial on its use for a completely different purpose – treating incontinence – there were apparently several suicides.[70] This is important and interesting information, and the FDA holds the relevant data: it conducted a review on this issue, and came to a view on whether the risk was significant. But you cannot see any of that on the FDA website, because duloxetine never got a licence for use in treating incontinence.[71] The trial data was only used by the FDA to inform its internal ruminations. This is an everyday situation.

But even when you are allowed to see trial results held by regulators, getting this information from their public websites is supremely tricky. The search functions on the FDA website are essentially broken, while the content is haphazard and badly organised, with lots missing, and too little information to enable you to work out if a trial was prone to bias by design. Once again – partly, here, through casual thoughtlessness and incompetence – it is impossible to get access to the basic information that we need. Drug companies and regulators deny this:

they say that if you search their websites, everything is there. So let's walk, briefly, through the process in all its infuriating glory. The case I will use was published three years ago in *JAMA* as a useful illustration of how broken the FDA site has become:[72] replicating it today, in 2012, nothing has changed.

So: let's say we want to find the results from all the trials the FDA has, on a drug called pregabalin, in which the drug is used to treat pain for diabetics whose nerves have been affected by their disease (a condition called 'diabetic peripheral neuropathy'). You want the FDA review on this specific use, which is the PDF document containing all the trials in one big bundle. But if you search for 'pregabalin review', say, on the FDA website, you get over a hundred documents: none of them is clearly named, and not one of them is the FDA review document on pregabalin. If you type in the FDA application number – the unique identifier for the FDA document you're looking for – the FDA website comes up with nothing at all.

If you're lucky, or wise, you'll get dropped at the Drugs@FDA page: typing 'pregabalin' there brings up three 'FDA applications'. Why three? Because there are three different documents, each on a different condition that pregabalin can be used to treat. The FDA site doesn't tell you which condition each of these three documents is for, so you have to use trial and error to try to find out. That's not as easy as it sounds. I have the correct document for pregabalin and diabetic peripheral neuropathy right here in front of me: it's almost four hundred pages long, but it doesn't tell you that it's about diabetic peripheral neuropathy until you get to page 19. There's no executive summary at the beginning – in fact, there's no title page, no contents page, no hint of what the document is even about, and it skips randomly from one sub-document to another, all scanned and bundled up in the same gigantic file.

If you're a nerd, you might think: these files are electronic; they're PDFs, a type of file specifically designed to make sharing electronic documents convenient. Any nerd will know that if you want to find something in an electronic document, it's easy: you just use the 'find' command: type in, say, 'peripheral neuropathy', and your computer will find the phrase straight off. But no: unlike almost any other serious government document in the world, the PDFs from the FDA are a *series of photographs* of pages of text, rather than the text itself. This means you cannot search for a phrase. Instead, you have to go through it, searching for that phrase, laboriously, by eye.

I could go on. I will. There's some kind of 'table of contents' on the seventeenth page, but it gets the page numbers wrong. I've finished now. There is simply no reason for this obfuscation and chaos. These problems aren't caused by technical issues specific to trials, and they would cost hardly any money at all to fix. This is plainly, simply, unhelpful, and the best we can hope is that it's driven by thoughtlessness.

That's a tragedy, because if you can unearth this document and decode it, you will find that it is full of terrifying gems: perfect examples of situations in which a drug company has used dodgy statistical methods to design and analyse a study, in such a way that it is predestined – from the outset – to exaggerate the benefits of the drug.

For example, in the five trials on pregabalin and pain, lots of people dropped out during the study period. This is common in medical trials, as you will shortly see, and it often happens because people have found a drug to be unhelpful, or have had bad side effects. During these trials you're measuring pain at regular intervals. But if some people drop out, you're left with an important question: what kind of pain score should you use for them in your results? We know, after all, that people dropping out are more likely to have done badly on the drug.

Pfizer decided to use a method called 'Last Observation Carried Forward', which means what you'd expect: you take the last measurement of pain severity while the patients were on the drug, from just before they dropped out, and then paste that in for all the remaining pain measures that they missed, after they stopped coming to follow-up appointments.

The FDA disapproved of this: it pointed out, quite correctly, that Pfizer's strategy would make the drug look better than it really is. For a fairer picture, we have to assume that the drop-outs stopped taking the drug because of side effects, so their pain score should reflect the reality, which is that they would never get any benefit from the drug in normal use. The correct level of pain to record for them is, therefore, their pain at the beginning of the study, before they had any kind of treatment (if you're interested, this is called 'Baseline Observation Carried Forward'). The analysis was duly redone, properly, and a more modest, more accurate view of the benefits of the drug was produced. In this case, it turns out that using the 'last observation' method overestimated the improvement in pain by about a quarter.

Here's the catch. Four out of five of these trials were then published in the peer-reviewed academic literature, the place where doctors look for evidence on whether a drug works or not (one trial wasn't published at all). Every single one of the published analyses used 'Last Observation Carried Forward', the dodgy method, the one that exaggerates the benefits of the drug. Not one of them acknowledges that 'last observation' is a technique that overstates these benefits.

You can see why it is important that we have access to all the information we can possibly get on every drug trial: not only are some whole trials withheld from us, but there are often hidden flaws in the methods used. The devil is in the detail, and there are many dodgy trials, as we shall soon see, with flaws that may not be clear even in the academic papers, let alone in the thin and uninformative summaries from regulators. Furthermore, as we shall also see very shortly, there are often worrying discrepancies between the regulators' summary documents and what actually happened in the trial.

This is why we need a more detailed document on each trial: a Clinical Study Report (CSR). These can be thousands of pages long, but they are complete enough for the reader to reconstruct exactly what happened to all the participants; and they will let you find out where the bodies are buried. Everywhere in the world, if they are relying on the trial to get a marketing authorisation, drug companies give this study report to the regulator – though still only for formally licensed uses of the drug – so both have a copy, and both should be happy to hand it over.

We will now see what happens when you ask them.

Three: Regulators withhold study reports that they do have

In 2007, researchers from the Nordic Cochrane Centre were working on a systematic review for two widely used diet drugs, orlistat and rimonabant. A systematic review, as you know, is the gold-standard summary of the evidence on whether a treatment is effective. These are life-saving, because they give us the best possible understanding of the true effects of a treatment, including its side effects. But doing this requires access to all of the evidence: if some is missing, especially if unflattering data is deliberately harder to obtain, we will be left with a distorted picture.

The researchers knew that the trial data they were able to find in the published academic literature was probably incomplete, because negative trials are routinely left unpublished. But they also knew that the European Medicines Agency (EMA) would have much of this information, since the manufacturers of drugs are obliged to give the study reports to the regulator when trying to get them onto the market, in any country. Since regulators are supposed to act in the interests of patients, they applied to the EMA for the protocols and the study reports. That was in June 2007.

In August, the EMA responded: it had decided not to give out the study reports for these trials, and was invoking the section of its rules which allows it to protect the commercial interests and intellectual property of drug companies. The researchers replied immediately, almost by return of post: there is nothing in the study reports that will undermine the protection of someone's commercial interests, they explained. But if there was, could the EMA please explain why it felt the commercial interests of the drug companies should override the welfare of patients?

We should pause, for just one moment, and think about what the EMA is doing here. It is the regulator that approves and monitors drugs for the whole of Europe, with the aim of protecting the public. Doctors and patients can only make meaningful decisions about treatments if they have access to all the data. The EMA has that data, but has decided that the drug companies' interests are more important. Having spoken to a lot of people in regulation, I can offer one small insight into what on earth they might be thinking. Regulators, in my experience, are preoccupied with the idea that they see all the data, and use it to make the decision about whether a drug should go on the market, and that this is enough: doctors and patients don't need to see the data, because the regulator has done all that work.

This misunderstands a crucial difference between the decisions made by regulators and the decisions made by doctors. Contrary to what some regulators seem to think, a drug is not either 'good' and therefore on the market, or 'bad' and therefore off it. A regulator makes a decision about whether it's in the interests of the population as a whole that the drug should be available for use, at all, ever – even if only in some very obscure circumstance, infrequently and cautiously. This bar is set pretty low, as we shall see, and lots of drugs that are on the market (in fact, the overwhelming majority) are hardly ever used.

A doctor needs to use the same information as that available to the regulator in order to make a very different decision: is this the right drug for the patient in front of me right now? The simple fact that a drug is approved for prescription doesn't mean it's particularly good, or the best. In fact, there are complex decisions to be made in each clinical situation about which drug is best. Maybe the patient has failed to get better on one drug, so you want to try another, from a different class of drugs; maybe the patient has mild kidney failure, so you don't want to use the most popular drug, as that causes very occasional problems in patients with dodgy kidneys; maybe you need a drug that won't interfere with other drugs the patient is taking.

These complex considerations are the reason we are OK with having a range of drugs on the market: even if some of them are less useful overall, they might be useful in specific circumstances. But we need to be able to see all of the information about them, in order to make these decisions. It is not enough for the regulators to grandly state that they have approved a drug, and therefore we should all feel happy to prescribe it. Doctors and patients need the data just as much as regulators do.

In September 2007 the EMA confirmed to the Cochrane researchers that it wasn't going to share the study reports on orlistat and rimonabant, and explained that it had a policy of never disclosing the data given as part of a marketing authorisation. A serious problem had emerged. These weight-loss drugs were being widely prescribed throughout Europe, but doctors and patients were unable to access important information about whether they worked, how bad the side effects were, which was more effective, or any of a whole host of other important questions. Real patients were being exposed to potential harm, in everyday prescribing decisions, through this lack of information that was enforced by the EMA.

The researchers went to the European Ombudsman with two clear allegations. First, the EMA had failed to give sufficient reasons for refusing them access to the data; and second, the EMA's brief, dismissive claim that commercial interests must be protected was unjustified, because there was no material of commercial interest in the trial results, other than the data on safety and effectiveness, which doctors and patients obviously need to access. They didn't know it at the time, but this was the beginning of a battle for data that would shame the EMA, and would last more than three years.

It took four months for the EMA to respond, and over the next year it simply reiterated its position: as far as it was concerned, any kind of information the disclosure of which would 'unreasonably undermine or prejudice the commercial interests of individuals or companies' was commercially confidential. The study reports, it said, might contain information about the commercial plans for the drug. The researchers responded that this was unlikely, but was in any case of marginal importance, as part of a much more important and pressing situation: 'As a likely consequence of [the] EMA's position, patients would die unnecessarily, and would be treated with inferior and potentially harmful drugs.' They regarded the EMA's position as ethically indefensible. More than that, they said, the EMA had a clear conflict of interest: this data could be used to challenge its summary views on the benefits and risks of these treatments. The EMA had failed to explain why doctors and patients having access to study reports and protocols should undermine anyone's reasonable commercial interests, and why these commercial interests were more important than the welfare of patients.

Then, almost two years into this process, the EMA changed tack. Suddenly, it began to argue that study reports contain personal data about individual patients. This argument hadn't been raised by the EMA before, but it's also not true. There may have been some information in some whole sections of the study reports that gave detail on some individual participants' odd presentations, or on possible side effects, but these were all in the same appendix, and could easily be removed.

The EU Ombudsman's conclusions were clear: the EMA had failed in its duty to give an adequate or even coherent explanation of why it was refusing access to this important information.

He made a preliminary finding of maladministration. After doing that, he was under no obligation to offer any further opinion on the weak excuses offered by the EMA, but he decided to do so anyway. His report is damning. The EMA had flatly failed to address a serious charge that its withholding information about these trials was against the public interest, and exposed patients to harm. The Ombudsman also described how he had gone through the study reports in detail himself, and had found that they contained neither any commercially confidential information, nor any details about the commercial development of the drugs. The EMA's claims that answering the request would have put it under an inappropriate administrative burden were untrue, and it had overestimated the work this would have involved: specifically, he explained, removing any personal data, where it did sometimes appear, would be easy.

The Ombudsman told the EMA to hand over the data, or produce a convincing explanation of why it shouldn't. Amazingly, the EMA, the drugs regulator covering the whole of Europe, still refused to hand over the documents. During this delay, people certainly suffered unnecessarily, and some possibly also died, simply for want of information. But the behaviour of the EMA then deteriorated even further, into the outright surreal. Any scrap of the company's thinking about how to run the trial, it argued, which could be intuited from reading the study reports and protocols, was commercially sensitive with respect to its thoughts and plans. This was true, the EMA said, even though the drugs were already on the market, and the information was from final clinical trials, at the very end of the commercial drug-development process. The researchers responded that this was perverse: they knew that withheld data is often negative, so if anything, any other company seeing negative data about these drugs would be *less* likely to try to get a competitor to market, if it appeared that the benefits of the drugs were more modest than had been thought.

That wasn't the end of it. The EMA also grandly dismissed the idea that lives were at risk, stating that the burden of proof was on the researchers to demonstrate this. To me this is – if you can forgive me – a slightly contemptuous attitude, especially given what happens in the next paragraph. It's plainly true that if doctors and patients are unable to see which is the best treatment, then they will make worse decisions, exposing patients to unnecessary harm. Furthermore, it is obvious that larger numbers of academics making transparent judgements about publicly accessible trial data are a much more sensible way of determining the risks and benefits of an intervention than a brief blanket 'yes or no' edict and summary from a regulator. This is all true of drugs like orlistat and rimonabant, but it's also true of any drug, and we will see many cases where academics spotted problems with drugs that regulators had missed.

Then, in 2009, one of the two drugs, rimonabant, was taken off the market, on the grounds that it increases the risk of serious psychiatric problems and suicide. This, at the same time that the EMA was arguing that researchers were wrong to claim that withholding information was harming patients.

And then the EMA suddenly claimed that the design of a randomised trial itself is commercially confidential information.

In case it needs reiterating, let me remind you that the first trial appears in the Bible, Daniel 1:12, and although the core ideas have certainly been refined over time, all trials are essentially identical experiments, generic across any field, with the basics of a modern trial sketched out at least half a century ago. There is absolutely no earthly sense in which anyone could realistically claim that the design of a randomised controlled trial is a commercially confidential or patentable piece of intellectual property.

This was now a farce. The researchers opened all barrels. The EMA was in breach of the Declaration of Helsinki, the international code of medical ethics, which states that everyone involved in research has a duty to make trial findings public. The researchers knew that published papers give a flattering subset of the trial data, and so did the EMA. Patients would die if the EMA continued to withhold data. There was nothing of serious commercial value in there. The EMA's brief public summaries of the data were inaccurate. The EMA was complicit in the exploitation of patients for commercial gain.

It was now August 2009, and the researchers had been fighting for more than two years for access to data on two widely prescribed drugs, from the very organisation that is supposed to be protecting patients and the public. They weren't alone. The French 'prescribers' bulletin' *Prescrire* was trying at the same time to get the EMA's documents on rimonabant. It was sent some unhelpful documents, including the rather remarkable 'Final Assessment Report', from the Swedish agency that had handled the drug's approval much earlier. You can read this PDF in full online. Or rather, you can't. On the following page you can see exactly what the scientific analysis of the drug looked like – the document the EMA sent to one of France's most respected journals for doctors.[73] I think it tells a rather clear story, and to add to the insult, there are sixty pages of this in total.

In the meantime, the Danish Medical Authority had handed over fifty-six study reports to Cochrane (though it still needed more from the EMA); a complaint from the drug company about this had been rejected by the Danish government, which had seen no problem with commercial information (there was none), nor administrative hassle (it was minimal), nor the idea that the design of a randomised trial is commercial information (which is laughable). This was chaos. The EMA – which you may remember was responsible for EudraCT, the transparency tool that is held in secret – was going out on a very peculiar limb. It seemed that it would do anything it could to withhold this information from doctors and patients. As we shall see, sadly, this level of secrecy is its stock in trade.

LÄKEMEDELSVERKET
MEDICAL PRODUCTS AGENCY

Date:

But now we come to the end of this particular road for the EMA. It handed the full, final study reports over to the Ombudsman, reminding him that even the table of contents for each was commercial. Once he had had these documents in his hand, his final opinion came swiftly. There was no commercial data here. There was no confidential patient information here. Hand it over to the public, now. The EMA, at glacial pace, agreed to set a deadline for delivering the data to the researchers, doctors and patients who needed it. The Ombudsman's final ruling was published at the end of November 2010.[74] The initial complaint was made in June 2007. That is a full three and a half years of fighting, obstruction and spurious arguments from the EMA, during which one of the drugs had to be removed from the market because it was harming patients.

After this precedent had been set, things had to shift at the EMA: they were forced to revise their approach, and study reports were briefly made more widely available under a new policy (this window has been shut down, in 2013, as you will read in the Afterword). But even this policy of sharing reports did not solve the problem of access to all information about a trial: the EMA often don't hold all the modules of all the reports for all the uses of all the medicines they have approved. And they only hold records for the most recent few years, which is hopelessly incomplete, since in medicine we use treatments that have come on the market over the past few decades. In fact, this loophole is illustrated by the very first request made by the Cochrane researchers who won this great breakthrough with the Ombudsman. They applied for documents on antidepressants: a good place to start, since these drugs have been the focus of some particularly bad behaviour over the years (though we should remember that missing trial data pervades every corner of medicine). What happened next was even more bizarre than the EMA's three-year battle to withhold information on orlistat and rimonabant.

The researchers put in their request to the EMA, but were told that the drugs had been approved back in the era when marketing authorisations were given out by individual countries, rather than by the EMA centrally. These local authorisations were then 'copied' to all other nations. The MHRA, the UK drugs regulator, held the information the researchers wanted, so they would have to approach it for a copy. The researchers dutifully wrote to the MHRA, asked for the reports on a drug called fluxetine, and then waited patiently. Finally the answer came: the MHRA explained that it would be happy to hand over this information, but there was a problem.

The documents had all been shredded.[75]

This was in keeping, it explained, with the agency's retention policy, which was that such documents were only kept if they were of particular scientific, historical or political interest, and the files didn't meet those criteria. Let's just take a moment to think through what the criteria must be. The SSRI antidepressant drugs have seen many scandals on hidden data, and that should be enough on its own, but if you think back to the beginning of this chapter, one of them – paroxetine – was involved in an unprecedented, four-year-long investigation into whether criminal charges could be brought against GSK. That investigation into paroxetine was the largest investigation that the MHRA has ever conducted into drug safety: in fact, it was the largest investigation the MHRA has ever conducted, of any kind, ever. Quite apart from that, these original study reports contain vitally important data on safety and efficacy. But the MHRA shredded them all the same, feeling that they were not of sufficient scientific, historical or political interest.[76]

I can only leave you there.

Where are we so far?

The missing data saga is long and complex, and involves patients around the world being put at risk, and some major players which have let us down to an extraordinary extent. Since we are almost at the end of it, this is a good moment to gather together what we've seen so far.

Trials are frequently conducted, but then left unpublished, and so are unavailable to doctors and patients. Only half of all trials get published, and those with negative results are twice as likely to go missing as those with positive ones. This means that the evidence on which we base decisions in medicine is systematically biased to overstate the benefits of the treatments we give. Since there is no way to compensate for this missing data, we have no way of knowing the true benefits and risks of the drugs doctors prescribe.

This is research misconduct on a grand, international scale. The reality of the problem is universally recognised, but nobody has bothered to fix it:

- Ethics committees allow companies and individual researchers with a track record of unpublished data to do more trials on human participants.
- University administrators and ethics committees permit contracts with industry that explicitly say the sponsor can control the data.
- Registers have never been enforced.
- Academic journals continue to publish trials that were unregistered, despite pretending that they won't.
- Regulators have information that would be vital to improve patient care, but are systematically obstructive:
- they have poor systems to share poor summaries of the information that they do have;

- and they are bizarrely obstructive of people asking for more information.
- Drug companies hold trial results that even regulators don't see.
- Governments have never implemented laws compelling companies to publish data.
- Doctors' and academics' professional bodies have done nothing.

What to do about all this?

You will need to wait, because in the next section there are even greater horrors to come.

Trying to get trial data from drug companies: the story of Tamiflu

Governments around the world have spent billions of pounds on stockpiling a drug called Tamiflu. In the UK alone we have spent several hundred million pounds – the total figure is not yet clear – and so far we've bought enough tablets to treat 80 per cent of the population if an outbreak of bird flu occurs. I'm very sorry if you have the flu, because it's horrid being ill; but we have not spent all this money to reduce the duration of your symptoms in the event of a pandemic by a few hours (though Tamiflu does do this, fairly satisfactorily). We have spent this money to reduce the rate of 'complications': a medical euphemism, meaning pneumonia and death.

Lots of people seem to think Tamiflu will do this. The US Department of Health and Human Services said it would save lives and reduce hospital admissions. The European Medicines Agency said it would reduce complications. The Australian drugs regulator said so too. Roche's website said it reduces complications by 67 per cent.[77] But what is the evidence that Tamiflu really will reduce complications? Answering questions like this is the bread and butter of the Cochrane Collaboration, which you will remember is the vast and independent non-profit international collaboration of academics producing hundreds of systematic reviews on important questions in medicine every year. In 2009 there was concern about a flu pandemic, and an enormous amount of money was being spent on Tamiflu. Because of this, the UK and Australian governments specifically asked the Cochrane Respiratory Diseases Group to update its earlier reviews on the drug.

Cochrane reviews are in a constant review cycle, because evidence changes over time as new trials are published. This should have been a pretty everyday piece of work: the previous review, in 2008, had found some evidence that Tamiflu does indeed reduce the rate of complications. But then a Japanese paediatrician called Keiji Hayashi left a comment that would trigger a revolution in our understanding of how evidence-based medicine should work. This wasn't in a publication, or even a letter: it was a simple online comment, posted underneath the Tamiflu review on the Cochrane website.

You've summarised the data from all the trials, he explained, but your positive conclusion is really driven by data from just one of the papers you cite, an industry-funded meta-analysis led by an author called Kaiser. This, 'the Kaiser paper', summarises the findings of ten earlier trials, but from these ten trials, only two have ever been published in the scientific literature. For the remaining eight, your only information comes from the brief summary in this secondary, industry source. That's not reliable enough.

In case it's not immediately obvious, this is science at its best. The Cochrane review is readily accessible online; it explains transparently the methods by which it looked for trials, and then analysed them, so any informed reader can pull the review apart, and understand where the conclusions came from. Cochrane provides an easy way for readers to raise criticisms. And, crucially, these criticisms did not fall on deaf ears. Tom Jefferson is an editor at the Cochrane Acute Respiratory Infections Group, and the lead author on the 2008 review. He realised immediately that he had made a mistake in blindly trusting the Kaiser data. He said so, without any defensiveness, and then set about getting the information in a straight, workpersonlike fashion. This began a three-year battle, which is still not resolved, but which has thrown stark light on the need for all researchers to have access to clinical study reports on trials wherever possible.

First, the Cochrane researchers wrote to the authors of the Kaiser paper, asking for more information. In reply, they were told that this team no longer had the files, and that they should contact Roche, the manufacturer of Tamiflu. So naturally they wrote to Roche and asked for the data.

This is where the problems began. Roche said it would hand over some data, but the Cochrane reviewers would need to sign a confidentiality agreement. This was an impossibility for any serious scientist: it would prevent them from conducting a systematic review with any reasonable degree of openness and transparency. More than this, the proposed contract also raised serious ethical issues, in that it would have required the Cochrane team to actively withhold information from the reader: it included a clause saying that on signing it, the reviewers would never be allowed to discuss the terms of this secrecy agreement; and more than that, they would be forbidden from ever publicly acknowledging that it even existed. Roche was demanding a secret contract, with secret terms, requiring secrecy about trial data, in a discussion about the safety and efficacy of a drug that has been taken by hundreds of thousands of people around the world. Jefferson asked for clarification, and never received a reply.

Then, in October 2009, the company changed tack: they would like to hand the data over, they explained, but another meta-analysis was being conducted elsewhere. Roche had given them the study reports, so Cochrane couldn't have them. This was a simple non-sequitur: there is no reason why many groups shouldn't all work on the same question. In fact, quite the opposite: replication is the cornerstone of good science. Roche's excuse made no sense. Jefferson asked for clarification, but never received a reply.

Then, one week later, unannounced, Roche sent seven short documents, each around a dozen pages long. These contained excerpts of internal company documents on each of the clinical trials in the Kaiser meta-analysis. This was a start, but it didn't contain anything like enough information for Cochrane to assess the benefits, or the rate of adverse events, or fully to understand exactly what methods were used in the trials.

At the same time, it was rapidly becoming clear that there were odd inconsistencies in the information on this drug. First, there was considerable disagreement at the level of the broad conclusions drawn by different organisations. The FDA said there were no benefits on complications, while the Centers for Disease Control and Prevention (in charge of public health in the USA – some wear nice naval uniforms in honour of their history on the docks) said it did reduce complications. The Japanese regulator made no claim for complications, but the EMA said there was a benefit. In a sensible world, we might think that all these organisations should sing from the same hymn sheet, because all would have access to the same information. Of course, there is also room for occasional reasonable disagreement, especially where there are close calls: this is precisely why doctors and researchers should access to all the information about a drug, so that they can make their own judgements.

Meanwhile, reflecting these different judgements, Roche's own websites said completely different things in different jurisdictions, depending on what the local regulator had said. It's naïve, perhaps, to expect consistency from a drug company, but from this and other stories it's clear that industry utterances are driven by the maximum they can get away with in each territory, rather than any consistent review of the evidence.

In any case, now that their interest had been piqued, the Cochrane researchers also began to notice that there were odd discrepancies between the frequency of adverse events in different databases. Roche's global safety database held 2,466 neuropsychiatric adverse events, of which 562 were classified as 'serious'. But the FDA database for the same period held only 1,805 adverse events in total. The rules vary on what needs to be notified to whom, and where, but even allowing for that, this was odd.

In any case, since Roche was denying them access to the information needed to conduct a proper review, the Cochrane team concluded that they would have to exclude all the unpublished Kaiser data from their analysis, because the details could not be verified in the normal way. People cannot make treatment and purchasing decisions on the basis of trials if the full methods and results aren't clear: the devil is often in the detail, as we shall see in Chapter 4, on 'bad trials', so we cannot blindly trust that every study is a fair test of the treatment.

This is particularly important with Tamiflu, because there are good reasons to think that these trials were not ideal, and that published accounts were incomplete, to say the least. On closer examination, for example, the patients participating were clearly unusual, to the extent that the results may not be very relevant to normal everyday flu patients. In the published accounts, patients in the trials are described as typical flu patients, suffering from normal flu symptoms like cough, fatigue, and so on. We don't do blood tests on people with flu in routine practice, but when these tests are done – for surveillance purposes – then even during peak flu season only about one in three people with 'flu' will actually be infected with the influenza virus, and most of the year only one in eight will really have it. (The rest are sick from something else, maybe just a common cold virus.)

Two thirds of the trial participants summarised in the Kaiser paper tested positive for flu. This is bizarrely high, and means that the benefits of the drug will be overstated, because it is being tested on perfect patients, the very ones most likely to get better from a drug that selectively attacks the flu virus. In normal practice, which is where the results of these trials will be applied, doctors will be giving the drug to real patients who are diagnosed with 'flu-like illness', which is all you can realistically do in a clinic. Among these real patients, many will not actually have the influenza virus. This means that in the real world, the benefits of Tamiflu on flu will be diluted, and many more people will be exposed to the drug who don't actually have flu virus in their systems. This, in turn, means that the side effects are likely to creep up in significance, in comparison with any benefits. That is why we strive to ensure that all trials are conducted in normal, everyday, realistic patients: if they are not, their findings may not be relevant to the real world.

So the Cochrane review was published without the Kaiser data in December 2009, alongside some explanatory material about why the Kaiser results had been excluded, and a small flurry of activity followed. Roche put the short excerpts it had sent over online, and committed to make full study reports available (it still hasn't done so).

What Roche posted was incomplete, but it began a journey for the Cochrane academics of learning a great deal more about the real information that is collected on a trial, and how that can differ from what is given to doctors and patients in the form of brief, published academic papers. At the core of every trial is the raw data: every single record of blood pressure of every patient, the doctors' notes describing any unusual symptoms, investigators' notes, and so on. A published academic paper is a short description of the study, usually following a set format: an introductory background; a description of the methods; a summary of the important results; and then finally a discussion, covering the strengths and weaknesses of the design, and the implications of the results for clinical practice.

A clinical study report, or CSR, is the intermediate document that stands between these two, and can be very long, sometimes thousands of pages.[78] Anybody working in the pharmaceutical industry is very familiar with these documents, but doctors and academics have rarely heard of them. They contain much more detail on things like the precise plan for analysing the data statistically, detailed descriptions of adverse events, and so on.

These documents are split into different sections, or 'modules'. Roche has shared only 'module 1', for only seven of the ten study reports Cochrane has requested. These modules are missing vitally important information, including the analysis plan, the randomisation details, the study protocol (and the list of deviations from that), and so on. But even these incomplete modules were enough to raise concerns about the universal practice of trusting academic papers to give a complete story about what happened to the patients in a trial.

For example, looking at the two papers out of ten in the Kaiser review which were published, one says: 'There were no drug-related serious adverse events,' and the other doesn't mention adverse events. But in the 'module 1' documents on these same two studies, there are ten serious adverse events listed, of which three are classified as being possibly related to Tamiflu.[79]

Another published paper describes itself as a trial comparing Tamiflu against placebo. A placebo is an inert tablet, containing no active ingredient, that is visually indistinguishable from the pill containing the real medicine. But the CSR for this trial shows that the real medicine was in a grey and yellow capsule, whereas the placebos were grey and ivory. The 'placebo' tablets also contained something called dehydrocholic acid, a chemical which encourages the gall bladder to empty.[80] Nobody has any clear idea of why, and it's not even mentioned in the academic paper; but it seems that this was not actually an inert dummy pill placebo.

Simply making a list of all the trials conducted on a subject is vitally important if we want to avoid seeing only a biased summary of the research done on a subject; but in the case of Tamiflu even this proved to be almost impossible. For example, Roche Shanghai informed the Cochrane group of one large trial (ML16369), but Roche Basel seemed not to know of its existence. But by setting out all the trials side by side, the researchers were able to identify peculiar discrepancies: for example, the largest 'phase 3' trial – one of the large trials that are done to get a drug onto the market – was never published, and is rarely mentioned in regulatory documents.*

There were other odd discrepancies. Why, for example, was one trial on Tamiflu published in 2010, ten years after it was completed?[82] Why did some trials report completely different authors, depending on where they were being discussed?[83] And so on.

The chase continued. In December 2009 Roche had promised: 'full study reports will also be made available on a password-

---

* Setting out a simple list of trials is also important for other reasons, including something called 'duplicate publication'. A British anaesthetist called Martin Tramèr conducted a review on the efficacy of a nausea drug called ondansetron, and noticed that lots of the data seemed to be replicated. On closer inspection, it turned out that lots of trials had been conducted in lots of different places, and then bundled up into multi-centre trials.[81] But the results for many individual patients had been written up several times over, bundled up with other data, in different journals and different papers. Crucially, data which showed the drug in a better light were more likely to be duplicated than the data which showed it to be less impressive; overall this led to a 23 per cent overestimate of the drug's efficacy.

protected site within the coming days to physicians and scientists undertaking legitimate analyses'. This never happened. Then an odd game began. In June 2010 Roche said: Oh, we're sorry, we thought you had what you wanted. In July it announced that it was worried about patient confidentiality (you may remember this from the EMA saga). This was an odd move: for most of the important parts of these documents, privacy is no issue at all. The full trial protocol, and the analysis plan, are both completed before any single patient is ever touched. Roche has never explained why patient privacy prevents it from releasing the study reports. It simply continued to withhold them.

Then in August 2010 it began to make some even more bizarre demands, betraying a disturbing belief that companies are perfectly entitled to control access to information that is needed by doctors and patients around the world to make safe decisions. First, it insisted on seeing the Cochrane reviewers' full analysis plan. Fine, they said, and posted the whole protocol online. Doing so is completely standard practice at Cochrane, as it should be for any transparent organisation, and allows people to suggest important changes before you begin. There were few surprises, since all Cochrane reports follow a pretty strict manual anyway. Roche continued to withhold its study reports (including, iron-ically, its own protocols, the very thing it demanded Cochrane should publish, and that Cochrane had published, happily).

By now Roche had been refusing to publish the study reports for a year. Suddenly, the company began to raise odd personal concerns. It claimed that some Cochrane researchers had made untrue statements about the drug, and about the company, but refused to say who, or what, or where. 'Certain members of Cochrane Group involved with the review of the neuraminidase inhibitors,' it announced, 'are unlikely to approach the review with the independence that is both necessary and justified.' This is an astonishing state of affairs, where a company feels it should be allowed to prevent individual researchers access to data that should be available to all; but still Roche refused to hand over the study reports.

Then it complained that the Cochrane reviewers had begun to copy journalists in on their emails when responding to Roche staff. I was one of the people copied in on these interactions, and I believe that this was exactly the correct thing to do. Roche's excuses had become perverse, and the company had failed to keep its promise to share all study reports. It's clear that the modest pressure exerted by researchers in academic journals alone was having little impact on Roche's refusal to release the data, and this is an important matter of public health, both for the individual case of this Tamiflu data, and for the broader issue of companies and regulators harming patients by withholding information.

Then things became even more perverse. In January 2011 Roche announced that the Cochrane researchers had already been given all the data they need. This was simply untrue. In February it insisted that all the studies requested were published (meaning academic papers, now shown to be misleading on Tamiflu). Then it declared that it would hand over nothing more, saying: 'You have all the detail you need to undertake a review.' But this still wasn't true: it was still withholding the material it had publicly promised to hand over 'within a few days' in December 2009, a year and a half earlier.

At the same time, the company was raising the broken arguments we have already seen: it's the job of regulators to make these decisions about benefit and risk, it said, not academics. Now, this claim fails on two important fronts. First, as with many other drugs, we now know that not even the regulators had seen all the data. In January 2012 Roche claimed that it 'has made full clinical study data available to health authorities around the world for their review as part of the licensing process'. But the EMA never received this information for at least fifteen trials. This was because the EMA had never requested it.

And that brings us on to our final important realisation: regulators are not infallible. They make outright mistakes, and they make decisions which are open to judgement, and should be subject to second-guessing and checking by many eyes around the world. In the next chapter we will see more examples of how regulators can fail, behind closed doors, but here we will look at one story that illustrates the benefit of 'many eyes' perfectly.

Rosiglitazone is a new kind of diabetes drug, and lots of researchers and patients had high hopes that it would be safe and effective.[84] Diabetes is common, and more people develop the disease every year. Sufferers have poor control of their blood sugar, and diabetes drugs, alongside dietary changes, are supposed to fix this. Although it's nice to see your blood sugar being controlled nicely in the numbers from lab tests and machines at home, we don't control these figures for their own sake: we try to control blood sugar because we hope that this will help reduce the chances of real-world outcomes, like heart attack and death, both of which occur at a higher rate in people with diabetes.

Rosiglitazone was first marketed in 1999, and from the outset it was a magnet for disappointing behaviour. In that first year, Dr John Buse from the University of North Carolina discussed an increased risk of heart problems at a pair of academic meetings. The drug's manufacturer, GSK, made direct contact in an attempt to silence him, then moved on to his head of department. Buse felt pressured to sign various legal documents. To cut a long story short, after wading through documents for several months, in 2007 the US Senate Committee on Finance released a report describing the treatment of Dr Buse as 'intimidation'.

But we are more concerned with the safety and efficacy data. In 2003 the Uppsala Drug Monitoring Group of the World Health Organization contacted GSK about an unusually large number of spontaneous reports associating rosiglitazone with heart problems. GSK conducted two internal meta-analyses of its own data on this, in 2005 and 2006. These showed that the risk was real, but although both GSK and the FDA had these results, neither made any public statement about them, and they were not published until 2008.

During this delay, vast numbers of patients were exposed to the drug, but doctors and patients only learned about this serious problem in 2007, when cardiologist Professor Steve Nissen and colleagues published a landmark meta-analysis. This showed a 43 per cent increase in the risk of heart problems in patients on rosiglitazone. Since people with diabetes are already at increased risk of heart problems, and the whole point of treating diabetes is to reduce this risk, that finding was big potatoes. His findings were confirmed in later work, and in 2010 the drug was either taken off the market or restricted, all around the world.

Now, my argument is not that this drug should have been banned sooner, because as perverse as it sounds, doctors do often need inferior drugs for use as a last resort. For example, a patient may develop idiosyncratic side effects on the most effective pills, and be unable to take them any longer. Once this has happened, it may be worth trying a less effective drug, if it is at least better than nothing.

The concern is that these discussions happened with the data locked behind closed doors, visible only to regulators. In fact, Nissen's analysis could only be done at all because of a very unusual court judgement. In 2004, when GSK was caught out withholding data showing evidence of serious side effects from paroxetine in children, the UK conducted an unprecedented four-year-long investigation, as we saw earlier. But in the US, the same bad behaviour resulted in a court case over allegations of fraud, the settlement of which, alongside a significant payout, required GSK to commit to posting clinical trial results on a public website.

Professor Nissen used the rosiglitazone data, when it became available, found worrying signs of harm, and published this to doctors, which is something that the regulators had never done, despite having the information years earlier. (Though before doctors got to read it, Nissen by chance caught GSK discussing a copy of his unpublished paper, which it had obtained improperly.[85])

If this information had all been freely available from the start, regulators might have felt a little more anxious about their decisions, but crucially, doctors and patients could have disagreed with them, and made informed choices. This is why we need wider access to full CSRs, and all trial reports, for all medicines, and this is why it is perverse that Roche should be able even to contemplate deciding which favoured researchers should be allowed to read the documents on Tamiflu.

Astonishingly, a paper published in April 2012 by regulators from the UK and Europe suggests that they might agree to more data sharing, to a limited extent, within limits, for some studies, with caveats, at the appropriate juncture, and in the fullness of time.[86] Before feeling any sense of enthusiasm, we should remember that this is a cautious utterance, wrung out after the dismal fights I have already described; that it has not been implemented; that it must be set against a background of broken promises from all players across the whole field of missing data; and that in any case, regulators do not have all the trial data anyway. But it is an interesting start.

Their two main objections – if we accept their goodwill at face value – are interesting, because they lead us to the final problem in the way we tolerate harm to patients from missing trial data. First, they raise the concern that some academics and journalists might use study reports to conduct histrionic or poorly conducted reviews of the data: to this, again, I say, 'Let them,' because these foolish analyses should be conducted, and then rubbished, in public.

When UK hospital mortality statistics first became easily accessible to the public, doctors were terrified that they would be unfairly judged: the crude figures can be misinterpreted, after all, because one hospital may have worse figures simply because it is a centre of excellence, and takes in more challenging patients than its neighbours; and there is random variation to be expected in mortality rates anyway, so some hospitals might look unusually good, or bad, simply through the play of chance. Initially, to an extent, these fears were realised: there were a few shrill, unfair stories, and people overinterpreted the results. Now, for the most part, things have settled down, and many lay people are quite able to recognise that crude analyses of such figures are misleading. For drug data, where there is so much danger from withheld information, and so many academics desperate to conduct meaningful analyses, and so many other academics happy to criticise them, releasing the data is the only healthy option.

But second, the EMA raises the spectre of patient confidentiality, and hidden in this concern is one final prize.

So far I have been talking about access to trial reports, summaries of patients' outcomes in trials. There is no good reason to believe that this poses any threat to patient confidentiality, and where there are specific narratives that might make a patient identifiable – a lengthy medical description of one person's idiosyncratic adverse event in a trial, perhaps – these can easily be removed, since they appear in a separate part of the document. These CSRs should undoubtedly, without question, be publicly available documents, and this should be enforced retrospectively, going back decades, to the dawn of trials.

But all trials are ultimately run on individual patients, and the results of those individual patients are all stored and used for the summary analysis at the end of the study. While I would never suggest that these should be posted up on a public website – it would be easy for patients to be identifiable, from many small features of their histories – it is surprising that patient-level data is almost never shared with academics.

Sharing data of individual patients' outcomes in clinical trials, rather than just the final summary result, has several significant advantages. First, it's a safeguard against dubious analytic practices. In the VIGOR trial on the painkiller Vioxx, for example, a bizarre reporting decision was made.[87] The aim of the study was to compare Vioxx against an older, cheaper painkiller, to see if it was any less likely to cause stomach problems (this was the hope for Vioxx), and also if it caused more heart attacks (this was the fear). But the date cut-off for measuring heart attacks was much earlier than that for measuring stomach problems. This had the result of making the risks look less significant, relative to the benefits, but it was not declared clearly in the paper, resulting in a giant scandal when it was eventually noticed. If the raw data on patients was shared, games like these would be far easier to spot, and people might be less likely to play them in the first place.

Occasionally – with vanishing rarity – researchers are able to obtain raw data, and re-analyse studies that have already been conducted and published. Daniel Coyne, Professor of Medicine at Washington University, was lucky enough to get the data on a key trial for epoetin, a drug given to patients on kidney dialysis, after a four-year-long fight.[88] The original academic publication on this study, ten years earlier, had switched the primary outcomes described in the protocol (we will see later how this exaggerates the benefits of treatments), and changed the main statistical analysis strategy (again, a huge source of bias). Coyne was able to analyse the study as the researchers had initially stated they were planning to in their protocol; and when he did, he found that they had dramatically overstated the benefits of the drug. It was a peculiar outcome, as he himself acknowledges: 'As strange as it seems, I am now the sole author of the publication on the predefined primary and secondary results of the largest outcomes trial of epoetin in dialysis patients, and I didn't even participate in the trial.' There is room, in my view, for a small army of people doing the very same thing, re-analysing all the trials that were incorrectly analysed, in ways that deviated misleadingly from their original protocols.

Data sharing would also confer other benefits. It allows people to conduct more exploratory analyses of data, and to better investigate – for example – whether a drug is associated with a particular unexpected side effect. It would also allow cautious 'subgroup analyses', to see if a drug is particularly useful, or particularly useless, in particular types of patients.

The biggest immediate benefit from data sharing is that combining individual patient data into a meta-analysis gives more accurate results than working with the crude summary results at the end of a paper. Let's imagine that one paper reports survival at three years as the main outcome for a cancer drug, and another reports survival at seven years. To combine these two in a meta-analysis, you'd have a problem. But if you were doing the meta-analysis with access to individual patient data, with treatment details and death dates for all of them, you could do a clean combined calculation for three-year survival.

This is exactly the kind of work being done in the area of breast cancer research, where a small number of charismatic and forceful scientists just happen to have driven a pioneering culture of easier collaboration. The summaries they are publishing represent real collaboration, between vast numbers of people, and on vast numbers of patients, producing highly reliable guidance for doctors and patients.

The process sheds a stark light on the reality of data collaboration on such a large scale. Here, for example, is the author list on an academic paper from the *Lancet* in November 2011: it's reporting an immense, definitive, and incredibly useful meta-analysis of breast cancer treatment outcomes, using individual patient data pooled from seventeen different trials. The author list is printed in four-point font size (though I suspect that might go wrong in the e-book edition . . .) because there are

Darby S, McGale P, Correa C, Taylor C, Arriagada R, Clarke M, Cutter D, Davies C, Ewertz M, Godwin J, Gray R, Pierce L, Whelan T, Wang Y, Peto R.Albain K, Anderson S, Arriagada R, Barlow W, Bergh J, Bliss J, Buyse M, Cameron D, Carrasco E, Clarke M, Correa C, Coates A, Collins R, Costantino J, Cutter D, Cuzick J, Darby S, Davidson N, Davies C, Davies K, Delmestri A, Di Leo A, Dowsett M, Elphinstone P, Evans V, Ewertz M, Gelber R, Gettins L, Geyer C, Goldhirsch A, Godwin J, Gray R, Gregory C, Hayes D, Hill C, Ingle J, Jakesz R, James S, Kaufmann M, Kerr A, MacKinnon E, McGale P, McHugh T, Norton L, Ohashi Y, Paik S, Pan HC, Perez E, Peto R, Piccart M, Pierce L, Pritchard K, Pruneri G, Raina V, Ravdin P, Robertson J, Rutgers E, Shao YF, Swain S, Taylor C, Valagussa P, Viale G, Whelan T, Winer E, Wang Y, Wood W, Abe O, Abe R, Enomoto K, Kikuchi K, Koyama H, Masuda H, Nomura Y, Ohashi Y, Sakai K, Sugimachi K, Toi M, Tominaga T, Uchino J, Yoshida M, Haybittle JL, Leonard CF, Calais G, Geraud P, Collett V, Davies C, Delmestri A, Sayer J, Harvey VJ, Holdaway IM, Kay RG, Mason BH, Forbes JF, Wilcken N, Bartsch R, Dubsky P, Fesl C, Fohler H, Gnant M, Greil R, Jakesz R, Lang A, Luschin-Ebengreuth G, Marth C, Mlineritsch B, Samonigg H, Singer CF, Steger GG, Stöger H, Canney P, Yosef HM, Focan C, Peek U, Oates GD, Powell J, Durand M, Mauriac L, Di Leo A, Dolci S, Larsimont D, Nogaret JM, Philippson C, Piccart MJ, Masood MB, Parker D, Price JJ, Lindsay MA, Mackey J, Martin M, Hupperets PS, Bates T, Blamey RW, Chetty U, Ellis IO, Mallon E, Morgan DA, Patnick J, Pinder S, Olivotto I, Ragaz J, Berry D, Broadwater G, Cirrincione C, Muss H, Norton L, Weiss RB, Abu-Zahra HT, Portnoj SM, Bowden S, Brookes C, Dunn J, Fernando I, Lee M, Poole C, Rea D, Spooner D, Barrett-Lee PJ, Mansel RE, Monypenny IJ, Gordon NH, Davis HL, Cuzick J, Lehingue Y, Romestaing P, Dubois JB, Delozier T, Griffon B, Mace Lesec'h J, Rambert P, Mustacchi G, Petruzelka, Pribylova O, Owen JR, Harbeck N, Jänicke F, Meisner C, Schmitt M, Thomssen C, Meier P, Shan Y, Shao YF, Wang X, Zhao DB, Chen ZM, Pan HC, Howell A, Swindell R, Burrett JA, Clarke M, Collins R, Correa C, Cutter D, Darby S, Davies C, Delmestri A, Elphinstone P, Evans V, Gettins L, Godwin J, Gray R, Gregory C, Hermans D, Hicks C, James S, Kerr A, MacKinnon E, Lay M, McGale P, McHugh T, Sayer J, Taylor C, Wang Y, Albano J, de Oliveira CF, Gervásio H, Gordilho J, Johansen H, Mouridsen HT, Gelman RS, Harris JR, Hayes D, Henderson C, Shapiro CL, Winer E, Christiansen P, Ejlertsen B, Ewertz M, Jensen MB, Møller S, Mouridsen HT, Carstensen B, Palshof T, Trampisch HJ, Dalesio O, de Vries EG, Rodenhuis S, van Tinteren H, Comis RL, Davidson NE, Gray R, Robert N, Sledge G, Solin LJ, Sparano JA, Tormey DC, Wood W, Cameron D, Chetty U, Dixon JM, Forrest P, Jack W, Kunkler I, Rossbach J, Klijn JG, Treurniet-Donker AD, van Putten WL, Rotmensz N, Veronesi U, Viale G, Bartelink H, Bijker N, Bogaerts J, Cardoso F, Cufer T, Julien JP, Rutgers E, van de Velde CJ, Cunningham MP, Huovinen R, Joensuu H, Costa A, Tinteri C, Bonadonna G, Gianni L, Valagussa P, Goldstein LJ, Bonneterre J, Fargeot P, Fumoleau P, Kerbrat P, Luporsi E, Namer M, Eiermann W, Hilfrich J, Jonat W, Kaufmann M, Kreienberg R, Schumacher M, Bastert G, Rauschecker H, Sauer R, Sauerbrei W, Schauer A, Schumacher M, Blohmer JU, Costa SD, Eidtmann H, Gerber G, Jackisch C, Loibl S, von Minckwitz G, de Schryver A, Vakaet L, Belfiglio M, Nicolucci A, Pellegrini F, Pirozzoli MC, Sacco M, Valentini M, McArdle CS, Smith DC, Stallard S, Dent DM, Gudgeon CA, Hacking A, Murray E, Panieri E, Werner ID, Carrasco E, Martin M, Segui MA, Galligioni E, Lopez M, Erazo A, Medina JY, Horiguchi J, Takei H, Fentiman IS, Hayward JL, Rubens RD, Skilton D, Scheurlen H, Kaufmann M, Sohn HC, Untch M, Dafni U, Markopoulos C, Dafni D, Fountzilas G, Mavroudis D, Klefstrom P, Saarto T, Gallen M, Margreiter R, de Lafontan B, Mihura J, Roché H, Asselain B, Salmon RJ, Vilcoq JR, Arriagada R, Bourgier C, Hill C, Koscielny S, Laplanche A, Lê MG, Spielmann M, A'Hern R, Bliss J, Ellis P, Kilburn L, Yarnold JR, Benraadt J, Kooi M, van de Velde AO, van Dongen JA, Vermorken JB, Castiglione M, Coates A, Colleoni M, Collins J, Forbes J, Gelber RD, Goldhirsch A, Lindtner J, Price KN, Regan MM, Rudenstam CM, Senn HJ, Thuerlimann B, Bliss JM, Chilvers CE, Coombes RC, Hall E, Marty M, Buyse M, Possinger K, Schmid P, Untch M, Wallwiener D, Foster L, George WD, Stewart HJ, Stroner P, Borovik R, Hayat H, Inbar MJ, Robinson E, Bruzzi P, Del Mastro L, Pronzato P, Sertoli MR, Venturini M, Camerini T, De Palo G, Di Mauro MG, Formelli F, Valagussa P, Amadori D, Martoni A, Pannuti F, Camisa R, Cocconi G, Colozza A, Passalacqua R, Aogi K, Takashima S, Abe O, Ikeda T, Inokuchi K, Kikuchi K, Sawa K, Sonoo H, Korzeniowski S, Skolyszewski J, Ogawa M, Yamashita J, Bastiaanet E, van de Velde CJ, van der Water W, van Nes JG, Christiaens R, Neven P, Paridaens R, Van den Bogaert W, Braun S, Janni W, Martin P, Romain S, Janauer M, Seifert M, Sevelda P, Zielinski CC, Hakes T, Hudis CA, Norton L, Wittes R, Giokas G, Kondylis D, Lissaios B, de la Huerta R, Sainz MG, Altemus R, Camphausen K, Cowan K, Danforth D, Lichter A, Lippman M, O'Shaughnessy J, Pierce LJ, Steinberg S, Venzon D, Zujewski JA, D'Amico C, Lioce M, Paradiso A, Chapman JA, Gelmon K, Goss PE, Levine MN, Meyer R, Parulekar W, Pater JL, Pritchard KI, Shepherd LE, Tu D, Whelan T, Nomura Y, Ohno S, Anderson A, Bass G, Brown A, Bryant J, Costantino J, Dignam J, Fisher B, Geyer C, Mamounas EP, Paik S, Redmond C, Swain S, Wickerham L, Wolmark N, Baum M, Jackson IM, Palmer MK, Perez E, Ingle JN, Suman VJ, Bengtsson NO, Emdin S, Jonsson H, Del Mastro L, Venturini M, Lythgoe JP, Swindell R, Kissin M, Erickson B, Hannisdal E, Jacobsen AB, Varhaug JE, Erikstein B, Gundersen S, Hauer-Jensen M, Høst H, Jacobsen AB, Nissen-Meyer R, Blamey RW, Mitchell AK, Morgan DA, Robertson JF, Ueo H, Di Palma M, Mathé G, Misset JL, Levine M, Pritchard KI, Whelan T, Morimoto K, Sawa K, Takatsuka Y, Crossley E, Harris A, Talbot D, Taylor M, Martin AL, Roché H, Cocconi G, di Blasio B, Ivanov V, Paltuev R, Semiglazov V, Brockschmidt J, Cooper MR, Falkson CI, A'Hern R, Ashley S, Dowsett M, Makris A, Powles TJ, Smith IE, Yarnold JR, Gazet JC, Browne L, Graham P, Corcoran N, Deshpande N, di Martino L, Douglas P, Hacking A, Høst H, Lindtner A, Notter G, Bryant AJ, Ewing GH, Firth LA, Krushen-Kosloski JL, Nissen-Meyer R, Anderson H, Killander F, Malmström P, Rydén L, Arnesson LG, Carstensen J, Dufmats M, Fohlin H, Nordenskjöld B, Söderberg M, Carpenter JT, Murray N, Royle GT, Simmonds PD, Albain K, Barlow W, Crowley J, Hayes D, Gralow J, Green S, Hortobagyi G, Livingston R, Martino S, Osborne CK, Adolfsson J, Bergh J, Bondesson T, Celebioglu F, Dahlberg K, Fornander T, Fredriksson I, Frisell J, Göransson E, Iiristo M, Johansson U, Lenner E, Löfgren L, Nikolaidis P, Perbeck L, Rotstein S, Sandelin K, Skoog L, Svane G, af Trampe E, Wadström C, Castiglione M, Goldhirsch A, Maibach R, Senn HJ, Thürlimann B, Hakama M, Holli K, Isola J, Rouhento K, Saaristo R, Brenner H, Hercbergs A, Martin AL, Roché H, Yoshimoto M, Paterson AH, Pritchard KI, Fyles A, Meakin JW, Panzarella T, Pritchard KI, Bahi J, Reid M, Spittle M, Bishop H, Bundred NJ, Cuzick J, Ellis IO, Fentiman IS, Forbes JF, Forsyth S, George WD, Pinder SE, Sestak I, Deutsch GP, Gray R, Kwong DL, Pai VR, Peto R, Senanayake F, Boccardo F, Rubagotti A, Baum M, Forsyth S, Hackshaw A, Houghton J, Ledermann J, Monson K, Tobias JS, Carlomagno C, De Laurentiis M, De Placido S, Williams L, Hayes D, Pierce LJ, Broglio K, Buzdar AU, Love RR, Ahlgren J, Garmo H, Holmberg L, Liljegren G, Lindman H, Wärnberg F, Asmar L, Jones SE, Gluz O, Harbeck N, Liedtke C, Nitz U, Litton A, Wallgren A, Karlsson P, Linderholm BK, Chlebowski RT, Caffier H.

seven hundred individual researchers named in it. I typed each of them in by hand for you.

This is what medicine should look like. An honest list of all the people involved, free access to information, and all the data pooled together, giving the most accurate information we can manage, to inform real decisions, and so prevent avoidable suffering and death.

We are a very, very long way away from there.

What can be done?

We urgently need to improve access to trial data. In addition to the previous suggestions, there are small changes that would vastly improve access to information, and so improve patient care.

1. The results of all trials conducted on humans must be made available within one year of completion, in summary table form if academic journal publication has not occurred. This requires the creation of a body that is charged with publicly auditing whether or not trials have withheld results at twelve months; and primary legislation that is enforced, as a matter of urgency, internationally, with stiff penalties for transgression. In my view these penalties should include fines, but also prison terms for those who are found to be responsible for withholding trial data, as patients are harmed by this process.

2. All systematic reviews – such as Cochrane reviews – that draw together trial results on any clinical question should also include a section on the trials which they know have been conducted, but whose results are being withheld. This should state: which completed trials have not reported results; how many patients' worth of information there is in each unreported trial; the names of the organisations and the individuals who are withholding the data; the efforts the reviewers have made to get the information from them. This is trivial extra work, as review teams already attempt to access this kind of data. Documenting this will draw attention to the problem, and make it easier for doctors and the public to see who is responsible for harming patient care in each area of medicine.

3. All clinical study reports should also be made publicly available, for all the trials that have ever been conducted on humans. This will be cheap, as the only costs are in finding one paper copy, scanning it and placing it online, perhaps with a check to remove confidential patient information. There is a vast mountain of highly relevant data on drugs that is currently being withheld, distorting what we know about treatments in widespread current

use. Many of these documents will be sitting in the dry paper archives of drug companies and regulators. We need legislation compelling the industry to hand them over. Our failure to fix this is costing lives.

4. We need to work on new methods for academics to extract summary information from these documents, as they are more detailed than published academic papers. The Cochrane group working on Tamiflu have made great progress here, learning as they go, and this field will need manuals.

5. We should work towards all trialists having an obligation to share patient-level data wherever possible, with convenient online data warehouses,[89] and streamlined systems whereby legitimate senior researchers can make requests for access, in order to conduct pooled analyses and double-check the results reported in published trials.

None of this is difficult, or impossible. Some of it is technical, for which I apologise. The field of missing data is a tragic and strange one. We have tolerated the emergence of a culture in medicine where information is routinely withheld, and we have blinded ourselves to the unnecessary suffering and death that follows from this. The people we should have been able to trust to handle all this behind the scenes – the regulators, the politicians, the senior academics, the patient organisations, the professional bodies, the universities, the ethics committees – have almost all failed us. And so I have had to inflict the details on you, in the hope that you can bring some pressure to bear yourself.

If you have any ideas about how we can fix this, and how we can force access to trial data – politically or technically – please write them up, post them online, and tell me where to find them.

2

Where Do New Drugs Come From?

Where do drugs come from?

That is the tale of hidden trial data. In the remainder of this book we will see how the pharmaceutical industry distorts doctors' beliefs about drugs through misleading and covert marketing; we will also see how trials can be flawed by design, and how regulators fail to regulate. But first, we must see how drugs are invented in the first place, and how they come to be available for prescription at all. This is a dark art, and generally remains a mystery for both doctors and patients. There are hidden traps at every turn, odd incentives, and frightening tales of exploitation. This is where new drugs come from.

From laboratory to pill

A drug is a molecule that does something useful, somewhere in the human body,[1] and luckily, there's no shortage of such molecules. Some are found in nature, in particular from plants: this makes sense, because we share a lot of molecular make-up with plants. Sometimes you just extract the molecule, but more commonly you add a few bits onto it here and there, through some elaborate chemical process, or take a few bits away, in the hope of increasing the potency, or reducing the side effects.

Often you'll have some idea about the mechanism you're targeting, and usually that's because you're copying the mechanism by which an existing drug works. For example, there's an enzyme in the body called cyclooxygenase, which helps to make molecules that signal inflammation. If you stop that enzyme working, it helps to reduce pain. Lots of drugs work like this, including aspirin, ibuprofen, ketoprofen, fenoprofen and so on. If you can find a new molecule that stops cyclooxygenase working in a laboratory dish, then it's probably going to stop it working in an animal, and if it does that, then it's probably going to help reduce pain in a person. If nothing disastrous has happened to animals or humans in the past when they've taken a drug that stops that enzyme working, then your new drug is fairly likely (though not certain) to be safe.

A new drug that operates in a completely new way is much more of a development risk, because it's unpredictable, and much more likely to fail at every step described above. But that kind of new drug would also be a more significant move forward in medical science. We'll discuss the tension between copying and innovating later.

One way that drugs are developed is by a process called screening, one of the most boring jobs imaginable for a young laboratory scientist. Hundreds, maybe thousands, of molecules, all of slightly different shapes and sizes, will be synthesised in the hope that they will operate on a particular target in the body. Then you come up with a lab method that lets you measure whether the drug is inducing the change you hope for – stopping an enzyme from working properly, for example – and then you try every drug out, one after the other, measuring their effects until you come up with a good one. Lots of great data is created during this period, and then thrown away, or locked in one drug company's vault.

Once you've got something that works in a dish, you give it to an animal. At this point you're measuring lots of different things. How much of the drug do you find in the blood after the animal eats the pill? If the answer is 'very little', your patients will need to eat giant horse pills to get an active dose, and that isn't practical. How long does the drug stay in the blood before it's broken down in the body? If the answer is 'one hour', your patients will need to take a pill twenty-four times a day, and that's not useful either. You might look at what your drug molecule gets turned into when it's broken down in the body, and worry about whether any of those breakdown products are harmful themselves.

At the same time you'll be looking at toxicology, especially very serious things that would rule a drug out completely. You want to find out if your drug causes cancer, for example, fairly early on in the development process, so you can abandon it. That said, you might be OK if it's a drug that people only take for a few days; by the same token, if it harms the reproductive system but is a drug for – let's say – Alzheimer's, you might be less worried (I only said less worried: old people do have sex). There are lots of standard methods at this stage. For example, it can take several years to find out if your drug has given living animals cancer, so even though you need to do this for regulatory approval, you'll also do early tests in a dish. One example is the Ames test, which lets you see if a drug has caused mutation in bacteria very quickly, by looking at what kinds of food they need to survive in a dish.

It's worth noting at this point that almost all drugs with desirable effects will also have unintended toxic effects at some higher dose. That's a fact of life. We're very complicated animals, but we only have about 20,000 genes, so lots of the building blocks of the human body are used several times over, meaning that something which interferes with one target in the body might also affect another, to a greater or lesser extent, at a higher dose.

So, you'll need to do animal and lab studies to see if your drug interferes with other things, like the electrical conductivity of the heart, that won't make it popular with humans; and various screening tests to see if it has any effect on common drug receptors, rodents' kidneys, rodents' lungs, dogs' hearts, dogs' behaviour; and various blood tests. You'll look at the breakdown products of the drug in animal and human cells, and if they give very different results you might try testing it in another species instead.

Then you'll give it in increasing doses to animals, until they are dead, or experiencing very obvious toxic effects. From this you'll find out the maximum tolerable dose in various different species (generally a rat or other rodent, and a non-rodent, usually a dog), and also get a better feel for the effects at doses below the lethal ones. I'm sorry if this paragraph seems brutal to you. It's my view – broadly speaking, as long as suffering is minimised – that it's OK to test whether drugs are safe or not on animals. You might disagree, or you might agree but prefer not to think about it.

If your patients are going to take the drug long-term, you'll be particularly interested in effects that emerge when animals have been taking it for a while, so you'll generally dose animals for at least a month. This is important, because when you come to give your drug to humans for the first time, you can't give it to them for longer than you've given it to animals.

If you're very unlucky, there'll be a side effect that animals don't get, but humans do. These aren't hugely common, but they do happen: practolol was a beta-blocker drug, very useful for various heart problems, and the molecule looks almost exactly the same as propranolol (which is widely used and pretty safe). But out of the blue, practolol turned out to cause something called multi-system oculomucocutaneous syndrome, which is horrific. That's why we need good data on all drugs, to catch this kind of thing early.

As you can imagine, this is all very time-consuming and expensive, and you can't even be sure you've got a safe, effective drug once you've got this far, because you haven't given it to a single human yet. Given the improbability of it all, I find it miraculous that any drug works, and even more miraculous that we developed safe drugs in the era before all this work had to be done, and was technically possible.

Early trials

So now you come to the nerve-racking moment where you give your drug to a human for the first time. Usually you will have a group of healthy volunteers, maybe a dozen, and they will take the drug at escalating doses, in a medical setting, while you measure things like heart function, how much drug there is in the blood, and so on.

Generally you want to give the drug at less than a tenth of the 'no adverse effects' dose in the animals that were most sensitive to it. If your volunteers are OK at a single dose, you'll double it, and then move up the doses. You're hoping at this stage that your drug only causes adverse effects at a higher dose, if at all, and certainly at a much higher dose than the one at which it does something useful to the expected target in the body (you'll have an idea of the effective dose from your animal studies). Of all the drugs that make it as far as these phase 1 trials, only 20 per cent go on to be approved and marketed.

Sometimes – mercifully rarely – terrible things happen at this stage. You will remember the TGN1412 story, where a group of volunteers were exposed to a very new kind of treatment which interfered with signalling pathways in the immune system, and ended up on intensive care with their fingers and toes rotting off. This is a good illustration of why you shouldn't give a treatment simultaneously to several volunteers if it's very unpredictable and from an entirely new class.

Most new drugs are much more conventional molecules, and generally the only unpleasantness they cause comes from nausea, dizziness, headache and so on. You might also want a few of your test subjects to have a dummy pill with no medicine in it, so you can try to determine if these effects are actually from the drug, or are just a product of dread.

At this moment you might be thinking: what kind of reckless maniac gives their only body over for an experiment like this? I'm inclined to agree. There is, of course, a long and noble tradition of self-experimentation in science (at the trivial end, I have a friend who got annoyed with feeding his mosquitoes the complicated way, and started sticking his arm in the enclosure, rearing a PhD on his own blood). But the risks might feel more transparent if it's your own experiment. Are the subjects in first-in-man trials reassured by blind faith in science, and regulations?

Until the 1980s, in the US, these studies were often done on prisoners. You could argue that since then such outright coercion has been softened, rather than fully overturned. Today, being a guinea pig in a clinical trial is a source of easy money for healthy young people with few better options: sometimes students, sometimes unemployed people, and sometimes much worse. There's an ongoing ethical discussion around whether such people can give meaningful consent, when they are in serious need of money, and faced with serious financial inducements.[2] This creates a tension: payments to subjects are supposed to be low, to reduce any 'undue inducement' to risky or degrading experiences, which feels like a good safety mechanism in principle; but given the reality of how many phase 1 subjects live, I'd quite like them to be paid fairly well. In 1996 Eli Lilly was found recruiting homeless alcoholics from a local shelter.[3] Lilly's director of clinical pharmacology said: 'These individuals want to help society.'

That's an extreme case. But even at best, volunteers come from less well-off groups in society, and this means that drugs taken by all of us are tested – to be blunt – on the poor. In the US, this means people without medical insurance, and that raises another interesting issue: the Declaration of Helsinki, the ethics code which frames modern medical activity, says that research is justified if the population from whom participants are drawn would benefit from the results. A new AIDS drug shouldn't be tested on people in Africa, for example, who could never afford to buy it. But uninsured unemployed people in the US do not have access to expensive medical treatments either – and they may still not have access to all expensive treatments after the coming changes – so it's not clear that they could benefit from this research. On top of that, most agencies don't offer free treatment to injured subjects, and none give them compensation for suffering or lost wages.

This is a strange underworld that has been brought to light for the academic community by Carl Elliot, an ethicist, and Robert Abadie, an anthropologist who lived among phase 1 participants for his PhD.[4] The industry refers to these participants by the oxymoron 'paid volunteers', and there is a universal pretence that they are not paid for their work, but merely reimbursed for their time and travel expenses. The participants themselves are under no such illusions.

Payment is often around $200 to $400 a day, studies can last for weeks or more, and participants will often do several studies in a year. Money is central to the process, and payment is often back-loaded, so you only receive full payment if you complete the study, unless you can prove your withdrawal was due to serious side effects. Participants generally have few economic alternatives, especially in the US, and are frequently presented with lengthy and impenetrable consent forms, which are hard to navigate and understand.

You can earn better than the minimum wage if you 'guinea

pig' full-time, and many do: in fact, for many of them it's a job, but it's not regulated as any other job might be. This is perhaps because we feel uncomfortable regarding this source of income as a profession, so new problems arise. Participants are reluctant to complain about poor conditions, because they don't want to miss out on future studies, and they don't go to lawyers for the same reason. They may be disinclined to walk away from studies that are unpleasant or painful, too, for fear of sacrificing income. One participant describes this as 'a mild torture economy': 'You're not being paid to do a job . . . you're being paid to endure.'

If you really want to rummage in this underworld, I recommend a small photocopied magazine called *Guinea Pig Zero*. For anyone who likes to think of medical research as a white-coated exercise, with crisp protocols, carried out in clean glass-and-metal buildings, this is a rude awakening.

> The drugs are hitting the boys harder than the girls. The ephedrine is not so bad, it's like . . . over the counter speed. Then they increased our dosage and things got funky. This is when the gents took to the mattresses . . . We women figured we had more endurance . . . No. 2 was feeling so bad that he hid the pills under his lounge during the dosing procedure. The coordinator even checked his mouth and he still got away with it . . . this made No. 2 twice as sick after the next dosing – he couldn't fake it for the rest of the study.[5]

Guinea Pig Zero published investigations into deaths during phase 1 trials, advice for participants, and long, thoughtful discursions on the history of guinea-pigging (or, as the subjects themselves call it, 'our bleeding, pissing work'). Illustrations show rodents on their backs with thermometers in their anuses, or cheerfully offering up their bellies to scalpels. This wasn't just idle carping, or advice on how to break the system. The volunteers developed 'research unit report cards' and discussed unionising: 'The need exists for a set of standard expectations to be set down in an independently controlled, guinea-pig based forum so we volunteers can rein in the sloppy units in a way that doesn't bring ourselves harm.'

These report cards were informative, heartfelt and entertaining, but as you might expect, they were not welcomed by the industry. When three of them were picked up by *Harper's* magazine, it resulted in libel threats and apologies. Similarly, following a Bloomberg news story from 2005 – in which more than a dozen doctors, government officials and scientists said the industry failed to adequately protect participants – three illegal immigrants from Latin America said they were threatened with deportation by the clinic they had raised concerns about.

We cannot rely solely on altruism to populate these studies, of course. And even where altruism has provided, historically, it has been in extreme or odd circumstances. Before prisoners, for example, drugs were tested on conscientious objectors, who also wore lice-infested underpants in order to infect themselves with typhus, and participated in 'the Great Starvation Experiment' to help Allied doctors understand how we should deal with malnourished concentration camp victims (some of the starvation subjects committed acts of violent self-mutilation).[6]

The question is not only whether we feel comfortable with the incentives and the regulation, but also whether this information is all new to us, or simply brushed under the carpet. You might imagine that research all takes place in universities, and twenty years ago you'd have been correct. But recently, and very rapidly, almost all research activity has been outsourced, often far away from universities, into small private clinical research organisations, which sub-contract for drug companies, and run their trials all around the world. These organisations are atomised and diffuse, but they are still being monitored by frameworks devised to cope with the ethical and procedural problems arising in large institutional studies, rather than small businesses. In the US, in particular, you can shop around for Institutional Review Board approval, so if one ethics committee turns you down, you simply go to another.

This is an interesting corner of medicine, and phase 2 and 3 trials are being outsourced too. First, we need to understand what those are.

Phase 2 and 3

So, you've established that your drug is broadly safe, in a few healthy people referred to by popular convention as 'volunteers'. Now you want to give it to patients who have the disease you're aiming to treat, so you can try to understand whether it works or not.

This is done in 'phase 2' and 'phase 3' clinical trials, before a drug comes to market. The line between phase 2 and 3 is flexible, but broadly speaking, in phase 2 you give your drug to a couple of hundred patients, and try to gather information on short-term outcomes, side effects and dosage. This will be the first time you get to see if your blood-pressure drug does actually lower blood pressure in people who have high blood pressure, and it might also be the first time you learn about very common side effects.

In phase 3 studies you give your drug to a larger group of patients, usually somewhere between three hundred and 2,000, again learning about outcomes, side effects and dosage. Crucially, all phase 3 trials will be randomised controlled trials, comparing your new treatment against something else. (All of these pre-marketing trials, you will notice, are in fairly small numbers of people, which means that rarer side effects are very unlikely to be picked up. I'll come back to this later.)

Here again, you may be wondering: who are these patients, and where do they come from? It's clear that trial participants are not representative of all patients, for a number of different reasons. First, we need to consider what drives someone to participate in a trial. It would be nice to imagine that we all recognise the public value of research and that all research had public value. Unfortunately, many trials are conducted on drugs that are simply copies of other companies' products, and are therefore an innovation designed merely to make money for a drugs company, rather than a significant leap forward for patients. It's hard for participants to work out whether a trial really does represent a meaningful clinical question, so to an extent we can understand people's reluctance to take part. But in any case, wealthy patients from the developed world have become more reluctant to participate in trials across the board, and this raises interesting issues, both ethical and practical.

In the US, where for decades many millions of people have been unable to pay for health care, clinical trials have often been marketed as a way to access free doctors' appointments, scans, blood tests and treatment. One study compared insurance status in people who agreed to participate in a clinical trial with those who declined;[7] participants are a diverse population, but still, those agreeing to be in a trial were seven times more likely to have no health insurance. Another study looked at strategies to improve targeted recruitment among Latinos, a group with lower wages and poorer health care than the average:[8] 96 per cent agreed to participate, a rate far higher than would normally be expected.

These findings echo what we saw in phase 1 trials, where only the very poor were offering themselves for research. They also raise the same ethical question: trial participants are supposed to come from the population of people who could realistically benefit from the answers provided by that trial. If participants are the uninsured, and the drugs are only available to the insured, then that is clearly not the case.

But selective recruitment of poor people for trials in the USA is trivial compared to another new development, about which many patients – but also many doctors and academics – are entirely ignorant. Drug trials are increasingly outsourced around the world, to be conducted in countries with inferior regulation, inferior medical care, different medical problems and – in some cases – completely different populations.

'CROs' and trials around the world

Clinical research organisations are a very new phenomenon. Thirty years ago, hardly any existed: now there are hundreds, with a global revenue of $20 billion in 2010, representing about a third of all pharma R&D spending.[9] They conduct the majority of clinical trials research on behalf of industry, and in 2008 CROs ran more than 9,000 trials, with over two million participants, in 115 countries around the world.

This commercialisation of trials research raises several new concerns. First, as we have already seen, companies often bring pressure to bear on academics they are funding, discouraging them from publishing unflattering results and encouraging them to put spin on both the methods and the conclusions of their work. When academics have stood up to those pressures, the threats have turned into grim realities. What employee or chief executive of a CRO is likely to stand up to a company which is directly paying the bills, when the staff all know that the CRO's hope for future business rides on how it manages each demanding client?

It's also interesting to note that the increasing commercialisation of research has driven many everyday clinicians away from trials, even when the trials come from the more independent end of the spectrum. Three British academics have written recently of their difficulty in getting doctors to help them recruit patients for a study requested by the European medicines regulator, but paid for by Pfizer: 'Academics wrote the protocols, collaborators are academic, and the study data are owned by the steering committees (on which industry has no say), which also control analyses and publications. A university is the sponsor. Funding is from industry, which has no role in study conduct, data collection, or data interpretation.'[10] UK doctors and primary care trusts regarded this study as commercial, and were reluctant to hand over their patients. They are not alone. The Danish Board of Medicine regards these kinds of studies as commercial, which means that any practice participating in them must declare the interest, further reducing recruitment. In the US, meanwhile, the use of private community doctors to conduct trials has expanded enormously, with incentives approaching $1 million a year for the most enterprising medics.[11]

For a window into the commercial reality of the CRO world, you can look at the way these services are presented when they are being promoted to pharmaceutical companies, and see how far this reality is from the needs of patients, and any spirit of neutral enquiry. Quintiles, the largest company, offers to help its industry customers to 'better identify, promote, and prove the value of a particular drug to key stakeholders'.[12] 'You've spent hundreds of millions of dollars and years bringing your product through the drug development process,' it says. 'Now you face multiple opportunities – and possibly more requirements – to demonstrate safety and effectiveness in larger populations.' There are also cases of CROs and drug companies with contracts that share the risk of a poor outcome between them, increasing the chances of a conflict of interest even further.

These aren't smoking guns. They simply illustrate the banal commercial reality of what these companies do: they find stuff out, of course, but their main objective is to make a company's drug look good so that regulators, doctors and patients will swallow it. That's not ideal in science. It's not fraud either. It's just not ideal.

It would be wrong to imagine that this shift in culture has been driven by a hope that CROs would produce more flattering findings than other options. They are attractive because they're fast, efficient, focused and cheap. And they're especially cheap because, like many other industries, they can also outsource their work to poorer countries. As the former chief executive of GSK explained in a recent interview, running a trial in the US costs $30,000 per patient, while a CRO can do it in Romania for $3,000.[13] That is why GSK aims to move half of its trials to low-cost countries, and it is part of a global trend.

In the past, only 15 per cent of clinical trials were conducted outside the USA. Now it's more than half. The average rate of growth in the number of trials in India is 20 per cent a year, in China 47 per cent, in Argentina 27 per cent, and so on, simply because they are better at attracting CRO business, at lower cost. At the same time, trials in the US are falling by 6 per cent a year (and in the UK by 10 per cent a year).[14] As a result of these trends, many trials are now being conducted in developing countries, where regulatory oversight is poorer, as is the normal standard of clinical care. This raises a huge number of questions about data integrity, the relevance of findings to developed world populations, and ethics – all issues with which regulators around the world are currently struggling.

There are many anecdotal accounts of bad behaviour with trial data in poorer countries, and it's clear that the incentives for massaging results are greater in a country where a clinical trial is paying its subjects vastly more than local wages could match. There are also difficulties with regulatory requirements that fall across two countries or two languages, as well as translation issues in patient reports, especially for unexpected side effects. Site visits for monitoring may be of variable quality, and countries differ in how much corruption is routine in public life. There may also be less local familiarity with administrative requirements concerning data integrity (which have been a bone of contention between industry and regulators in the developed world).[15]

These are, we should be clear, only hints of problems with data. There have been cases where trials from developing countries have produced positive results, while those conducted elsewhere showed no benefit, but to the best of my knowledge, very little quantitative research has so far been conducted comparing the results of trials in poorer countries with those in the US and Western Europe. This means we can draw no firm conclusions on data integrity; it also means – I would suggest – that this is an open field for highly publishable research by anyone reading this book. A barrier to such research, though, would be access to the most basic information. One review of articles in leading medical journals reporting on multi-centre trials found that less than 5 per cent gave any information on recruitment numbers from each individual country.[16]

There is also the issue of publication bias, where whole trials disappear. We saw in the previous chapter how unflattering data can go missing in action. European and US researchers with stable academic positions have had difficulties retaining the right to publish, sometimes engaging in heated confrontations with drug companies. It's hard to imagine that such problems wouldn't be exacerbated in a developing country setting, where commercial research has introduced unprecedented investment for individuals, institutions and communities. This is especially problematic because the trials registers, where protocols should be posted before a trial begins, are often poorly policed around the world; and trials from the developing world – or simply data from new sites – may only come to the attention of the international community after they are completed.

But there is a more interesting problem with running trials in such diverse populations: people, and medicine, are not the same the world over. It is known – we shall see more later – that trials are generally conducted in very unrepresentative 'ideal' patients, who are often less sick than real-world patients, and on fewer other drugs. In trials conducted in developing countries, these problems can be exacerbated. A typical patient in Berlin or Seattle with high blood pressure may have been on several drugs for several years. Now, you might collect data on the benefits of a new blood-pressure drug in Romania, or India, where the patients may not have been on any other medication, because access to what would be regarded as normal medical treatment in the West is much less common. Are those findings really transferable, and relevant, to American patients, on all their tablets?

Beyond differences in routine treatment, there will also be a different social context. Are patients diagnosed with depression in China really the same as patients diagnosed with depression in California? And then there are genetic differences. You might know, from drinking with friends, that many Oriental people metabolise drugs, especially alcohol, differently from Western-ers: if a drug has few side effects at a particular dose in Botswana, can you really rely on that data for your patients in Tokyo?[17]

There are other cultural considerations. Trials are not simply a one-way street: they are also a way to create new markets, in countries like Brazil, say, by reshaping norms of clinical prac-tice, and modifying patients' expectations. Sometimes this may be a good thing, but trials can also create expectations for drugs that cannot be afforded. And they can even, by distorting local employment markets, draw good physicians away from clinical work in their own communities, and off into research jobs (just as Europe has taken expensively trained doctors and nurses away from developing countries by emigration).

But more than anything, these trials raise enormous issues about the ethics of research, and meaningful informed consent.[18] The incentives offered to participants in developing countries can exceed the average annual wage. Some countries have a culture of 'doctor knows best', where patients will be more likely to accept unusual or experimental treatments simply because their doctor has offered them (a doctor with a significant personal financial interest, we should note, since they are paid for each recruited patient). The background and risks – that a drug is new, that they may actually be taking a dummy placebo pill – may not be clearly communicated to patients. Informed consent may not be adequately policed. Standards of ethics oversight can also vary: in a survey of researchers in developing countries, half said their research was not reviewed at all by an institutional review board.[19] A review of Chinese trial publications found that only 11 per cent mentioned ethics approval, and only 18 per cent discussed informed consent.[20] This is a very different ethical context for research to that in Europe or the US; and while international regulators have tried to keep up, it's not yet clear whether the steps they've taken will be successful.[21] What's more, oversight is especially problematic, since these trials are often used to bolster the marketing case for a drug after it has come to market, and are not included in the bundle of documents presented for regulatory approval, meaning they are less subject to Western regulatory control.

Trials in the developing world also present the issue of fairness that we met earlier in our discussion of phase 1 trials: the people participating in trials are supposed to come from a population that could reasonably expect to benefit from their results. In several damning cases, especially from Africa, it is very clear that this was not the case. In some cases, more horrifically, it seems that effective treatment which was available may have been withheld as part of the drug company's effort to conduct a trial.

The most notorious story is the Trovan antibiotic study conducted by Pfizer in Kano, Nigeria, during a meningitis epidemic. An experimental new antibiotic was compared, in a randomised trial, with a low dose of a competing antibiotic that was known to be effective. Eleven children died, roughly the same number from each group. Crucially, the participants were apparently not informed about the experimental nature of the treatments, and moreover, they were not informed that a treatment known to be effective was available, immediately, from Médecins sans Frontières next door at the very same facility.

Pfizer argued in court – successfully – that there was no international norm requiring it to get informed consent for a trial involving experimental drugs in Africa, so the cases relating to the trial should be heard in Nigeria only. That's a chilling thing to hear a company claim about experimental drug trials, and it was knocked back in 2006 when the Nigerian Ministry of Health released its report on the trial. This stated that Pfizer had violated Nigerian law, the UN Convention on the Rights of the Child and the Declaration of Helsinki.

This all took place in 1996, and was the inspiration for John le Carré's novel The Constant Gardener. You may think 1996 was a long time ago, but the facts in these matters are always on a delay, and in contentious or litigated issues the truth can move very slowly. In fact, Pfizer only settled the case out of court in 2009, and several disturbing new elements of what is clearly an ongoing saga emerged in the WikiLeaks diplomatic cables made public in 2010.[22] One US diplomatic cable describes a meeting in April 2009 between Pfizer's country manager and US officials at the American embassy in Abuja, where smears of a Nigerian official involved in the litigation are casually discussed.

According to [Pfizer's country manager], Pfizer had hired investigators to uncover corruption links to Federal Attorney General Michael Aondoakaa to expose him and put pressure on him to drop the federal cases. He said Pfizer's investigators were passing this information to local media. A series of damaging articles detailing Aondoakaa's 'alleged' corruption ties were published in February and March. Liggeri contended that Pfizer had much more damaging information on Aondoakaa and that Aondoakaa's cronies were pressuring him to drop the suit for fear of further negative articles.[23]

Pfizer deny any wrongdoing in the Trovan trials, and say the statements contained in the cable are false.[24] Its $75 million settlement was subject to a confidentiality clause.

These issues are disturbing in themselves, but they must also be seen against the wider context of trials in the developing world on drugs that are not available for normal clinical use in those countries. It's a classic ethics essay quandary, but very much set in real life: imagine you're in a country where modern AIDS medicines cannot be afforded. Is it reasonable to run a trial of an expensive new AIDS drug in that setting? Even if it has been broadly demonstrated to be safe? What if the control group in your trial are simply receiving dummy placebo pills, which is to say, effectively, nothing? In the USA, no patient would receive dummy sugar pills for AIDS. In this African country, perhaps 'nothing' is a commonplace treatment.

This is an area of considerable obfuscation and embarrassment, tangled up in complex regulatory frameworks which are starting to change in a worrying direction. In 2009, three researchers wrote in the *Lancet* drawing attention to one very notable shift.[25] For years, they explained, the FDA had insisted that when a company applied for a marketing authorisation for a drug in the US, all foreign trials given as evidence had to show that they were compliant with the Declaration of Helsinki.[26] In 2008 this requirement changed, but only for foreign trials, and the FDA shifted to the International Conference on Harmonisation (ICH) Good Clinical Practice (GCP) guidelines. These are not terrible, but they are only voted on by members from the EU, the USA and Japan. They are also more focused on procedures, while Helsinki clearly articulates moral principles. But most concerning are the differences in detail, when you consider that GCP is now the main ethical regulation for trials in the developing world.

Helsinki says that research must benefit the health needs of the populations where it is conducted. GCP does not. Helsinki discusses the moral need for access to treatment after a trial has finished. GCP does not. Helsinki restricts the use of dummy placebo sugar pills in trials, where there are effective treatments available. GCP does not. Helsinki also, incidentally, encourages investigators to disclose funding and sponsors, post the study design publicly, publish negative findings, and report results accurately. GCP does not. So this was not a reassuring regulatory shift, specifically for trials conducted outside the US, and specifically in 2008, at a time when studies were moving outside the US and the EU at a very rapid pace.

It's also worth mentioning that the pharmaceutical industry plays hardball with developing countries over the price of medicines. Like much else we are discussing, this is worthy of a book in its own right, but here is just one illustrative story. In 2007 Thailand tried to take a stand against the drug company Abbott over its drug Kaletra. There are more than half a million people living with HIV in Thailand (many of them can thank Western sex tourists for that), and 120,000 have AIDS. The country can afford first-line AIDS drugs, but many become ineffective with time, through acquired resistance. Abbott had been charging $2,200 a year for Kaletra in Thailand, which was – by morbid coincidence – roughly the same as the gross income per capita.

We give drug companies exclusive rights to manufacture the treatments they have discovered for a limited period of time – usually about eighteen years – in order to incentivise innovation. It's unlikely that the revenue available from selling drugs in poorer countries will ever incentivise innovation of new treatments to any great extent (we can see this very clearly from the fact that so many medical conditions that occur mainly in developing countries are neglected by the pharmaceutical industry). Because of this there are various international treaties, such as the Doha Declaration of 2001, under which a government can declare a public-health emergency and start manufacturing or buying copies of a patented drug. One memorable use of these 'compulsory licences' was when the US government insisted in the aftermath of the 9/11 attacks that it should be allowed to buy large amounts of cheap ciprofloxacin to treat anthrax when it was worried that spores were being sent to politicians by terrorists.

So in January 2007 the Thai government announced that it was going to copy Abbott's drug, only for the country's poor, to save lives. Abbott's response was interesting: it retaliated by completely withdrawing its new heat-stable version of Kaletra from the Thai market, and six other new drugs for good measure, and then announced that it would not bring these drugs back to the Thai market until the government promised not to use a 'compulsory licence' on its drugs again. It's hard to think of anything less in keeping with the Doha Declaration. If you want more moral context, the World Health Organization estimates that half of HIV transmissions in Thailand come from contact between sex workers and their clients. There are said to be two million women and 800,000 children under eighteen working in the Thai sex trade, much of which is servicing Western men, some of whom you may know personally.

So those are phase 1, 2 and 3 clinical trials: both the science of them and, I hope, some colour about the reality beyond the protocols, in the clinics and on the streets. It may have made you nervous. The story from there on is simple: the medicines regulator, whether it's the FDA, or the EMA, or some other country's body, looks at the results from these phase 1, 2 and 3 trials, works out if the drug is effective and the side effects are acceptable, then either asks for some more trials, tells the company to bin the drug, or lets it go on the market for prescription by any willing doctor. That's the theory, at any rate.

In reality, things are much messier.

3

Bad Regulators

Getting your drug approved

After all the effort and expense of discovering a molecule and conducting trials, still nobody can prescribe your drug: first you must go to regulators and get them to approve it for marketing in their territory. Like so much in medicine, this field has been protected from public scrutiny by the complicated nature of the process, and in general even doctors don't fully understand what regulators do: as just one illustration of this reality, a 2006 survey by Ipsos MORI found that 55 per cent of hospital doctors and 37 per cent of GPs[1] had never even heard of the MHRA, the UK medicines regulator.[2]

In principle, the job of a regulator is simple: it approves a drug in the first place, after seeing trials that show it works; it monitors a drug's safety once it's on the market; it communicates risks and hazards to doctors; and it removes unsafe and ineffective medications from sale. Unfortunately, as we shall see, while there are many good people trying to do their best for patients, regulators are beset with problems: there are pressures from industry; pressures from government; problems with funding; issues of competence; conflicts of interest within their own ranks; and worst of all, again, a dangerous obsession with secrecy.

Pressures on regulators

Sociologists of regulation – such people exist – talk about something called 'regulatory capture'.[3] This is the process whereby a state regulator ends up promoting the interests of the industry it is supposed to monitor, at the expense of the public interest. It can happen for a number of reasons, and many of them are very human. For example, if you work in the technical business of drug approval and pharmacovigilance, who can you chat to about your day's work? To your partner, it's all baffling and pedantic; but the people in the regulatory affairs division of the companies you work with every day, they understand. You have so much in common with them. So industry bodies – not even necessarily the pharmaceutical companies themselves – might offer things as intangible as friendship, and opportunities to socialise.

That is how regulatory capture works, and it has been discussed at length in the academic literature, as well as by those who seek to influence regulators. A disarmingly honest illustration of how industries view this process can be found in a book called *The Regulation Game: Strategic Use of the Administrative Process*:

> Effective lobbying requires close personal contact between the lobbyists and government officials. Social events are crucial to this strategy. The object is to establish long-term personal relationships transcending any particular issue. Company and industry officials must be 'people' to the agency decision-makers, not just organisational functionaries. A regulatory official contemplating a decision must be led to think of its impact in human terms. Officials will be much less willing to hurt long-time acquaintances than corporations. Of course, there are also important tactical elements of lobbying . . . This is most effectively done by identifying the leading experts in each relevant field and hiring them as consultants or advisors, or giving them

research grants and the like. This activity requires a modicum of finesse; it must not be too blatant, for the experts themselves must not recognise that they have lost their objectivity and freedom of action. At a minimum, a programme of this kind reduces the threat that the leading experts will be available to testify or write against the interests of the regulated firms.[4]

Then there is free movement of staff between regulators and drug companies, a revolving door which creates problems that are very hard to monitor and contain. Government regulators tend not to pay very well, and after some time working for the MHRA you might start to notice that the people in the regulatory affairs department at the companies you work with – the ones you socialise with – all seem to have much smarter cars than you. They live in much more expensive areas, and their children go to much better schools. But they do basically the same job as you, just on the other side of the fence. In fact, as somebody with inside knowledge of the regulator, you can be very valuable to a pharmaceutical company, especially since it is a field where the written rules are often lengthy but vague, and many of the details on 'what you can get away with' are effectively an oral tradition.

This movement of personnel then creates a further problem: what if people working for a regulator, while still in post, are thinking of their future at a drug company? It's possible, after all, that they might be reluctant to make decisions that would alienate a potential future employer. This is a very difficult conflict of interest to monitor and regulate, since no declarable current arrangement exists; it's also hard to predict who will move; and there is little opportunity to impose sanctions retrospectively. Furthermore, if individuals working at a regulator are changing their behaviour based on vague thoughts of future employment, it probably won't be on the basis of a specific job plan, or a specific exchange of favours, so clear evidence of corruption will be hard to spot. The whole process may not even be conscious, and in any case, all large organisations are like container ships, taking a long time to change direction. Rather, you might see a shift in the spirit of the workers, and a slow, gradual reorientation of priorities and implicit organisational goals.

The clearest illustration of how these problems are managed by the European Medicines Agency comes from the head of the organisation itself. The EMA regulates the pharmaceutical industry throughout the whole of Europe, and has taken over the responsibilities of the regulators in individual member countries. In December 2010 Thomas Lonngren stepped down as its executive director. On the 28th of that month he sent a letter telling the EMA management board that he was going to start working as a private consultant to the pharmaceutical industry, starting in just four days' time, on 1 January 2011.[5]

Some places, and some fields, have clear regulations on this kind of thing. In the USA, for example, you have to wait a year after leaving the Defense Department before you can work for a defence contractor. After ten days the chairman of the EMA wrote back to Lonngren saying that his plans were fine.[6] He didn't impose any further restrictions, and nor, remarkably, did he ask for any information on what kind of work Lonngren planned to do.[7] Lonngren had said in his letter that there would be no conflict of interest, and that was enough for everyone concerned.

My concern here is not Thomas Lonngren – although I think we would all struggle to be impressed by his behaviour, since we are all, in Europe at any rate, his former employers. This story is interesting, rather, for what it tells us about the EMA and its casual approach to these kinds of problems. A man who was previously overseeing the approval of new medicines now advises companies on how to get their drugs approved, having given the EMA four days' notice of this plan, between Christmas and New Year, and *nobody* in the organisation regards that as a problem, even though it represents a staggering conflict of interest. In fact, this is not so unusual: the Corporate Europe

Observatory recently produced a report detailing fifteen similar cases of senior EU officials passing through the revolving door between government and industry.[8]

But regulator employees aren't the only people operating with a conflict of interest (a notion I'll examine more in Chapter 6). Many of the patient representatives who sit on boards at the EMA, including the two on their management board, come from organisations heavily funded by pharmaceutical companies. This is despite EMA rules stating that 'members of the Management Board … shall not have financial or other interests in the pharmaceutical industry which could affect their impartiality'.

The same problem exists among the scientific and medical experts advising the agencies, and sitting on their boards. In the USA, at the FDA meeting on how to manage controversial cox-2 painkiller drugs like Vioxx, ten out of thirty-two members of the panel had some kind of conflict of interest, and nine of those ten voted to keep the drugs on the market, compared with the 60/40 split among the remainder of the committee. Research reviewing a series of FDA votes found that experts are slightly more likely to vote in a company's interest if they have a financial tie to that company (though their exclusion would not have affected the overall outcome of the meetings studied).[9]

There are endless stories of conflicts of interest at the FDA, and regulatory decisions that have been distorted by political pressure. I don't find these kinds of stories very interesting (though I'm glad others document them), because they're often more soap opera than science, but there's plainly a problem,[10] and it's not a new one. In the 1950s, US Senator Estes Kefauver ran a series of hearings into the activities of the FDA. He noted that drugs were often approved despite offering no new benefits. Along with many other changes, he recommended that drugs should be licensed, and then be subject to rigorous subsequent reviews for renewal once they were on the market, but FDA officials opposed him, and medical officers there complained extensively of industry influence. One possible explanation for this odd situation comes from the payments that were uncovered: the head of one division had received $287,000 from drug companies, which is over $2 million in today's money.[11]

Today, a troubling sense of distorted priorities can still be detected in anonymous surveys of people working at the regulators, though the influences appear to be political rather than financial. The fabulously named 'Union of Concerned Scientists' recently surveyed 1,000 scientists working at the FDA and found that 61 per cent said they knew of cases where 'Department of Health and Human Services or FDA political appointees have inappropriately injected themselves into FDA determinations or actions'. A fifth said that they themselves 'have been asked, for non-scientific reasons, to inappropriately exclude or alter technical information or their conclusions in an FDA scientific document'. Only 47 per cent thought the FDA 'routinely provides complete and accurate information to the public'.[12] If you're worried about a think tank conducting a survey, the US Department of Health and Human Services had conducted a similar one two years earlier: again, a fifth of those surveyed said they had been pressured to approve a drug despite reservations about efficacy and safety.[13]

Then there are testimonies from within the organisation. During the withdrawal of Vioxx from the market, several questionable decisions were made by the FDA. Afterwards, David Graham, Associate Director for Science and Medicine in the Office of Drug Safety, told the US Senate Committee on Finance: 'The FDA has become an agent of industry. I have been to many, many internal meetings, and as soon as a company says it is not going to do something, the FDA backs down. The way it talks about industry is "our colleagues in industry".'

Various suggestions have been made, over the years, on how to manage the problem of regulatory experts with industry ties.

One, of course, is to exclude them from the decision-making process completely, though this can introduce new problems, if it is hard to find *anyone* with no such ties. This is not because academics are all corrupt or money-grabbing, but rather because university staff have been energetically encouraged to work collaboratively with industry for over two decades, by governments around the world, in the belief that this will stimulate innovation and reduce costs for the public sector. Having actively created this situation, it would be bizarre if we now had to seriously consider excluding some of our best academics from being able to inform us on issues of efficacy and safety. The question, then, becomes how we can monitor and contain the conflicts of interest that arise.

A second suggestion is that the membership and voting of these panels should be open. On this, the FDA is far ahead of the EMA, where membership, voting and comments have been secret since its inception, with some promises of greater transparency being made in the past year (you will know from what you have already read not to judge a promise from the EMA at face value, but to wait and see what actually happens). It's worth saying here that while I don't think it should change anyone's mind about transparency, there is one argument for secret meetings with unattributed comments: people may be more candid if they know they are speaking off the record, unattributably. 'I shouldn't tell you this,' a professor might say to a room full of people he trusts, 'but everyone at MGB knows that drug is rubbish, and the latest unreported trial is looking bad too.'

There are other quirks that lead regulators to feel – perhaps – disoriented as to where their allegiances lie. Up until 2010, for example, the EMA sat in the European Commission's Enterprise and Industry Directorate, rather than under health, which might make you worry that political oversight was more focused on the economic benefits of a friendly relationship with the $700 billion pharmaceutical industry than on the interests of patients.[14]

In both the United States and the European Union, regulators are paid for almost entirely by drug companies, through the fees that are paid to pass regulatory hurdles. Until a few years ago, when approval was centralised to the EMA, this was a particular source of concern in Europe, because drug companies could choose which country they would seek approval in, and this created something of a competition. Overall this payment model has created an impression that the companies are the customer, but this is not simply because they write the cheques: this change in funding was brought in specifically to improve approval times for industry.

Approving a drug

So what do regulators mean by 'effective' when they measure the benefits of a new drug? The specific details for each drug are often a matter of ad hoc negotiation, and in the dark arts of getting a drug approved, inside knowledge and oral tradition are often as valuable as knowing the rules: for example, research has shown that applications from large companies, which have greater experience of the regulatory process, pass through to approval faster than those from smaller companies. In general, however, a company would expect to have to provide two or three trials, with a thousand or more participants, showing that its drug works.

This is where the smoke and mirrors begin. Although the notion of a simple randomised trial should be straightforward, in reality there are all kinds of distortions and perversions that can come into play, in the comparisons that are made, and the outcomes that are measured for success. For me, 'What works?' is the most basic practical question that every patient faces, and the answer isn't complicated. Patients want to know: what's the best treatment for my disease?

The only way to answer this question when a new drug comes along is by comparing it against the best currently available treatment. But that is not what drug regulators require of a treatment for it to get onto the market. Often, even when there are effective treatments around already, regulators are happy for a company simply to show that its treatment is better than nothing – or rather, better than a dummy placebo pill with no medicine in it – and the industry is happy to clear that low bar.

'Better than nothing'

This raises several serious problems, the first of which is ethical. It's obviously wrong to put patients in a trial where half of them will be given a placebo, if there is a currently available option which is known to be effective, because you are actively depriving half of your patients of treatment for their disease. Remember, these are not healthy volunteers, giving their bodies over for financial reward: these are real patients, often with serious medical problems, hoping for treatment and exposing themselves to some inconvenience (but hopefully no more than that) in order to advance the state of medical knowledge for other sufferers in the future.

What's more, if patients participate in a trial that uses a placebo instead of a currently available effective treatment, they are suffering a double sting. In all likelihood, the trial they're taking part in is not attempting to answer a clinically meaningful question, relevant to medical practice. Because doctors and patients aren't interested in whether a new drug is better than nothing, except as a matter of the most abstract and irrelevant science. We're interested in the practical question of whether it's better than the best currently available option, and when a drug is approved, at the very least we'd expect to see trials which answer this question.

That is not what we get. A paper from 2011 looked at the evidence supporting every single one of the 197 new drugs approved by the FDA between 2000 and 2010, at the time they were approved.[15] Only 70 per cent had data to show they were better than other treatments (and that's after you ignore drugs for conditions for which there was no current treatment). A full third had no evidence comparing them with the best currently available treatment, even though that's the only question that matters to patients.

As we have seen, the Declaration of Helsinki is pretty hot on patients not being exposed to unnecessary harm in trials. It started to get hot on the misuse of placebos in a 2000 amendment which says that the use of dummy pills is only acceptable when

> for compelling and scientifically sound methodological reasons [placebo] is necessary to determine the efficacy or safety of an intervention and the patients who receive placebo . . . will not be subject to any risk of serious or irreversible harm. Extreme care must be taken to avoid abuse of this option.

You will be interested to note that this amendment marked the beginning of the process by which the FDA distanced itself from Helsinki as its main source of regulatory guidance, especially for trials conducted outside the USA (as we discussed earlier, in the section on CROs).[16]

This same perverse problem of inadequate comparators also exists in the EU.[17] To get a licence to market your drug, the EMA does not require you to show that it is better than the best currently available treatment, even if that treatment is universally used: you simply have to show that it is better than nothing. A study from 2007 found that only half the drugs approved between 1999 and 2005 had been studied in comparison with other treatments at the time they were allowed onto the market (and, shamefully, only one third of those trials were published and publicly accessible to doctors and patients).[18]

Many researchers have argued for this problem of 'Better than what?' to be flagged up as prominently as possible, ideally on the leaflet that patients receive in the packet, since this is the only part of the marketing and communications process on which the regulators can exert clear, unambiguous control. One recent paper suggested some plain, simple wording: 'Although this drug has been shown to lower blood pressure more effectively than placebo, it has not been shown to be more effective than other members of the same drug class.'[19] It has been ignored.

Surrogate outcomes

Placebo controls are not the only problem with the trials that are used to obtain marketing approval. Often, drugs are approved despite showing no benefit at all on real-world outcomes, such as heart attacks or death: instead, they are approved for showing a benefit on 'surrogate outcomes', such as a blood test, that is only weakly or theoretically associated with the real suffering and death that we are trying to avoid.

This is best understood with an example. Statins are drugs that lower cholesterol, but you don't take them because you want to change your cholesterol figures on a blood test print-out: you take them because you want to lower your risk of having a heart attack, or dying. Heart attack and death are the real outcomes of interest here, and cholesterol is just a surrogate for those, a process outcome, something that we hope is associated with the real outcome, but it might not be, either not at all, or perhaps not very well.

Often there is a fair reason for using a surrogate outcome, not as your only indicator, but at least for some of the data. People take a long time to die (it's one of the great problems of research, if you can forgive the thought), so if you want an answer quickly, you can't wait around for them to have a heart attack and die. In these circumstances, a surrogate outcome like a blood test is a reasonable thing to measure, as an interim arrangement. But you still have to do long-term follow-up studies at some stage, to find out if your hunch about the surrogate outcome was right after all. Unfortunately, the incentives for companies – which are by far largest funders of trials – are all focused on short-term gains, either to get their drug on the market as soon as possible, or to get results before the drug comes off patent, while it still belongs to them.

This is a major problem for patients, because benefits on surrogate endpoints often don't translate into real-life benefits. In fact, the history of medicine is full of examples where quite the opposite was true.

Probably the most dramatic and famous comes from the Cardiac Arrhythmia Suppression Trial (CAST), which tested three anti-arrhythmic drugs to see if they prevented sudden death in patients who were at higher risk because they had a certain kind of abnormal heart rhythm.[20] The drugs prevented these abnormal rhythms, so everyone thought they must be great: they were approved onto the market to prevent sudden death in patients with abnormal rhythms, and doctors felt pretty good about prescribing them. When a proper trial measuring death was conducted, everyone felt a bit embarrassed: the drugs increased the risk of death to such a huge extent that the trial had to be stopped early. We had been cheerfully handing out tablets that killed people (it's been estimated that well over a hundred thousand people died as a result).

Even when they don't actively increase your risk of death, sometimes drugs which work well to change surrogate outcomes simply don't make any difference to the real outcomes that we're most interested in. Doxazosin was an expensive branded blood-pressure drug, and it worked extremely well for lowering the blood-pressure reading in a doctor's office – about as well as chlorthalidone, a simple old-fashioned blood-pressure drug that had been off patent for many years. Eventually, a trial was done comparing the two on real-world outcomes like heart failure (using government funding, since it was in nobody's financial interest); and it had to stop early, because patients on doxazosin were doing so much worse.[21] The manufacturer of doxazosin, Pfizer, mounted a magnificent marketing campaign, and there was barely any change in the use of the drug.[22] I will discuss this kind of campaigning later.

There are endless examples of drugs for which the only evidence available uses surrogate outcomes. If you have diabetes, the thing you're worried about is death, and horrible problems in your feet, your kidneys, your eyes and so on. You worry about your blood-sugar level and your weight because they are useful guides to whether your diabetes is under control, but they're nothing compared to the basic important question: will this drug actually reduce my risk of dying? Right now, there are all sorts of new diabetes drugs on the market. The 'glucagon-like peptide-1 receptor drugs', for example, are pretty exciting to a lot of doctors. If you look at the latest systematic review of their benefits, published in December 2011 (this one just happens to be open in front of me – it could have been any number of drugs) you will see that they lower blood sugar, lower blood pressure, lower cholesterol, and all this great stuff;[23] but nobody's ever checked to see if they actually stop you dying, which is all that the people taking them really care about.

The same goes for side effects. Depo-Provera is a reasonably good contraceptive, but there's some concern about whether it makes you more vulnerable to fractures. The research into this looks at bone mineral density, rather than actual fractures.[24]

When you come to get your drug approved to go on the market, regulators will often permit you to show proof of effectiveness only on surrogate outcomes. For 'accelerated approval', for drugs that are the first in a new class, or are treating a condition that has no current treatment, they may let you get away with a surrogate outcome that has barely been validated, which means there is very little research into how well it really is associated with the real-world outcomes of the disease. For context here, it's worth remembering that the examples above, where we were misled, came from surrogate outcomes that are regarded as 'well validated'. This would be fine if getting onto the market was just the beginning of the story, a starting gun for cautious prescription, in the context of larger monitoring of real-world outcomes. Unfortunately, as we will now see, things aren't like that.

Accelerated approval

Gathering and assessing trial evidence takes a long time, but regulators have to balance several opposing forces. Doctors with an eye on public health are often keen to make sure that the evidence for a new product is as good as possible, partly because many new drugs are only trivially useful in comparison with what already exists; but also because the pre-approval period is the time when demands on a drug company for compelling research are most likely to be met.

Drug companies, meanwhile, would like to get their drug on the market as swiftly and cheaply as possible. This isn't just impatience for revenue; it's also a fear of losing revenue outright, because the clock is already ticking for patent expiry even before the approval process starts. That strong commercial incentive is communicated in muscular fashion to governments, which push regulators to approve rapidly, and often measure speed of approval as a key outcome for the regulator.

This can have worrying effects, which might lead you to believe that quality of evidence is not the only factor affecting a drug's approval. For many decades, for example, the FDA's performance was measured by how many drugs it managed to approve in each calendar year.[25] This led to a phenomenon known as the 'December Effect', whereby a very large proportion of the year's approvals were rushed through in a panic during the last few weeks around Christmas. By graphing the proportion of approvals that were made in December over the course of thirty years (below, from Carpenter 2010), we can see the size of this effect, and also trace the arrival of a more aggressive pro-industry stance during Ronald Reagan's presidency (1981–89). If approvals were evenly distributed throughout the year, we'd expect to only see 8 per cent in each month: during the late eighties, the proportion passing in December rose to more than half, and it's hard to believe that this was simply when the assessments were complete.

These kinds of pressures can also be seen in the approval times for medicines, which have fallen hugely around the world: in the US they have reduced by half since 1993, on the back of previous cuts before that; and in the UK they dropped even more dramatically, from 154 working days in 1989 to forty-four days just a decade later.

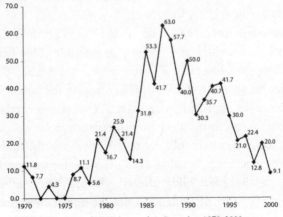

Percentage of NME Approvals in December, 1970–2000

It would be a mistake to imagine that drug companies are the only people applying pressure for fast approvals. Patients can also feel they are being deprived of access to drugs, especially if they are desperate. In fact, in the 1980s and 1990s the key public drive for faster approvals came from an alliance forged between drug companies and AIDS activists such as ACT UP.

At the time, HIV and AIDS had suddenly appeared out of nowhere, and young, previously healthy gay men were falling ill and dying in terrifying numbers, with no treatment available. We don't care, they explained, if the drugs that are currently being researched for effectiveness might kill us: we want them, because we're dying anyway. Losing a couple of months of life because a currently unapproved drug turned out to be dangerous was nothing, compared to a shot at a normal lifespan. In an extreme form, the HIV-positive community was exemplifying the very best motivations that drive people to participate in clinical trials: they were prepared to take a risk, in the hope of finding better treatments for themselves or others like them in the future. To achieve this goal they blocked traffic on Wall Street, marched on the FDA headquarters in Rockville, Maryland, and campaigned tirelessly for faster approvals.

As a result of this campaign, a series of new regulations were implemented to allow accelerated approval of certain new drugs. This legislation was intended for use on lifesaving drugs, in situations where there was no currently available medical treatment. Unfortunately, now that they have been in place for more than a decade, we can see that this is not how they have been used.

Midodrine

Once a drug has been approved it is very rare for a regulator to remove it from the market, especially if the only issue is lack of efficacy, rather than patients actively dying because of side effects. Where they do finally make such a move, it is usually after phenomenal delay.

Midodrine is a drug used to treat 'orthostatic hypotension', a drop in blood pressure causing dizziness when you stand up.[26] While this is doubtless unpleasant for those who experience it, and there may be an increased risk of, say, falls when feeling dizzy, this condition is generally not what most people would regard as serious or life-threatening. Furthermore, the extent to which it is regarded as a singular medical problem varies between countries and cultures. But because there were no previous drugs available to treat it, midodrine was able to be approved through the accelerated programme, in 1996, with weak evidence, but the promise of better studies to follow.

Specifically, midodrine was approved on the basis of three very small, brief trials (two of them only two days long) in which many of the people receiving the drug dropped out of the study completely. These trials showed a small benefit on a surrogate outcome – changes in blood-pressure recordings when the participants stood up – but no benefit on real-world outcomes like dizziness, quality of life, falls, and so on. Because of this, after midodrine was approved through the urgent approval scheme, the manufacturer, Shire, had to promise it would do more research once the drug was on the market.

Year after year, no satisfactory trials appeared. In August 2010, *fourteen years later*, the FDA announced that unless Shire finally produced some compelling data showing that midodrine improved actual symptoms and day-to-day function, rather than some numbers on a blood-pressure machine after one day, it would take the drug off the market for good.[27] This seemed like an assertive move which should finally provoke compliance, but the result was quite the opposite. Effectively, the company said: 'Fine.' The drug was off patent: anyone could make it, and indeed Shire now only made 1 per cent of the midodrine sold, with Sandoz, Apotex, Mylan and other companies making the rest. In such a crowded market there was very little money to be made from selling this medicine, and certainly no incentive to invest in research that would only help other companies sell a hundred times more of the same product. Fourteen years on from midodrine's original approval, the FDA found that there is such a thing as being simply too late.

But that was not the end of the story. Suddenly a vast army of midodrine users and special-interest patient groups appeared, with politicians at their helm: 100,000 patients had filed prescriptions for the drug in 2009 alone. To them, this pill was a life-saver, and the only drug available to treat their condition. If all companies were going to be banned from making it, with the drug taken off the market, this would be a disaster. The fact that no trial had ever demonstrated any concrete benefit was irrelevant: quack remedies such as homeopathy continue to maintain a viciously loyal fan base, despite the fact that homeopathic pills by definition contain no active ingredients at all, and despite research overall showing them to perform no better than placebo.* These midodrine patients didn't care about what trials found: they 'knew' that their drug worked, with the certainty of true believers. And now the government was planning to take it away because of some complicated administrative transgression. Surrogate what? This must have sounded like irrelevant word-play from where they were trying to stand.

The FDA was forced to backtrack and leave the drug on the market. Slow negotiations have continued over the post-marketing trials, but the FDA now has very little leverage with any company over this drug. Almost two decades after midodrine was first approved as an urgent, exceptional case, the drug companies are still making promises to do proper trials. As of 2012, these trials were nowhere to be seen.

---

* If this is of particular interest to you, it's covered at length in my previous book, *Bad Science*.

This is a serious problem, and it goes way beyond this single, rather trivial medicine. The General Accounting Office is the investigative audit branch of the US Congress. In 2009, it examined the FDA's failure to chase these kinds of post-approval studies, and its findings were damning: between 1992 and 2008, ninety drugs had been given accelerated approval on the basis of surrogate endpoints alone, with the drug companies making a commitment to conduct 144 trials in total. As of 2009, one in every three of those trials was still outstanding.[28] No drug had *ever* been taken off the market because its manufacturer had failed to hand over outstanding trial data.

The British academic John Abraham is a social scientist who has done more than anyone to shine a light into the traditions and processes of regulators around the world. He has concluded that accelerated approval is simply part of a consistent trend towards deregulation, for the benefit of the industry. It's useful to walk through just one of the case studies he has worked on, with colleague Courtney Davis, to see how regulators around the world dealt with the *best* possible candidates for these urgent assessments.[29]

Gefitinib (brand name Iressa) is a cancer drug made by AstraZeneca, for desperate patients who have reached the end of the line. It's approved for non-small-cell lung cancer, which is a serious life-threatening diagnosis, and it's approved for use as 'third-line' treatment, after all else has failed. Its accelerated approval was driven partly by patient campaigning, just like the AIDS campaigners who drove the introduction of accelerated approval legislation in the first place. It's also a good case study, because the manufacturer did actually conduct its follow-up studies, which is fairly unusual (only 25 per cent of the cancer drugs being studied by Abraham have done so).

For standard approval of a lung-cancer treatment you need to show meaningful improvement in either survival or symptoms. But 'tumour response' – a reduction in tumour size seen on a body scan – is a fairly typical surrogate outcome for cancer drugs, which can be used to get accelerated approval; after doing so, you will then need to do more trials to find out if this translates into benefits that actually matter to patients.

Initially, AstraZeneca provided evidence from a small trial showing a 10 per cent drop in tumour size on Iressa. This was regarded by the FDA as unimpressive, especially since the patients in the trial were unusual, with slower-growing tumours than you'd usually see. But the company pressed on, and began much larger trials measuring the impact on survival. It had expected to finish these studies after the drug was rapidly approved, but in fact they were completed before then. The trials on real-world outcomes found no survival benefit. What's more, contradicting the smaller, earlier study, they found no improvement in tumour size. One FDA scientist summarised the findings fairly bluntly: 'It's 2,000 patients saying Iressa doesn't work versus 139 saying it works marginally.'

At the same time, the company was also giving the drug to 12,000 desperate, dying patients with no other option, through something known as an 'expanded access programme'. This is common when patients have shown no response to any other medication, and are regarded as too unfit for clinical trials (although I would argue that trials should ideally include every-one eligible for treatment, since we only do them to try to answer the question of whether a drug works in real-world patients). These programmes can cost companies money, but they also generate a huge amount of goodwill from desperate people, their families and their organised patient groups.

Regulators now, like so many public bodies, set high store by 'public engagement', and this is an admirable goal, if done well. But what we see here is not an example of good public engage-ment. Large, well-conducted, fair tests of Iressa had shown that it was no better than a dummy sugar pill containing no medicine. Yet many dying patients from the expanded access programmes, who had been given the drug for free, travelled with advocacy groups to give compelling evidence to the FDA. From their perspective this was 'a wonderful drug', they explained, 'light-years better than previous treatment'. It 'began to eliminate cancer symptoms in seven days'. Tumours were '90 per cent gone in three months', said one. Whether that was exaggeration or fluke, the reality is that fair tests showed no benefit. But the desperate patients disagreed, and asserted their case plainly and simply: Iressa 'will save lives'. This personal testimony was in all likelihood a combination of the placebo effect and the natural fluctuation in symptoms that all patients experience. That didn't seem to matter.

When the committee charged with approving the drug cast their votes, they went 11–3 in favour.

It's hard to know what to make of this process, since the vote went against not only the surrogate outcome data, but also the evidence from very large trials showing no benefit on real-world outcomes or survival. But we are all human, and it is hard to reject a drug when you're faced with moving life-and-death testimony. One FDA scientist told John Abraham during his field work: '[Patient testimonials] definitely have an influence over advisory committees. That's what Iressa proves.' Several of these patients had been funded to attend the FDA advisory committee meeting by AstraZeneca. We can only wonder if individuals who had not been successfully treated with Iressa would have been flown across the country to speak their personal truth. Perhaps not. Perhaps they might be dead.

The FDA could have rejected the view of its expert committee, and that might have been wise. Not only was there no evidence of benefit: there were reports from Japan of fatal pneumonia associated with Iressa, affecting 2 per cent of patients, a third of whom died within a fortnight. But the FDA approved the drug all the same. AstraZeneca was compelled to conduct a further 1,700-patient study, which again found no benefit over placebo. Iressa stayed on the market. Another treatment appeared, and this one was effective in third-line treatment of non-small-cell lung cancer. Iressa stayed on the market.

The FDA did send out a letter saying that no new patients should be started on Iressa, but drugs that are on the market get used by doctors, often quite haphazardly, driven by marketing, habit, familiarity, rumour and lack of clear current information. Iressa continued to be prescribed for new patients. And still it stayed on the market.

We can see from the percentages in surveys that post-marketing trials requested by regulators are often neglected; and cynical doctors will often tell you that ineffective drugs are commonly marketed. But midodrine and Iressa are, I think, two cases that really put flesh on those bones. Accelerated approval is *not* used to get urgent drugs to market for emergency use and rapid assessment. Follow-up studies are not done. These accelerated approval programmes are a smokescreen.

The impact on innovation

As we have seen, drugs regulators don't require that new drugs are particularly good, or an improvement on what came before; they don't even require that drugs are particularly effective. This has interesting consequences on the market more broadly, because it means that the incentives for producing new drugs that improve patients' lives are less intense. One thing is clear from all the stories in this book: drug companies respond rationally to incentives, and when those incentives are unhelpful, so are drug companies.

To explore whether new drugs represent any kind of forward movement in a field, we would have to examine all those approved over a period of time. This is exactly what a recent paper by some Italian researchers achieved.[30] They took every drug acting on the central nervous system that had been approved since the first day the European regulator began approving drugs, and looked to see whether they represented any degree of innovation.

As you would expect by now, they found several serious problems with the data given to support these drugs' applications. All the approved drugs had only been shown to be better than a placebo. Important information was missing from the trial reports: for example, there was no clear data about the number of people dropping out of each trial, which is important information, as it helps to show whether a drug is intolerable because of side effects. Then there were serious problems in the design of the trials. The majority (seventy-five out of eighty-three) were very brief. They were also small: not one of the submitted studies had enough participants to accurately detect a difference between a currently available treatment and a new one, on the rare occasions when they tried to do so.

The researchers concluded that the problem was straightforward: if the rules don't require a company to show a new drug's superiority to current treatments, they are unlikely to develop better medicines.

This is well demonstrated by the phenomenon of 'me-too' drugs. If you think back to the previous chapter, you will remember that developing a completely new molecule, with a completely new mechanism of action in the body, is a very risky and difficult business. Because of that, once a company has an established drug on the market, others will often try to produce their own version of that drug: so there are a great many antidepressants around from the class known as 'selective serotonin reuptake inhibitors', or SSRIs, for example. Developing a drug like this is much more of a safe bet.

Often these me-too drugs don't represent a significant therapeutic benefit, so many people regard them as wasteful, an unnecessary use of development money, potentially exposing trial participants to unnecessary harm for individual companies' commercial gain rather than medical advancement. I'm not entirely sure this is correct: among a class of drugs, one may be better than the others, or have fewer idiosyncratic side effects, so in that sense these copycats can be useful, sometimes. On the other hand, we have no way of knowing what amazing new drugs might have been created if we incentivised companies by insisting that they demonstrate superiority. These are not easy counterfactuals to unpick, and I never feel fully satisfied with the economists' models of the impact on innovation, on either side.

However, we can see, by tracing the life of these me-too drugs, that the market does not work entirely as we might wish it to for us, the people who collectively pay for health services.[31] For example, you might expect that multiple competing drugs in the same market would bring prices down, but an economic analysis, using Swedish data, showed that drugs regarded by the FDA as showing no advantage over the existing options enter the market at the same price. Another followed the price of an ulcer drug called cimetidine, and found that it became more expensive when ranitidine, from the same class, came on the market; and both drugs' prices kept on rising when the competitors famotidine and nizatidine came out.

Perhaps the clearest story is told by tracing the recent history of another class of drugs, the 'proton pump inhibitors' used to treat reflux and heartburn. These are common medical problems, so it's a lucrative area, and omeprazole was one of the most lucrative drugs in the class. Around a decade ago it was making AstraZeneca $5 billion every year, which was about a third of its total revenue from all drugs. But it was about to come out of patent, and once generic drug manufacturers could make their own pills containing the same drug, the price would plummet, and the revenue would disappear. So AstraZeneca introduced something called a 'me-again' drug.

This is an interesting new twist on the original idea. Me-too drugs are entirely new molecules that work in a similar way to the old ones; with a me-again drug, the *same* molecule is relaunched onto the same market for the same disease, but with one clever difference.

Complex molecules like drugs can exist in right- and left-handed forms, called 'enantiomers'. The chemical formula for each of the different molecules is the same, and you will find the same atoms in the same order attached to the same parts of the same rings. The only difference is that a particular twist in a chain goes one way in one enantiomer, and the other way in the other, in the same way that your left and right gloves are identical mirror-images, made of the same material, the same weight, and so on. But left- and right-handed versions of drugs can have slightly different properties. Maybe the molecule only fits neatly into the receptor where it exerts its influence if it's the right-hand version. Maybe it only fits nicely into the jaws of the enzyme that will break it down if it's the left-hand form. This will affect what it does in your body. Recently, with increasing frequency, drug companies have started to release 'single enantiomer preparations', where you only get the right-handed version, say, of an existing treatment in your pill. The companies claim this as a new drug, and so add a whole new patent lifetime to their profits.

This can be a good financial bet. It's usually easy to get market-ing authorisation, because the mixed form of the drug has already been licensed, and you have lots of trials showing that form of the drug to be better than nothing. The second job, of convincing people that the single enantiomer is better than the mixture, is down to your marketing department, and may not be subject to much detailed formal scrutiny from a regulator at all.

So if they can have genuinely different properties in the body, why do people often think it's dubious (or 'evergreen-ing') for a drug company to put out a version of an existing drug that is only the single enantiomer form? First, these different properties are often slight, so all the concerns about me-too drugs are also applicable for me-again drugs. Then there's the issue of timing: it's striking that companies often put out a me-again drug towards the end of the patent life of the original. It's also worth bearing in mind that there's no free ride, as ever, and that treatments with beneficial effects can also have side effects. The right-handed version of fluoxetine (Prozac) seemed like a great idea: it had a longer half-life than the original mixture, and this raised the possibility of an anti-depressant tablet that could be taken once a week, rather than once a day; but it also turned out to cause something called 'QT-prolongation', a change in the electric patterns in the heart which is associated with things like an increased risk of sudden death. But last, most strikingly, alongside these possible extra risks, the new 'single enantiomer' pills often don't seem to work much better than the mixed ones, despite being very much more expensive.

Let's look at omeprazole, our heartburn drug. Come 2002, AstraZeneca knew it was about to lose $5 billion a year, a third of its revenue, which would be a disaster for its profits and its stock price. But in 2001 the company launched esomeprazole, and this was a great success: in fact, AstraZeneca still takes $5 billion a year from the drug today. In the US it's huge, a top-three blockbuster. In the UK it takes £44 million a year, but the amount of esomeprazole we get for that large amount of money is tiny, because brand-new esomeprazole costs ten times as much as old-fashioned omeprazole.

Here's the kicker: new esomeprazole, just the left-handed version of the molecule rather than a mixture of both forms, is really no better than the plain old mixture in omeprazole tablets. The evidence is mixed, but it's clear that there is no dramatic difference between any of the various members of this class of drug, and certainly no special amazing, unique benefit from esomeprazole.[32]

So why do doctors prescribe it? This is the power of the

marketing machine in medicine, as we will shortly see. The direct-to-consumer campaign in the US was vast: AstraZeneca spent $260 million on ads in 2003,[33] and purplepill.com, its website to promote the drug, eventually pulled over a million visitors every quarter.[34] Against this, there was a considerable backlash. Kaiser Permanente, the American medical insurance giant, kept esomeprazole off its list of prescribable drugs, after deciding it was pointlessly expensive. Thomas Scully, the head of Medicare and Medicaid, gave speeches explaining that the drug was a waste of money; but with no final control over what gets prescribed in the two organisations, he sat and watched as they spent $800 million on this vastly expensive drug every year. When I say Scully gave speeches, he said: 'Any doctor that prescribes Nexium [the brand name for esomeprazole] should be ashamed of himself.' AstraZeneca complained about him to the White House and on Capitol Hill. Scully says that he was put under pressure to 'shut up'.[35] He hasn't.

Comparative effectiveness research

These are not stories that can leave anyone feeling actively impressed. But there is a more important problem behind our ethical concerns about how companies behave in situations like this: we have allowed ourselves to be left – as prescribers, as patients, and as the people paying for health care – without clear evidence comparing different treatments against each other. We have no idea which treatments are best, and by extension, we have no idea which are harmful. If you die from getting the third-best treatment, then you have died needlessly and avoidably, and have every right to be angry in your grave.

This would seem to be a simple matter: we need to do more trials after drugs come to market, comparing them against each other in head-to-head tests. Health care is a huge financial burden on all societies around the world, and in most countries outside the USA this burden is shouldered by the state. If we are so weak that we can't force drug companies, through our regulators, to conduct meaningful trials, then surely it makes sense for governments to fund them? This seems especially sensible when you consider that the cost of irrational prescribing, in most cases, is vastly more expensive than the cost of research to prevent it.

A clear early illustration of this comes from the ALLHAT trial, which started in 1994 and cost \$125 million. This project looked at high blood pressure, a condition that affects about a quarter of the adult population, with half of that number receiving medication for it. The trial compared chlorthalidone, a very cheap, old-fashioned blood-pressure pill, against amlodipine, a very expensive new one that was being very widely prescribed. We knew from trials comparing them head-to-head that the two drugs were equally good at controlling blood pressure, but these numbers aren't what matters to patients: a trial was needed that gave some patients the old drug, and some patients the new drug, and then measured how many people had heart attacks, and died. When ALLHAT finally did this comparison, measuring the real-world outcome that matters to patients, it found – to everyone's amazement – that the old drug was much better. The savings from this one project alone vastly outweigh the cost of the trial itself, even though it was a huge project: this study started in 1994, when I was still an undergraduate, and finished in 2002, when I had finished training (I tell you that, because later we will see how important and difficult it is for doctors to keep up).

So 'comparative effectiveness research', as this is called, is vitally important, but it has only recently been embraced. To give you an illustration of how slow and arduous this journey is: in 2008, shortly after being elected President, Barack Obama demonstrated to many academics and doctors that he had a clear understanding of the deep problems in health care, by committing to spend $1 billion on head-to-head trials of commonly used treatments, in order to find out which is best. In return he was derided by right-wing critics as 'anti-industry'.

On this issue, since people with resources often defend the pharmaceutical companies from deserved criticism, it's worth remembering one thing: health care really is one of those areas where we are all, in a very real sense, in it together. If you're super-rich, in the top 0.2 per cent of the population, you can buy pretty much anything you want. But however rich you are, if you become sick you can't innovate new medicines overnight, because that takes time, and more money than even you have. And you can't know the true effects of the medicines we have today, because nobody does, if they've not been properly tested, and if some results go missing in action. The most expensive doctors in the world don't know any better than anyone else, since any trained person can critically read the best systematic reviews on a given drug, what it will do to your life expectancy, and there is no hack, no workaround, for this broken system. Even if you are super-rich, even if you make $10 million a year, you are right here in it with the rest of us all.

So comparative effectiveness research is a vitally important field, for everyone, and in many cases the value of finding out what works best among the drugs we already have would hugely exceed the value of developing entirely new ones. This research is a very rational area in which to spend more money.[36]

Monitoring side effects

But effectiveness is only part of the story. Alongside issues of which drug is most effective, there is also the problem of safety. Like many doctors, I am constantly amazed by the enthusiasm with which doctors embrace prescribing new drugs. When a new medicine comes on the market, with no proven benefits over an existing one, doctors and patients are being offered a simple choice: do you want to use an old drug, a known quantity, for which we have many years of experience in monitoring side effects? Or do you want to take a completely new drug, with no demonstrated advantages, which may, for all we know, have some horrific idiosyncratic side effect quietly waiting to emerge?

I was taught at medical school that in this situation, a doctor should regard the rest of the medical profession as unpaid stunt doubles: let them make the risky mistakes on your behalf, sit back, watch, learn, and then come back out when it's safe. In some respects you could argue that this is useful advice for life more generally. But how are side effects monitored?

Once a drug is approved, its safety needs to be assessed. This is a complex business, with – to be fair – genuine methodological challenges, and glaring, unnecessary holes. The flaws are driven by unnecessary secrecy, poor communication, and an institutional reluctance to take drugs off the market. To understand these, we need to understand the basics of the field known as 'pharmacovigilance'.

It is important to recognise, before we even begin, that drugs will always come onto the market with unforeseen side effects. This is because you need data on lots of patients to spot rare side effects, but the trials that are used to get a drug approved are usually small, totalling between five hundred and 3,000 people. In fact, we can quantify how common a side effect must be in order to be detected in such a small number of people, by using a simple piece of maths called 'the rule of three'. If five hundred patients are studied in pre-approval trials, that is only enough patients to spot the side effects which occur more frequently than one in every 166 people; if 3,000 patients are studied, that is still only enough to spot side effects which affect more than one in every 1,000 people. The overall rule here is easy to apply: if a side effect hasn't yet occurred in n patients, then you can be 95 per cent confident that it will happen in fewer than one in $3/n$

patients (there's a mathematical explanation of why this is true in the footnote below, if you want one, but it makes my head hurt*). You can also use the rule of three in real life: if three hundred of your parachutes have opened just fine, for example, then assuming no other knowledge, the chance of one *not* opening, and sending you to certain death, is at least less than one in a hundred. This may or may not be reassuring for you.

Putting this in context: your drug might make one in every 5,000 people literally explode – their head blows off, their intestines fly out – through some idiosyncratic mechanism that nobody could have foreseen. But at the point when the drug is approved, after only 1,000 people have taken it, it's very likely that you'll *never* have witnessed one of these spectacular and unfortunate deaths. After 50,000 people have taken your drug, though, out there in the real world, you'd expect to have seen

---

* This is tough, but here's how to think through the derivation of the $3/n$ rule, if you're statistically inclined. Let's say we're eating week-old chicken, and the probability of death is 0.2, so the probability of no death is $(1-0.2)$, which is 0.8. If we have two observations – I eat mouldy chicken twice – then the probability of 'no death' is less: it's 0.8 x 0.8, or 0.64 (so, my chances of death are rising with every meal of mouldy chicken I eat). If I eat mouldy chicken n times, the probability of no death is 0.8^n, or, to go back to where the 0.8 came from, that's $(1-0.2)^n$, or more generally $(1-risk)^n$. Now we want to sit at the other end of the telescope. We want to know the *maximum* possible risk of something happening that is compatible with never having seen it happen, after n observations (or mouldy-chicken meals), with at most a 5 per cent margin of error. In equation terms, we would say that $(1-risk)^n$ equals 0.05, or rather, since we're not interested in $(1-risk)$, but in $(1-maximum\ risk)$, we'd say $(1-maximum\ risk)^n=0.05$. Now we just have to rearrange that equation, to make it give us maximum risk when we know n. The calculus-wrestling goes: $1-maximum\ risk = 0.05^{(1/n)}$, and for n greater than 30 that's approximately the same as $1-maximum\ risk = 1-(3/n)$. We're nearly there: take away the 'one minus' on both sides and you have maximum risk = $3/n$. That may have been a bit tougher than your average Vorderman maths session, but it is much more useful. 'I've Never Met a Nice South African' is a racist song about racists. Now you know to ask: 'How many have you met?'

about ten people explode overall (since, on average, it makes one in every 5,000 people explode).

Now, if your drug is causing a very rare adverse event, like exploding, you're actually quite lucky, because *weird* adverse events really stand out, as there's nothing like them happening already. People will talk about patients who explode, they'll write them up in short reports for academic journals, probably notify various authorities, coroners might be involved, alarm bells will generally ring, and people will look around for what is suddenly causing patients to explode very early on, probably quite soon after the first one goes off.

But many of the adverse events caused by drugs are things that happen a lot anyway. If your drug increases the chances of someone getting heart failure, well, there are a lot of people around with heart failure already, so if doctors see one more case of heart failure in their clinic, then they're probably not going to notice, especially if this drug is given to older people, who already experience a lot of heart failure anyway. Even detecting any signal of increased heart failure in a large group of patients might be tricky.

This helps us to understand the various different mechanisms that are used to monitor side effects by drug companies, regulators and academics. They fall into roughly three groups:

1. Spontaneous reports of side effects, from patients and doctors, to the regulator
2. 'Epidemiology' studies looking at the health records of large groups of patients
3. Reports of data from drug companies

Spontaneous reports are the simplest system. In most territories around the world, when a doctor suspects that a patient has developed some kind of adverse reaction to a drug, they can notify the relevant local authority. In the UK this is via something called the 'Yellow Card System': these freepost cards are given out to all doctors, making the system easy to use, and patients can also report suspected adverse events themselves, online at yellowcard.mhra.gov.uk (please do).

These spontaneous reports are then categorised by hand, and collated into what is effectively a giant spreadsheet, with one row for every drug on the market, and one column for every imaginable type of side effect. Then you look at how often each type of side effect is reported for each drug, and try to decide whether the figure is higher than you'd expect to see simply from chance. (If you're statistically minded, the names of the tools used, such as 'proportional reporting ratios' and 'Bayesian confidence propagation neural networks', will give you a clue as to how this is done. If you're not statistically minded, then you're not missing out; at least, no more here than elsewhere in your life.)

This system is good for detecting unusual side effects: a drug that made your head and abdomen literally explode, for example, would be spotted fairly easily, as discussed. Similar systems are in place internationally, most of the results from around the world are pooled together by WHO in Uppsala, and academics or companies can apply for access, with varying success.*

---

* If you're very interested in this issue, there is a serious problem in the detail. The International Society of Drug Bulletins, which represents the academics and pharmacists producing friendly summaries of data for doctors, has been campaigning for five years to have access to something called *Signal*, a WHO publication which reports drug-safety issues raised by the raw case reports data. WHO has consistently refused, insisting that only 'National Health Authorities' are allowed to see it, but in 2012 they changed their mind, and are now planning wider access for other independent groups. This, sadly, does not end the saga. Drug companies are still allowed early access, to 'read and comment' prior to publication. And *Signal* is just a publication describing findings; it is not the individual case reports, which are held in something called VigiBase, and which still remain secret. The UK and US data are more easily available, but the EU data, as you might expect, is not. You can read more in the ISDB press release and website.[37]

But this approach suffers from an important problem: not all adverse events are reported. The usual estimate is that in Britain, only around one in twenty gets fed back to the MHRA.[38] This is not because all doctors are slack. It would actually be perfect if that was the cause, because then at least we would know that all side effects on all drugs had an equal chance of not being reported, and we could still usefully compare the proportions of side-effect reports between each other, and between different drugs.

Unfortunately, different side effects from different drugs are reported at very different rates. A doctor might be more likely to be suspicious of a symptom being a side effect if the patient is on a drug that is new on the market, for example, so those cases may be reported more than side effects for older drugs. Similarly, if a patient develops a side effect that is already well known to be associated with a drug, a doctor will be much *less* likely to bother reporting it, because it's not an interesting new safety signal, it's just a boring instance of a well-known phenomenon. And if there are rumours or news stories about problems with a drug, doctors may be more inclined to spontaneously report adverse events, not out of mischief, but simply because they're more likely to remember prescribing the controversial drug when a patient comes back with an odd medical problem.

Also, a doctor's suspicions that something is a side effect at all will be much lower if it is a medical problem that happens a lot anyway, as we've already seen: people often get headaches, for example, or aching joints, or cancer, in the everyday run of life, so it may not even occur to a doctor that these problems are anything to do with a prescription they've given. In any case, these adverse events will be hard to notice against the high background rate of people who suffer from them, and this will all be especially true if they occur a long time after the patient starts on a new drug.

Accounting for these problems is extremely difficult. So spontaneous reporting can be useful if the adverse events are extremely rare without the drug, or are brought on rapidly, or are the kind of thing that is typically found as an adverse drug reaction (a rash, say, or an unusual drop in the number of white blood cells). But overall, although these systems are important, and contribute to a lot of alarms being usefully raised, generally they're only used to identify suspicions.[39] These are then tested in more robust forms of data.

Better data can come from looking at the medical records of very large numbers of people, in what are known as 'epidemiological' studies. In the US this is tough, and the closest you can really get are the administrative databases used to process payments for medical services, which miss most of the detail. In the UK, however, we're currently in a very lucky and unusual position. This is because our health care is provided by the state, not just free at the point of access, but also through one single administrative entity, the NHS. As a result of this happy accident, we have large numbers of health records that can be used to monitor the benefits and risks of treatments. Although we have failed to realise this potential across the board, there is one corner called the General Practice Research Database, where several million people's GP records are available. These records are closely guarded, to protect anonymity, but researchers in pharmaceutical companies, regulators and universities have been able to apply for access to specific parts of anonymised records for many years now, to see whether specific medicines are associated with unexpected harms. (Here I should declare an interest, because like many other academics I am doing some work on analysing this GPRD data myself, though not to look at side effects.)

Studying drug safety in the full medical record of patients who receive a prescription in normal clinical practice has huge advantages over spontaneous report data, for a number of reasons. First, you have all of a patient's medical notes, in coded form, as they appear on the clinic's computer, without any doctor having to make a decision about whether to bother flagging up a particular outcome.

You also have an advantage over those small approval trials, because you have a *lot* of data, allowing you to look at rare outcomes. And more than that, these are real patients. The people who participate in trials are generally unusual 'ideal patients': they're healthier than real patients, with fewer other medical problems, they're on fewer other medications, they're less likely to be elderly, very unlikely to be pregnant, and so on. Drug companies like to trial their drugs in these ideal patients, as healthier patients are more likely to get better and to make the drug look good. They're also more likely to give that positive result in a briefer, cheaper trial. In fact, this is another way in which database studies can have an advantage: approval trials are generally brief, so they expose patients to drugs for a shorter period of time than the normal duration of a prescription. But database studies give us information on what drugs do in real-world patients, under real-world conditions (and as we shall see, this isn't just restricted to the issue of side effects).

With this data, you can look for an association between a particular drug and an increased risk of an outcome that is already common, like heart attacks, for example. So you might compare heart-attack risk between patients who have received three different types of foot-fungus medication, for example, if you were worried that one of them might damage the heart. This is not an entirely straightforward business, of course, partly because you have to make important decisions about what you compare with what, and this can affect your outcomes. For example, should you compare people getting your worrying drug against other people getting a similar drug, or against people matched for age but not getting any drug? If you do the latter, are foot-fungus patients definitely comparable with age-matched healthy patients on your database? Or are patients with foot fungus, perhaps, more likely to be diabetic?

You can also get caught out by a phenomenon called 'channelling': this is where patients who have reported problems on previous drugs are preferentially given a drug with a solid reputation for being safe. As a result, the patients on the safe drug include many of the patients who are sicker to start with, and so are more likely to report adverse events, for reasons that have nothing to do with the drug. That can end up making the safe drug look worse than it really is; and by extension, it can make a riskier drug look better in comparison.

But in any case, short of conducting massive drug trials in routine care – not an insane idea, as we will see later – these kinds of studies are the best shot we have for making sure that drugs aren't associated with terrible harms. So they are conducted by regulators, by academics, and often by the manufacturer at the request of the regulator.

In fact, drug companies are under a number of obligations to monitor side effects, both general and specific, and report them to the relevant authority, but in reality these systems often don't work very well. In 2010, for example, the FDA wrote a twelve-page letter to Pfizer complaining that it had failed to properly report adverse events arising after its drugs came to market.[40] The FDA had conducted a six-week investigation, and found evidence of several serious and unexpected adverse events that had not been reported: Viagra causes serious visual problems, for example, and even blindness. The FDA said Pfizer failed to report these events in a timely fashion, by 'misclassifying and/or downgrading reports to non-serious, without reasonable justification'. You will remember the paroxetine story from earlier, where GSK failed to report important data on suicide. These are not isolated incidents.

Last, you can also get some data on side effects from trials, even though the adverse events we're trying to spot are rare, and therefore much less likely to appear in small studies. Here again, though, there have been problems. For example, sometimes companies can round up all kinds of different problems into one group, with a label that doesn't really capture the reality of what was happening to the patients. In antidepressant trials, adverse events like suicidal thoughts, suicidal behaviours and suicide attempts have been coded as 'emotional lability', 'admissions to hospital', 'treatment failures' or 'drop-outs'.[41] None of these really captures the reality of what was going on for the patient.

To try to manage these problems, for the past few years companies have been required by the EMA to produce something called a Risk Management Plan (RMP) on their drug, and here our problems begin again. These documents are written by the company, and explain the safety studies it has agreed with the regulator; but for absolutely no sane reason that I can imagine, the contents are kept secret, so nobody knows exactly what studies the companies have agreed to conduct, what safety issues they are prioritising, or how they are researching them.

A brief summary is available to doctors, academics and the public, and just recently academics have begun to publish papers assessing their contents, with damning findings.[42] After explaining that changes in risk identified from the RMP were communicated unpredictably and inadequately to doctors, one concludes: 'The main limitation of this study is the lack of publicly available data regarding the most significant aspects.' The researchers were simply deprived of information about the studies that were conducted to monitor drug safety. A similar study, given slightly better access, looked at the safety studies that *were* discussed in RMPs.[43] For about half of these studies, the RMP gave only a short description, or a commitment to conduct some kind of study, but no further information. In the full RMP document, where you would expect to have found full study protocols, the researchers found not one, for any of the eighteen drugs they looked at.

If these Risk Management Plans are drawn up in secret, and their contents are poorly communicated, but at the same time they are the tool used to get drugs to market with a lower threshold of evidence, then we have a serious and interesting new problem: it's possible that they are being used as a device to reassure the public, rather than to address a serious issue.[44]

When it comes to the secrecy of regulators, it is clear that there is an important cultural issue that needs to be resolved. I've spent some time trying to understand the perspective of public servants who are clearly good people, but still seem to think that hiding documents from the public is desirable. The best I can manage is this: regulators believe that decisions about drugs are best made by them, behind closed doors; and that as long as they make good decisions, it is OK for these to then be communicated only in summary form to the outside world.

This view, I think, is prevalent; but it is also misguided, in two ways. We have already seen many illustrations of how hidden data can be a cloak for mischief, and how many eyes are often valuable for spotting problems. But the regulators' apparent belief that we should have blind faith in their judgements also misses a crucial point.

A regulator and a doctor are trying to make two completely different decisions about a drug, even though they are using (or in doctors' case would *like* to use) the same information. A regulator is deciding whether it's in the interests of society overall that a particular drug should ever be available for use in its country, even if only in some very obscure circumstance, such as when all other drugs have failed. Doctors, meanwhile, are making a decision about whether they should use this drug right now, for the patient in front of them. Both are using the safety and efficacy data to which they have access, but they both need access to it in full, in order to make their very different decisions.

This crucial distinction is not widely understood by patients, who often imagine that an approved drug is a safe and effective one. In a 2011 US survey of 3,000 people, for example, 39 per cent believed that the FDA only approves 'extremely effective' drugs, and 25 per cent that it only approves drugs without serious side effects.[45] But that's not true: regulators frequently approve drugs that are only vaguely effective, with serious side effects, on the off-chance that they might be useful to someone, somewhere, when other interventions aren't an option. They are used by doctors and patients as second-best options, but we need all the facts to make safe and informed decisions.

Some would argue that cracks are appearing in this secrecy, with some new pharmacovigilance legislation coming into force for Europe in 2012 which is supposed to improve transparency.[46] But at best, this legislation is a very mixed bag. It does not give access to Risk Management Plans, but it does state that the EMA should publish the agendas, recommendations, opinions and minutes of various scientific committees, which are currently completely secret. We can only judge this small promised change on how it is implemented, if ever, and as we have seen, previous performance from the EMA does not inspire confidence. Even if we set aside the EMA's astonishing and perverse behaviour over CSRs for orlistat and rimonabant, which you will remember from Chapter 1, we should also recall that it has been mandated to provide an open clinical trials register for many years, but has simply failed to do so, still keeping much of that trials data secret to this very day.

In any case, this legislation has several serious flaws.[47] The EMA is being set up as the host of a single database for drug safety data, for example, yet this information will still be kept secret from health professionals, scientists and the public. But the most interesting shortcoming of this new legislation is an organisational one.

Many had called for a new 'drug safety agency' to be set up, monitoring risks after a drug came to market, as a stand-alone organisation, with its own powers and staff, completely separate from the organisation in charge of approving a drug when it first comes to market.[48] This may sound like a dull organisational irrelevancy, but in fact it speaks to one of the most disappointing problems that has been identified in the operations of regulators around the world: regulators that have approved a drug are often reluctant to take it off the market, in case that is seen as an admission of their failure to spot problems in the first place.

That is not idle pontification on my part. In 2004 the epidemiologist from the US Office of Drug Safety who led the review on Vioxx told the Senate Finance Committee: 'My experience with Vioxx is typical of how CDER [the FDA's Center for Drug Evaluation and Research] responds to serious drug safety issues in general . . . the new drug reviewing division that approved the drug in the first place, and that regards it as its own child, typically proves to be the single greatest obstacle to effectively dealing with serious drug safety issues.' Chillingly, in 1963, half a century ago, an FDA medical officer called John Nestor told Congress almost exactly the same thing: previous approval decisions were 'sacrosanct', he said. 'We were not to question decisions made in the past.'

This is a universal problem in the politics and management of regulators, and it can be seen in the organisational structures: around the world, the departments in charge of monitoring safety and removing drugs from the market are much smaller and less powerful than the departments that approve drugs, which makes institutions reluctant to impose suspensions. Since we are discussing matters of line management and organisational structure, and you might suspect that this is merely a vague, handwaving assertion, let me tell you that it is also the verdict of every serious study of regulators,[49] from the Institute of Medicine[50] to the semi-official biography of the FDA,[51] various academics,[52] and people from within the organisations.

That is the reason there were so many calls for the EU to create a new Drug Safety Agency, and that is why it's so concerning that these calls have been ignored. In fact, the same old models have been put back in place, only under different names. The EMA's Pharmacovigilance Risk Assessment Committee, which decides on whether to remove an approved drug from the market, still reports to the Committee for Medical Products for Human Use, which is the one that approves them in the first place. This perpetuates all of the old problems about removal being difficult, more lowly than approval, and an embarrassment to the approvers.

So what steps can a regulator take when it has established that there is a problem? In very extreme cases it can remove a drug from the market (although in the US, technically drugs usually stay on the market, with the FDA advising against their use). More commonly it will issue a warning to doctors through one of its drug safety updates, a 'Dear Doctor' letter, or by changing the 'label' (confusingly, in reality, a leaflet) that comes with the drug. Drug-safety updates are sent to most doctors, though it's not entirely clear whether they are widely read. But, amazingly, when a regulator decides to notify doctors about a side effect, the drug company can contest this, and delay the notice being sent out for months, or even years.

In February 2008, for example, the MHRA published a small piece in its bulletin *Drug Safety Update*, which is read by all too few people. The article stated that the agency was planning a change to the drug label for all statins, a class of drug given to reduce cholesterol and prevent heart attacks, following a review of clinical trial data, spontaneous reports of suspected adverse drug reactions, and the published literature. 'Product information for statins is being updated to reflect a number of different side-effects as class effects of all statins.' It explained: 'Patients should be made aware that treatment with any statin may sometimes be associated with depression, sleep disturbances, memory loss, and sexual dysfunction.' The agency also planned a new warning that – very rarely – statin therapy might be associated with interstitial lung disease, a serious medical condition.

The decision to add these new side effects to the label was

made in February 2008, but it took until November 2009 for an announcement that the change was finally being made. This is a delay of almost two years. Why did it take so long? The *Drugs and Therapeutics Bulletin* discovered the reason: 'One of the innovator MA [marketing authorisation] Holders was not in agreement with this wording.'[53] So, a drug company was able to delay the inclusion of safety warnings on a whole class of drugs prescribed to four million people in the UK for twenty-two months because it didn't agree with the wording.

But what good would have come of changing the label in any case?

This is the final component of our story. It's difficult for doctors and patients to get a clear, up-to-date picture of the risks and benefits of drugs, from any source, but since the regulators have privileged access to information, we should expect them to do a particularly clear job of communicating what they have, as there is by definition no competition for providing information here, and no opportunity to shop around: the regulators are the only people with access to all of the data.

Drug labels are lauded by regulators as a single, awesome repository of information, by which prescribers and patients alike can be educated and informed. In reality, they are chaotic and not very informative. They often discuss trials, but give no reference to enable you to find out more, or even to work out which trial they're discussing. Sometimes the basic elements of a trial are so bizarrely different in the regulator document and the published paper that it's hard to match them up even if you try, and even if the trial has been published. What's more, most labels feature long lists of hundreds of side effects, with poor information as to how common they are, even though most of them are very rare, and are not even confidently associated with the drug anyway. Too much information, communicated chaotically, is every bit as unhelpful as too little information.

Some US researchers have been campaigning for over a

decade to add a simple 'drug facts box' to the information given to doctors and patients alongside the rather dense and confusing 'label'. This box would be a summary document giving clear, quantitative information on the risks and benefits of the drug, using evidence-based strategies for communicating statistical information to lay people. There is randomised controlled trial evidence showing that patients given this drug facts box have better knowledge of the benefits and risks of their drugs.[54] The FDA has suggested that it will think about using them. I hope that one day it will, and that it will make these boxes itself.

So you can see the difference for yourself, on the following page is the drug facts box for a sleeping pill called 'Lunesta'.

This drugs facts box is briefer than the official label for the same drug, which appears opposite it: I think it's also much more informative. It doesn't solve all the problems of secrecy, or even all the problems of poor communication. But it does demonstrate very clearly that regulators have neither earned nor respected their special status when it comes to assessing and communicating risk.

Solutions

We have established that there are some very serious problems here, both in how we approve drugs, and in how we monitor their safety once they become available. Drugs are approved on weak evidence, showing no benefit over existing treatments, and sometimes no benefit at all. This gives us a market flooded with drugs that aren't very good. We then fail to collect better evidence on them once they're available, even when we have legislative power to force companies to do better trials, and even when they've promised to do so. Last, side-effects data is gathered in a slightly ad hoc fashion, behind closed doors, with secret documents and 'risk management plans' that are hidden

LUNESTA (compared to sugar pill) to reduce current symptoms for adults with insomnia

What is this drug for?   To make it easier to fall or to stay asleep

Who might consider taking it?   Adults age 18 and older with insomnia for at least 1 month

Recommended monitoring   No blood tests, watch out for abnormal behavior

Other things to consider doing   Reduce caffeine intake (especially at night), increase exercise, establish regular bedtime, avoid daytime naps

LUNESTA STUDY FINDINGS

788 healthy adults with insomnia for at least 1 month -- sleeping less than 6.5 hours per night and/or taking more than 30 minutes to fall asleep-- were given LUNESTA or a sugar pill nightly for 6 months.   Here's what happened:

What difference did LUNESTA make?	People given a sugar pill	People given LUNESTA (3 mg each night)
Did LUNESTA help?		
LUNESTA users fell asleep faster (15 minutes faster due to drug)	45 minutes to fall asleep	30 minutes to fall asleep
LUNESTA users slept longer (37 minutes longer due to drug)	5 hours 45 minutes	6 hours 22 minutes
Did LUNESTA have side effects?		
Life threatening side effects No difference between LUNESTA and a sugar pill	None observed	
Symptom side effects		
More had unpleasant taste in their mouth (additional 20% due to drug)	6% 6 in 100	26% 26 in 100
More had dizziness (additional 7% due to drug)	3% 3 in 100	10% 10 in 100
More had drowsiness (additional 6% due to drug)	3% 3 in 100	9% 9 in 100
More had dry mouth (additional 5% due to drug)	2% 2 in 100	7% 7 in 100
More had nausea (additional 5% due to drug)	6% 6 in 100	11% 11 in 100

How long has the drug been in use?
Lunesta was approved by FDA in 2005. As with all new drugs we simply don't know how its safety record will hold up over time. In general, if there are unforeseen, serious drug side effects, they emerge after the drug is on the market (when a large enough number of people have used the drug).

from doctors and patients for no good reason. The results of this safety monitoring are communicated inconsistently, through mechanisms that are uninformative and are therefore used infrequently, and which are, in any case, vulnerable to spectacular delays imposed by drug companies.

We could tolerate some of these problems, but enduring all of them at once creates a dangerous situation, in which patients are routinely harmed for lack of knowledge. It wouldn't matter, for example, that the market is flooded with drugs that are of little benefit, or are worse than their competitors, if doctors and patients knew this, could find out immediately and conveniently which are the best options, and could change their behaviour to reflect that. But this is not possible when we are deprived of existing information on risks and benefits by secretive regulators, or where good-quality trial data is not even collected.

There always have been – and always will be – programmes and measures in place to try and drive up these standards: NICE, in asking for more comparative data to decide on cost effectiveness, has helped. In my view, fixing this situation requires a significant cultural shift in how we approach new medicines; but before we get to that, there are several small, obvious steps which should go without saying.

1. Drug companies should be required to provide data showing how their new drug compares against the best currently available treatment, for every new drug, before it comes onto the market. It's fine that sometimes drugs will be approved despite showing no benefit over current treatments, because if a patient has an idiosyncratic reaction to the current common treatment, it is useful to have other inferior options available in your medical arsenal. But we need to know the relative risks and benefits, if we are to make informed decisions.

2. Regulators and healthcare funders should use their influence to force companies to produce more informative trials. The German government have led the field here, setting up an agency in 2010 called IQWiG, which looks at the evidence for all newly approved drugs, to decide if they should be paid for by Germany's healthcare providers. IQWiG has been brave enough to demand good quality trials, measuring real-world outcomes, and has already refused to approve payments for new drugs where the evidence provided is weak. As a result, companies have delayed marketing new drugs in Germany, while they try to produce better evidence that they really do work:[55] patients don't lose out, since there's no good evidence that these new drugs are useful. Germany is the largest market in Europe, at 80 million patients, and they're not a poor country. If all purchasers around the world held the line, and refused to buy drugs presented with weak evidence, then companies would be forced to produce meaningful trials much more quickly.

3. All information about safety and efficacy that passes between regulators and drug companies should be in the public domain, as should all data held by national and international bodies about adverse events on medications, unless there are significant privacy concerns on individual patient records. This has benefits that go beyond immediate transparency. Where there is free access to information about a treatment, we benefit from 'many eyes' on the problems around it, analysing them more thoroughly, and from more perspectives. Rosiglitazone, the diabetes drug, was removed from the market because of problems with heart failure, but those problems weren't identified and acted on by a regulator: they were spotted by an academic, working on data that was, unusually, made more publicly available as the result

of a court case. The problems with the pain drug Vioxx were spotted by independent academics outside the regulator. The problems with the diabetes drug benfluorex were spotted, again, by independent academics outside the regulator. Regulators should not be the only people who have access to this data.

4. We should aim to create a better market for communicating the risks and benefits of medications. The output of regulators is stuffy, legalistic and impenetrable, and reflects the interests of regulators, not patients or doctors. If all information is freely available, then it can be repurposed by those who have access to it, and précised into better forms. These could be publicly funded and given away, or privately funded and sold, depending on business models.

This is all simple. But there is a broader issue, that no government has ever satisfactorily addressed, bubbling under in the culture of medicine: we need more trials. Wherever there is true uncertainty about which treatment is best, we should simply compare them, see which is best at treating a condition, and which has worse side effects.

This is entirely achievable, and at the end of the next chapter I will outline a proposal for how we can carry out trials cheaply, efficiently and almost universally, wherever there is true uncertainty. It could be used at the point of approval of every new drug, and it could be used throughout all routine treatment.

But first, we need to see just how rubbish some trials can be.

4

Bad Trials

So far I've taken the idea of a clinical trial for granted, as if there was nothing complicated about it: you just take some patients; split them in half; give one treatment to one group, another to the other; and then, a while later, you see if there is any difference in outcomes between your two groups.

We're about to see the many different ways in which trials can be fundamentally flawed, by both design and analysis, in ways that exaggerate benefits and underplay harms. Some of these quirks and distortions are straightforward outrages: fraud, for example, is unforgivable, and dishonest. But some of them, as we will see, are grey areas. There can be close calls in hard situations, to save money or to get a faster result, and we can only judge each trial on its own merits. But it is clear, I think, that in many cases corners are cut because of perverse incentives.

We should also remember that many bad trials (including some of the ones discussed in the pages to follow) are conducted by independent academics. In fact, overall, as the industry is keen to point out, where people have compared the methods of independently sponsored trials against industry-sponsored ones, industry-sponsored trials often come out better. This may well be true, but it is almost irrelevant, for one simple reason: independent academics are bit players in this domain. Ninety per cent of published clinical trials are sponsored by the pharmaceutical industry. They dominate this field, they set the tone, and they create the norms.

Last, before we get to the meat, here is a note of caution. Some of what follows is tough: it's difficult science, that anyone can understand, but some examples will take more mental horsepower than others. For the complicated ones I've added a brief summary at the beginning, and then the full story. If you find it hard going, you could skip the details and take the summaries on trust. I won't be offended, and the final chapter of the book – on dodgy marketing – is filled with horrors that you mustn't miss.

To the bad trials.

Outright fraud

Fraud is an insult. In the rest of this chapter we will see wily tricks, close calls, and elegant mischief at the margins of acceptability. But fraud disappoints me the most, because there's nothing clever about it: nothing methodologically sophisticated, no plausible deniability, and no argument about whether it breaks the data. Somebody just made the results up, and that's that. Delete, ignore, start again.

So it's lucky – for me and for patients – that fraud is also fairly rare, as far as anyone can tell. The best current estimate of its prevalence comes from a systematic review in 2009, bringing together the results of survey data from twenty-one studies, asking researchers from all areas of science about malpractice. Unsurprisingly, people give different responses to questions about fraud depending on how you ask them. Two per cent admitted to having fabricated, falsified or modified data at least once, but this rose to 14 per cent when they were asked about

the behaviour of colleagues. A third admitted other questionable research practices, and this rose to 70 per cent, again, when
they were asked about colleagues.

We can explain at least part of the disparity between the
'myself' and 'others' figures by the fact that you are one person,
whereas you know lots of people, but since these are sensitive
issues, it's probably safe to assume that all responses are an
underestimate. It's also fair to say that sciences like medicine or
psychology lend themselves to fabrication, because so many
factors can vary between studies, meaning that picture-perfect
replication is rare, and as a result nobody will be very suspicious
if your results conflict with someone else's. In an area of science
where the results of experiments are more straightforwardly
'yes/no', failed replication would expose a fraudster much more
quickly.

All fields are vulnerable to selective reporting, however, and
some very famous scientists have manipulated their results in
this way. The American physicist Robert Millikan won a Nobel
Prize in 1923 after demonstrating with his oil-drop experiment
that electricity comes in discrete units – electrons. Millikan was
mid-career (the peak period for fraud) and fairly unknown. In
his famous paper from *Physical Review* he wrote: 'This is not a
selected group of drops, but represents all of the drops experi-
mented on during sixty consecutive days.' That claim was
entirely untrue: in the paper there were fifty-eight droplets, but
in his notebooks there are 175, annotated with phrases like
'publish this beautiful one' and 'agreement poor, will not work
out'. A debate has raged in the scientific literature for many
years over whether this constitutes fraud, and to an extent,
Millikan was lucky that his results could be replicated. But in
any case, his selective reporting – and his misleading description of it – lies on a continuum of all sorts of research activity
that can feel perfectly innocent, if it's not closely explored. What
should a researcher do with the outliers on a graph that is

otherwise beautifully regular? When they drop something on the floor? When the run on the machine was probably contaminated? For this reason, many experiments have clear rules about excluding data.

Then there is outright fabrication. Dr Scott Reuben was an American anaesthetist working on pain who simply never conducted at least twenty clinical trials published over the previous decade.[1] In some cases, he didn't even pretend to get approval for conducting studies on patients in his institution, and simply presented the results of trials that were conjured out of nothing. Data in medicine, as we should keep remembering, is not abstract or academic. Reuben claimed to have found that non-opiate medications were as effective as opiates for the management of pain after surgical operations. This pleased everyone, as opiates are generally addictive, and have more side effects. Practice in many places was changed, and now that field is in turmoil. Of all the corners in medicine where you could perpetrate fraud, and change the decisions that doctors and patients make together, pain is one area that really matters.

There are various ways that fraudsters can be caught, but constant vigilant monitoring by the medical and academic establishment is not one of them, as that doesn't happen to any sufficient extent. Often detection is opportunistic, accidental or the result of local suspicions. Malcolm Pearce, for example, was a British obstetric surgeon who published a case report claiming he had reimplanted an ectopic pregnancy, and furthermore that this had resulted in the successful delivery of a healthy baby. An anaesthetist and a theatre technician in his hospital thought this was unlikely, as they'd have heard if such a remarkable thing had happened; so they checked the records, found no matching records for any such event, and things collapsed from there.[2] Notably, in the same issue of the same journal, Pearce had also published a paper reporting a trial of two hundred women with polycystic ovary syndrome whom he treated for recurrent miscarriage. The trial never happened, and not only had Pearce invented the patients and the results, he had even concocted a fictitious name for the sponsoring drug company, a company that never existed. In the era of Google, that lie might not survive for very long.

There are other detection methods. The human brain is a fairly bad random-number generator, for example, and simple frauds have often been uncovered by forensic statisticians looking at last-digit frequency: if you're pencilling numbers into a column at random, you might have a slight unconscious preference for the number seven. To avoid this you might use a random-number generator, but here you would run into the odd problem of telltale uniformity in your randomness. The German physicist Jan Hendrik Schön co-authored roughly one paper every week in 2001, but his results were too accurate. Eventually someone noticed that two studies had the same amount of 'noise' superimposed on a perfect prototype result; it turned out that many of his figures had been generated by computer, using the very equations they were supposed to be checking, with supposedly realistic random variation built into the model.

There are all sorts of things we should be doing to catch outright fraud: better investigations, better routine monitoring, better communication from journal editors on suspicions about papers they reject, better protection of whistleblowers, random spot checks of primary data by journals, and so on. People talk about them, but they seldom do them, because responsibility for the problem is diffuse and unclear.

So, fraud: it happens, it's not clever, it's just criminal, and is perpetrated by bad people. But its total contribution to error in the medical literature is marginal when compared to the routine, sophisticated and – more than anything – plausibly deniable everyday methodological distortions which fill this book. Despite that fact, outright fraud is almost the only source of distortion that receives regular media coverage, simply because it's easy to understand. That's reason enough for me to leave it alone, and move on to the meat.

Test your treatment in freakishly perfect 'ideal' patients

As we have seen, patients in trials are often nothing like real patients seen by doctors in everyday clinical practice. Because these 'ideal' patients are more likely to get better, they exaggerate the benefits of drugs, and help expensive new medicines appear to be more cost effective than they really are.

In the real world, patients are often complicated: they might have many different medical problems, or take lots of different medicines, which all interfere with each other in unpredictable ways; they might drink more alcohol in a week than is ideal; or have some mild kidney impairment. That's what real patients are like. But most of the trials we rely on to make real-world decisions study drugs in unrepresentative, freakishly ideal patients, who are often young, with perfect single diagnoses, fewer other health problems, and so on.[3]

Are the results of trials in these atypical participants really applicable to everyday patients? We know, after all, that different groups of patients respond to drugs in different ways. Trials in an ideal population might exaggerate the benefits of a treatment, for example, or find benefits where there are none. Sometimes, if we're very unlucky, the balance between risk and benefit can even switch over completely, when we move between different populations. Anti-arrhythmic drugs, for example, were shown to be effective at prolonging life in patients with severe abnormal heart rhythms, but were also widely prescribed for patients after they'd had heart attacks, when they had only mild abnormal heart rhythms. When these drugs were finally trialled in this second population, we found – to everyone's horror – that they actively increased their risk of dying.[4]

Doctors and academics often ignore this problem, but when you start to stack up the differences between trial patients and real patients side by side, the scale of the problem is staggering.

One study from 2007 took 179 representative asthma patients from the general population and looked at how many would have been eligible to participate in a group of asthma treatment trials.[5] The answer was 6 per cent on average, and these weren't any old trials they were being rejected from: they were the trials that form the basis of the international consensus guidelines for treating asthma in GP clinics and hospitals. These guidelines are used around the world, and yet, as this study shows, they are based on trials that would have excluded almost every single real-world patient they're applied to.

Another study took six hundred patients being treated for depression in an outpatient clinic, and found that on average only a third of them would have been eligible to participate in thirty-nine recently published trials of treatments for depression.[6] People often talk about the difficulties in recruiting patients for research: but one study described how 186 patients with depression enquired about participating in two different trials of antidepressants, and more than seven out of every eight had to be turned away as they weren't eligible.[7]

To see what this looks like in reality, we can follow one group of patients with a particular medical problem. In 2011 some researchers in Finland took every patient who'd ever had a hip fracture, and worked out if they would have been eligible for the trials that have been done on bisphosphonate drugs, which are in widespread use for preventing fractures.[8] They started with 7,411 patients, but 2,134 were excluded straight off, because they were men, and all the trials have been done in women. Are there differences in how men and women respond to drugs? Sometimes, yes. Of the 5,277 patients remaining, 3,596 were excluded because they were the wrong age: patients in the trials had to be between sixty-five and seventy-nine. Then, finally, 609 patients were excluded because they didn't have osteoporosis. That only leaves 1,072 patients. So the data from the trials on these fracture-preventing drugs are only strictly applicable to about one of every seven patients with a fracture. They might still work in the people who've been excluded, but that's a judgement call you have to make; and even if they do work, the size of the benefit might be different in different people.

This problem goes way beyond simply measuring the effectiveness of drugs: it also distorts our estimates of their cost effectiveness (and in an era of escalating costs in health care, we need to worry about value). Here's one example from the new 'coxib' painkiller drugs. These are sold on the basis that they cause fewer gastrointestinal or 'GI' bleeds when compared with older, cheaper painkillers, like high-street ibuprofen.

Coxibs really do seem to reduce the risk of GI bleeds, which is good, because such bleeds can be extremely serious. In fact they lessened the risk by about a half in trials, which were conducted – of course – in ideal patients, who were at much higher risk of having a GI bleed. For the people running the trials this made perfect sense: if you want to show that a drug reduces the risk of having a bleed, it will be much easier and cheaper to show that in a population which is having lots of bleeds in the first place (because otherwise, if your outcome is really rare, you're going to need a huge number of patients in your trial).

But an interesting problem appears if you use these figures, on a change in the rate of GI bleeds in freakishly ideal trial patients, to calculate the cost of preventing a bleed in the real world. NICE estimated this cost at $20,000 per avoided bleed, but the real answer is more like $100,000.[9] We can see very easily how NICE got this wrong, by doing the maths on some simple rough round figures, though these are – pleasingly – almost exactly the same as the real ones (we must work in dollars here, by the way, because the analysis exposing this problem was published in a US academic journal).

The trial patients had a high risk of a bleed: over a year, fifty out of 1,000 had one. This was reduced to twenty-five out of 1,000 if they were on a coxib, because a coxib halves your risk of a bleed. A coxib drug costs an extra $500 a year for each patient. So, $500,000 spent on 1,000 patients buys you twenty-five fewer bleeds, and $500,000÷25 means the avoided bleeds cost you $20,000 each.

But if you look at the real patients getting coxibs in the GP records database, you can see that they have a much lower risk of bleeds: over a year, ten out of 1,000 had one. That goes down to five out of 1,000 if they were on a coxib, because a coxib halves your risk of a bleed. So, you still pay $500,000 for 1,000 patients to have a coxib for a year, but it only buys you five fewer bleeds, and $500,000÷5 means that these avoided bleeds now cost you $100,000 each. That is a lot more than $20,000.

This problem of trial patients being unrepresentative is called 'external validity', or 'generalisability' (in case you want to read more about it elsewhere). It can make a trial completely irrelevant to real-world populations, yet it is absolutely routine in research, which is conducted on tight budgets, to tight schedules, for fast results, by people who don't mind if their results are irrelevant to real-world clinical questions. This is a quiet, dismal scandal. There's no dramatic newspaper headline, and no single killer drug: just a slow and unnecessary pollution of almost the entire evidence base in medicine.

Test your drug against something useless

Drugs are often compared with something that's not very good. We've already seen this in companies preferring to test their drugs against a dummy placebo sugar pill that contains no medicine, as this sets the bar very low. But it is also common to see trials where a new drug is compared with a competitor that is known to be pretty useless; or with a good competitor, but at a stupidly low dose, or a stupidly high dose.

One thing that's likely to make your new treatment look good is testing it against something that doesn't work very well: this might sound absurd, or even cruel, so we're lucky that a researcher called Daniel Safer has pulled together a large collection of trials using odd doses specifically to illustrate this problem.[10] One study, for example, compares paroxetine against amitryptiline. Paroxetine is one of the newer antidepressants, and it is largely free from side effects like drowsiness. Amitryptyline is a very old drug, known to make people sleepy, so in real clinical practice it's often best to advise patients to take it only at night, because drowsiness doesn't matter so much when you're asleep. But in this trial amitryptyline was given twice a day, morning and night. The patients reported lots of daytime sleepiness on amitryptyline, making paroxetine look much better.

Alternatively, some trials compare the expensive new drug against an older one given at an unusually high dose, which means it has worse side effects by comparison. The world of antipsychotic medication gives an interesting illustration of this, and one that spans across several eras of research.

Schizophrenia is, like cancer, a disease for which treatments are not perfect, and the benefits of intervention must often be weighed against disadvantages. Each person with schizophrenia will have different goals. Some prefer to tolerate a higher risk of relapse because of their very strong desire to avoid side effects at any cost, and might choose a lower dose of medication; others may find that serious relapses damage their lives, costing them their home, job or friendships, and so they might choose to tolerate some side effects, in exchange for the benefits that go alongside them.

This is often a difficult decision, because side effects are common with schizophrenia medication: especially movement disorders (which are a little like the symptoms of Parkinson's disease) and weight gain. So the goal of drug innovation in this field has been to find tablets which treat the symptoms, but without causing side effects. A couple of decades ago there was a breakthrough: a new group of drugs were brought to market, the 'atypicals', which promised just that. A series of trials was set up to compare these new drugs with the old ones.

Safer found six trials comparing new-generation antipsychotic drugs with boring old-fashioned haloperidol – a drug well known to have serious side effects – at 20mg a day. This is not an insanely high dose of haloperidol: it wouldn't get you immediately struck off, and it's not entirely outside the maximum dose permitted in the *British National Formulary* (*BNF*), the standard reference manual for drug prescription. But it's an odd routine dose, and it's inevitable that patients receiving so much would report lots of side effects.

Interestingly, a decade later, history repeated itself: risperidone was one of the first of this new generation of antipsychotic drugs, so it came off patent first, immediately becoming very cheap, like the older generation of drugs. As a consequence, many drug companies wanted to show that their own expensive new-generation antipsychotic was better than risperidone, which was now suddenly cheap and old-fashioned: and so trials appeared comparing new drugs against risperidone at a dose of 8mg. Again, 8mg isn't an unimaginably high dose: but it's still pretty high, and patients on this dose of risperidone will be much more likely to report side effects, making the comparator drug look more attractive.

This – again – is a quiet and diffuse scandal. It doesn't mean that any of these specific drugs are outright, headline-grabbing killers: just that the evidence, overall, is distorted.

Trials that are too short

Trials are often brief, as we have seen, because companies need to get results as quickly as possible, in order to make their drug look good while it is still in patent, and owned by them. This raises several problems, including ones that we have already reviewed: specifically, people using 'surrogate outcomes', like changes in blood tests, instead of 'real-world outcomes', like changes in heart attack rates, which take longer to emerge. But brief trials can also distort the benefits of a drug simply by virtue of their brevity, if the short-term effects are different to the long-term ones.

An operation to remove a cancer, for example, has immediate short-term risks – you might die on the table in the operating theatre, or from an infection in the following week – but you hope that this short-term risk is offset by long-term benefits. If you do a trial to compare patients who have the operation with patients who don't, but only measure outcomes for one week, you might find that those having the operation die sooner than those who don't. This is because it takes months or years for people to die of the cancer you're cutting out, so the benefits of that operation take months and years to emerge, whereas the risks, the small number of people who die on the operating table, appear immediately.

The same problem presents itself with drug trials. There might be a sudden, immediate, short-term benefit from a weight-loss drug, for example, which deteriorates over time to nothing. Or there might be short-term benefits and long-term side effects, which only become apparent in longer trials. The weight-loss treatment Fenphen, for example, caused weight loss in the positive short-term trials, but when patients receiving it were observed over longer periods, it turned out that they also developed heart valve defects.[11] Benzodiazapine drugs like valium are very good for alleviating anxiety in the short term, and a trial lasting six weeks would show huge benefits; but over the months and years that follow, their benefits decrease, and patients become addicted. These adverse long-term outcomes would only be captured in a longer trial.

Longer trials are not, however, automatically always better: it's a question of the clinical question you are trying to answer, or perhaps trying to avoid. With an expensive cancer drug like Herceptin, for example, you might be interested in whether giving it for short periods is just as effective as giving it for long periods, in order to avoid paying for larger quantities of the drug unnecessarily (and exposing patients to a longer duration of side effects). For this you'd need short trials, or at the very least trials that reported outcomes over a long period, but after a short period of treatment. Roche applied for twelve-month treatment licences with Herceptin, presenting data from twelve-month-long trials. In Finland a trial was done with only a nine-week course of treatment, finding significant benefit, and the New Zealand equivalent of NICE decided to approve nine-week treatment. Roche rubbished this brief study, and commissioned new trials for a *two-year* period of treatment. As you can imagine, if we want to find out whether nine weeks of Herceptin are as good as twelve months of Herceptin, we need to run some trials comparing those two treatment regimes: funding trials like these is often a challenge.

Trials that stop early

If you stop a trial early, or late, because you were peeking at the
results as it went along, you increase the chances of getting a
favourable result. This is because you are exploiting the random
variation that exists in the data. It is a sophisticated version of
the way someone can increase their chances of winning in a coin
toss by using this strategy: 'Damn! OK, best of three . . . Damn!
Best of five? . . . Damn! OK, best of seven . . .'

Time and again in this book we have come back to the same
principle: if you give yourself multiple chances of finding a
positive result, but use statistical tests that assume you only had
one go, you hugely increase your chances of getting a misleading false positive. This is the problem with people hiding negative results. But it also creeps into the way people analyse studies
which haven't been hidden.

For example, if you flip a coin for long enough, then fairly
soon you'll get four heads in a row. That's not the same as saying
'I'm going to throw four heads in a row right now,' and then
doing so. We know that the time frame you put around some
data can allow you to pick out a clump of findings which please
you; and we know that this can be a source of mischief.

The CLASS trial compared a new painkiller called celecoxib
against two older pills over a six-month period. The new drug
showed fewer gastrointestinal complications, so lots more
doctors prescribed it. A year later it emerged that the original
intention of the trial had been to follow up for over a year. The
trial had shown no benefit for celecoxib over that longer period,
but when only the results over six months were included, the
drug shone. That became the published paper.

At this stage we should pause a moment, to recognise that it
can sometimes be legitimate to stop a trial early: for example, if

there is a massive, jaw-dropping difference in benefit between the two treatment groups; and specifically, a difference so great, so unambiguous and so informative that even when you factor in the risk of side effects, no physician of sound mind would continue to prescribe the losing treatment, and none will, ever again.

But you have to be very cautious here, and some terrible wrong results have been let through by people generously accepting this notion. For example, a trial of the drug biso-prolol during blood-vessel surgery stopped early, when two patients on the drug had a major cardiac event, while eighteen on placebo did. It seemed that the drug was a massive life-saver, and the treatment recommendations were changed. But when it began to seem that this trial might have overstated the benefits, two larger ones were conducted, which found that bisoprolol actually conferred no benefit.[12] The original finding had been incorrect, caused by researchers stopping the trial early after a fluke clump of deaths.

Here we should be clear that the ethics committee supervising a trial may themselves sometimes ask for a trial to be stopped early, and peeking at your data during a trial can raise a genuinely troubling ethical question. If you seem to have found evidence of harm for one or other treatment before the end of the study period (or, ethically similar, found that one treatment is radically better than the other), should you continue to expose the patients in your trial to what might be genuine harm in the interests of getting to the bottom of whether it's simply a chance finding? Or should you shut up shop and close the trial, potentially allowing that chance finding to pollute the medical literature, misinforming treatment decisions for larger numbers of patients in the future? This is particularly worrying when you consider that after a truncated trial, a larger one often has to be done anyway, exposing more people to risk, just to discover if your finding was an anomaly.

One way to restrict the harm that can come from early stopping is to set up 'stopping rules', specified before the trial begins, and carefully calculated to be extreme enough that they are unlikely to be triggered by the chance variation you'd expect to see, over time, in any trial. Such rules are useful because they restrict the intrusion of human judgement, which can introduce systematic bias.

But whatever we do about early stopping in medicine, it will probably pollute the data. A review from 2010 took around a hundred truncated trials, and four hundred matched trials that ran their natural course to the end: the truncated trials reported much bigger benefits, overstating the usefulness of the treatments they were testing by about a quarter.[13] Another recent review found that the number of trials stopped early has doubled since 1990,[14] which is probably not good news. At the very least, results from trials that stop early should be regarded with a large dose of scepticism. Particularly since these same systematic reviews show that trials which stop early often don't properly report their reasons for doing so.

And all of this, finally, becomes even more concerning when you look at which trials are being truncated early, who they're run by, and what they're being used for.

In 2008, four Italian academics pulled together all the randomised trials on cancer treatments that had been published in the preceding eleven years, and that were stopped early for benefit.[15] More than half had been published in the previous three years, suggesting once again that this issue has become more prevalent. Cancer is a fast-moving, high-visibility field in medicine, where time is money and new drugs can make big profits quickly. Eighty-six per cent of the trials that stopped early were being used to support an application to bring a new drug onto the market.

Trials that stop late

It would be a mistake to think that any of these issues illustrate transgressions of simple rules that should be followed thoughtlessly: a trial can be stopped too early, in ways that are foolish, but it can also be stopped early for sensible reasons. Similarly, the opposite can happen: sometimes a trial can be prolonged for entirely valid reasons, but sometimes, prolonging a trial – or including the results from a follow-up period after it – can dilute important findings, and make them harder to see.

Salmeterol is an inhaler drug used to treat asthma and emphysema. What follows[16] is – if you can follow the technical details to the end – pretty frightening, so, as always, remember that this is not a self-help book, and it contains no advice whatsoever about whether any one drug is good, or bad, overall. We are looking at flawed methods, and they crop up in trials of all kinds of drugs.

Salmeterol is a 'bronchodilator' drug, which means it works by opening up the airways in your lungs, making it easier for you to breathe. In 1996, occasional reports began to emerge of 'paradoxical bronchospasm' with salmeterol, where the opposite would happen, causing patients to become very unwell indeed. Amateur critics often like to dismiss anecdotes as 'unscientific', but this is wrong: anecdotes are weaker evidence than trials, but they are not without value, and are often the first sign of a problem (or an unexpected benefit).

Salmeterol's manufacturer, GSK, wisely decided to investigate these early reports by setting up a randomised trial. This compared patients on salmeterol inhalers against patients with dummy placebo inhalers, which contained no active medicine. The main outcome to be measured was carefully pre-specified as 'respiratory deaths and life-threatening experiences', combined together. The secondary outcomes were things like asthma-related deaths (which is a subset of all respiratory deaths), all-cause deaths, and 'asthma-related deaths or life-threatening experiences', again bundled up.

The trial was supposed to recruit 60,000 people and follow them up intensively for twenty-eight weeks, with researchers seeing them every four weeks to find out about progress and problems. For the six months after this twenty-eight-week period, investigators were asked to report any serious adverse events they knew of – but they weren't actively seeking them out.

What happened next is a dismal tale, told in detail in a *Lancet* paper some years later by Peter Lurie and Sidney Wolfe, working from the FDA documents. In September 2002 the trial's own monitoring board met and looked at the 26,000 patients who had been through so far. Judging by the main outcome – 'respiratory deaths and life-threatening experiences' – salmeterol was worse than placebo, although the difference wasn't quite statistically significant. The same was true for 'asthma-related deaths'. The trial board said to GSK: you can either run another 10,000 patients through to confirm this worrying hint, or terminate the trial, 'with dissemination of findings as quickly as possible'. GSK went for the latter, and presented this interim analysis at a conference (saying it was 'inconclusive'). The FDA got worried, and changed the drug's label to mention 'a small but significant increase in asthma-related deaths'.

Here is where it gets interesting. GSK sent its statistics dossier on the trial to the FDA, but the figures it sent weren't calculated using the method specified in the protocol laid down before the study began, which stipulated that the outcome figures for these adverse events should come from the twenty-eight-week period of the trial, as you'd imagine, when such events were being carefully monitored. Instead, GSK sent the figures for the full twelve-month period: the twenty-eight weeks when the adverse events were closely monitored, and also the six months after the trial finished, when adverse events weren't being actively sought out, so were less likely to be reported. This means that the high rate of adverse events from the first twenty-eight weeks of the trial was diluted by the later period, and the problem became much less prominent.

If you look at the following table, from the *Lancet* paper, you can see what a difference that made. Don't worry if you don't understand everything, but here is one easy bit of background, and one hard bit. 'Relative risk' describes how much more likely you were to have an event (like death) if you were in the salmeterol group, compared with the placebo group: so a relative risk of 1.31 means you were 31 per cent more likely to have that event (let's say, 'death').

The numbers in brackets after that, the '95 per cent CI', are the '95 per cent confidence interval'. While the single figure of the relative risk is our 'point estimate' for the difference in risk between the two groups (salmeterol and placebo), the 95 per cent CI tells us how certain we can be about this finding. Statisticians will be queuing up to torpedo me if I oversimplify the issue, but essentially, if you ran this same experiment, in patients from the same population, a hundred times, then you'd get slightly different results every time, simply through the play of chance. But ninety-five times out of a hundred the true relative risk would lie somewhere between the two extremes of the 95 per cent confidence interval. If you have a better way of explaining that in fifty-four words, my email address is at the back of this book.

GSK didn't tell the FDA which set of results it had handed over. In fact, it was only in 2004, when the FDA specifically asked, that it was told it was the twelve-month data. The FDA wasn't impressed, though this is expressed in a bland sentence: 'The Division presumed the data represented [only] the twenty-eight-week period as the twenty-eight-week period is

	Relative risk (95% CI)	
	28-week study	28-week study plus 6 months
Respiratory-related deaths and life-threatening experiences	1·39 (0·91–2·13)	1·16 (0·78–1·72)
Asthma-related deaths	4·33 (1·24–15·21)	2·50 (0·97–6·44)
Asthma-related deaths or life-threatening experiences	1·68 (0·99–2·85)	1·52 (0·92–2·52)
All-cause deaths	1·31 (0·83–2·08)	1·04 (0·70–1·55)

Table: Relative risks of primary outcome and major secondary outcomes in SMART study with and without inclusion of post-trial data clinically the period of interest.' It demanded the twenty-eight-week data, and said it was going to base all its labelling information on that. This data, as you can see, painted a much more worrying picture about the drug.

It took a couple of years from the end of the trial for these results to be published in an academic paper, read by doctors. Similarly, it took a long time for the label on the drug to explain the findings from this study.

There are two interesting lessons to be learnt from this episode, as Lurie and Wolfe point out. First, it was possible for a company to slow down the news of an adverse finding reaching clinicians and patients, even though the treatment was in widespread use, for a considerable period of time. This is something we have seen before. And second, we would never have known about any of this if the activities of the FDA Advisory Committees hadn't been at least partially open to public scrutiny, because 'many eyes' are often necessary to spot hidden flaws in data. Again, this is something we have seen before.

GSK responded in the *Lancet* that the twelve-month data was the only data analysed by the trial's board, which was independent of the company (the trial was run by a CRO).[17] It said that it communicated the risks urgently, sent a letter to doctors who'd prescribed salmeterol in January 2003, when the trial was formally stopped, and that a similar notice appeared on the GSK and FDA websites, stating that there was a problem.

Trials that are too small

A small trial is fine, if your drug is consistently life-saving in a condition that is consistently fatal. But you need a large trial to detect a small difference between two treatments; and you need a very large trial to be confident that two drugs are equally effective.

If there's one thing everybody thinks they know about research, it's that a bigger number of participants means a better study. That is true, but it's not the only factor. The benefit of more participants is that it evens out the random variation among them. If you've run a tiny trial of an amazing concentration-enhancing drug, with ten people in each group, then if only one person in one group had a big party the night before your concentration test, their performance alone could mess up your findings. If you have lots of participants, this sort of irritating noise evens itself out.

It's worth remembering, though, that sometimes a small study can be adequate, as the sample size required for a trial depends on a number of factors. For example, if you have a disease where everyone who gets it dies within a day, and you have a drug that you claim will cure this disease immediately, you won't need very many participants at all to show that your drug works. If the difference you're trying to detect between the two treatment groups is very subtle, though, you'll need many more participants to be able to detect this tiny difference against the natural background of everyday unpredictable variation in health for all the individuals in your study.

Sometimes you see a suspiciously large number of small trials being published on a drug, and when this happens it's reasonable to suspect that they might be marketing devices – a barrage of publications – rather than genuine acts of scientific enquiry. We'll also see an even more heinous example of marketing techniques in the section on 'marketing trials' shortly.

But there's a methodologically interesting problem hiding in here too. When you are planning a trial to detect a difference between two groups of patients, on two different treatments, you do something called a 'power calculation'. This tells you how many patients you will need if you're to have – say – an 80 per cent chance of detecting a true 20 per cent difference in deaths, given the expected frequency of deaths in your participants. If you complete your trials and find no difference in deaths between the two treatments, that means you cannot find evidence that one is better than the other.

This is not the same as showing that they are equivalent. If you want to be able to say that two treatments are equivalent, then for dismally complicated technical reasons (I had to draw a line somewhere) you need a much larger number of participants.

People often forget that. For example, the INSIGHT trial was set up to see if nifedipine was better than co-amilozide for treating high blood pressure. It found no evidence that it was. At the time, the paper said the two drugs had been found to be equivalent. They hadn't.[18] Many academics and doctors enjoyed pointing that out in the letters that followed.

Trials that measure uninformative outcomes

Blood tests are easy to measure, and often respond very neatly to a dose of a drug; but patients care more about whether they are suffering, or dead, than they do about the numbers printed on a lab report.

This is something we have already covered in the previous chapter, but it bears repeating, because it's impossible to overstate how many gaps have been left in our clinical knowledge through unjustified, blind faith in surrogate outcomes. Trials have been done comparing a statin against placebo, and these have shown that they save lives rather well. Trials have also compared one statin with another: but these all use cholesterol as a surrogate outcome. Nobody has ever compared the statins against each other to measure which is best at preventing death. This is a truly staggering oversight, when you consider that tens of millions of people around the world have taken these drugs, and for many, many years. If just one of them is only 2 per cent better at preventing heart attacks than the others, we are permitting a vast number of avoidable deaths, every day of the week. These tens of millions of patients are being exposed to unnecessary risk, because the drugs they are taking haven't been appropriately compared with each other; but each one of those patients is capable of producing data that could be used to compile new knowledge about which drug is best, in aggregate, if only it was systematically randomised, and the outcomes followed up. You will hear much more on this when we discuss the need for bigger, simpler trials in the next chapter, because this problem is not academic: lives are lost through our uncritical acceptance of trials that fail to measure real-world outcomes.

Trials that bundle their outcomes together in odd ways

Sometimes, the way you package up your outcome data can give misleading results. For example, by setting your thresholds just right, you can turn a modest benefit into an apparently dramatic one. And by bundling up lots of different outcomes, to make one big 'composite outcome', you can dilute harms; or allow freak results on uninteresting outcomes to make it look as if a whole group of outcomes are improved.

Even if you collect entirely legitimate outcome data, the way you pool these outcomes together over the course of a trial can be misleading. There are some simple examples of this, and then some slightly more complicated ones.

As a very crude example, many papers (mercifully, mostly, in the past) have used the 'worst-ever side-effects score' method.[19]

This can be very misleading, as it takes the worst side effects a patient has ever scored during a trial, rather than a sum of all their side-effects scores throughout its whole duration. In the graphs below, you can see why this poses such a problem, because the drug on the top is made to look as good as the drug on the bottom, by using this 'worst-ever side-effects score' method, even though the drug on the bottom is clearly better for side effects.

Another misleading summary can be created by choosing a cut-off for success, and pretending that this indicates a meaningful treatment benefit, where in reality there has been no such thing. For example, a 10 per cent reduction in symptom sever-

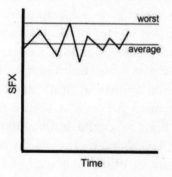

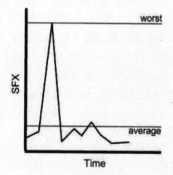

ity may be defined as success in a trial, even though it still leaves patients profoundly disabled.[20] This is particularly misleading if one treatment achieves a dramatic benefit if it works at all, and another a modest benefit if it works at all, but both get over the arbitrary and modest 10 per cent benefit threshold in the same number of patients: suddenly, a very inferior drug has been made to look just as good as the best in its class.

You can also mix lots of different outcomes together to produce one 'composite outcome'.[21] Often this is legitimate, but sometimes it can overstate benefits. For example, heart attacks are a fairly rare event in life generally, and also in most trials of cardiovascular drugs, which is why such trials often have to be very large, in order to have a chance of detecting a difference in the rate of heart attacks between the two groups. Because of this, it's fairly common to see 'important cardiovascular endpoints' all bundled up together. This 'composite outcome' will include death, heart attack and angina (angina, in case you don't know, is chest pain caused by heart problems: it's a worry, but not as much of a worry as heart attack and death). A massive improvement in that omnibus score can feel like a huge breakthrough for heart attack and death, until you look closely at the raw data, and see that there were hardly any heart attacks or deaths in the duration of the study at all, and all you're really seeing is some improvement in angina.

One particularly influential composite outcome came from a famous British trial called UKPDS, which looked to see whether intensively managing the blood-sugar levels of patients with diabetes made a difference to their real-world outcomes. This reported three endpoints: it found no benefit for the first two, which were death and diabetes-related death; but it did report a 12 per cent reduction in the composite outcome. This composite outcome consisted of lots of things:

- sudden death
- death from high or low blood sugar
- fatal heart attack
- non-fatal heart attack
- angina
- heart failure
- stroke
- renal failure
- amputation
- bleeding into the middle chamber of the eye
- diabetes-related damage to the arteries in the eye requiring laser treatment
- blindness in one eye
- cataracts requiring extraction

That's quite a list, and a 12 per cent reduction on all of it bundled up together certainly feels like 'patient-oriented evidence that matters', as we say in the business ('POEMs' if you prefer). But most of the improvement in this composite outcome was caused by a reduction in the number of people referred for laser treatment for damage to the arteries in their eyes. That's nice, but it's hardly the most important thing on that list, and it's very much a process outcome, rather than a concrete, real-world one. If you're interested in real-world outcomes, there wasn't even any change in the number of people experiencing visual loss, but in any case, it's clearly a much less important outcome than heart attacks, deaths, strokes or amputation. Similarly, the trial found a benefit for some blood markers suggestive of kidney problems, but no change in actual end-stage kidney disease.

This is only interesting because UKPDS has a slightly legendary status, among medics, as showing the benefit, on multiple outcomes, from intensive blood-sugar control for people with diabetes. How was this widespread belief created? One enterprising group of researchers decided to find every one of the thirty-five diabetes review papers citing the UKPDS study, and see what they said about it.[22] Twenty-eight said that the trial found a benefit for the composite outcome, but only one mentioned that most of this was down to improvements on the most trivial outcomes, and only six that it found no benefit for death, which is surely the ultimate outcome that matters. There is a terrifying reality revealed by this study: rumours, oversimplifications and wishful thinking can spread through the academic literature, just as easily as they do through any internet discussion forum.

Trials that ignore drop-outs

Sometimes patients leave a trial altogether, often because they didn't like the drug they were on. But when you analyse the two groups in your trial you have to make sure you analyse all the patients assigned to a treatment. Otherwise you overstate the benefits of your drug.

One classic failure at the analysis stage which can pervert your data horribly is to analyse patients according to the treatment they actually took, rather than the treatment they were assigned at the randomisation stage of the trial. At first glance, this seems perfectly reasonable: if 30 per cent of your patients dropped out and didn't take your new tablet, they didn't experience the benefit, and shouldn't be included in the 'new tablet' group at analysis.

But as soon as you start to think about why patients drop out of treatment in trials, the problems with this method start to become apparent. Maybe they stopped taking your tablets because they had horrible side effects. Maybe they stopped taking your tablets because they decided they didn't work, and just tipped them in the bin. Maybe they stopped taking your tablets, and coming to follow-up appointments, because they were dead, after your drug killed them. Looking at patients only by the treatment they took is called a 'per protocol' analysis, and this has been shown to dramatically overstate the benefits of treatments, which is why it's not supposed to be used.

If you keep all the patients prescribed your new treatment – including those who stopped taking it – in the 'new treatment' group when you do your final calculation, this is called an 'intention to treat' analysis. As well as being more conservative, this analysis makes much more sense philosophically. You're going to use the results of a trial to inform your decision about whether to 'give someone some tablets', not 'force some tablets down their throat compulsorily'. So you want the results to be from an analysis that looks at people according to what they were given by their doctor, rather than what they actually swallowed.

I've had the joy of marking sixty exam papers – a Groundhog Day experience if ever there was one – in which a fifth of the marks were to be earned by explaining 'intention to treat analysis'. This is at the absolute core of the evidence-based medicine curriculum, so it's utterly bizarre that there are still endless 'per protocol' analyses being reported by the drugs industry. One systematic review looked at all the trial reports submitted by companies to the Swedish drug regulator, and then the published academic papers relating to the same trials (if they even existed).[23] All but one of the submissions to the regulator featured both 'intention to treat' and 'per protocol' analyses, because regulators are, for all their faults and obsessive secrecy, at least a little sharper about methodological rigour than many academic journals. All but two of the academic papers, meanwhile, only reported one analysis, usually the 'per protocol' one that overstates the benefits. This is the version that doctors read. In the next section, we will see another example of how academic journals participate in the game of overstating results: often, for all their claims to be the gatekeepers for good-quality research, these journals do not do their job well.

Trials that change their main outcome after they've finished

If you measure a dozen outcomes in your trial, but cite an improvement in any one of them as a positive result, then your results are meaningless. Our tests for deciding if a result is statistically significant assume that you are only measuring one outcome. By measuring a dozen, you have given yourself a dozen chances of getting a positive result, rather than one, without clearly declaring that. Your study is biased by design, and is likely to find more positive results than there really are.

Imagine we're playing with dice, and we make a simple (albeit one-sided) arrangement: if I throw a double six, you have to give me £10. So I roll the dice, and they come up double three. But I still demand my £10, claiming that our original agreement was in fact that you give me £10 if I roll a double three; and you still pay me, with the cheerful encouragement of everyone around us. This exact scenario is played out in clinical academic research, as a matter of routine, every day, when we tolerate people doing something called 'switching the primary outcome'.

Before you begin a clinical trial, you write out the protocol. This is a document describing what you're going to do: how many participants you're going to recruit, where and how you're going to recruit them, what treatment each group will receive, and what outcomes you're going to measure. In a trial you'll measure all kinds of things as possible outcomes: perhaps a few different rating scales for 'pain', or 'depression', or whatever you're interested in; maybe 'quality of life', or 'mobility', that you'll measure with some kind of questionnaire; possibly 'death from all causes', and death from each of a number of specific causes too; and lots of other things.

Among all of these many outcomes, you will specify one (or perhaps a couple more, if you account for this in your analysis) as the main, primary outcome. You do this before the trial starts, because you're trying to avoid one simple problem: if you measure lots of things, some of them will come up as statistically significantly improved, simply from the natural random variation in all trial data. These are real people, remember, in the real world, and their pain, depression, mobility, quality of life and so on will all vary, for all kinds of reasons, many of which have nothing whatsoever to do with the intervention that you're testing in your trial.

If you're a pure-hearted researcher, you're using statistical tests specifically to identify genuine benefits of the treatment you're testing. You're trying to distinguish these real changes from the normal random variation of background noise that you would expect to see in your patients' results on various tests. More than anything, you want to avoid finding false positives.

The traditional cut-off for statistical significance is 'one in twenty'. Roughly speaking, clearing this bar means that if you repeated the same study over and over again, with the same methods, in participants taken from the same population, you'd expect to get the same positive finding you've observed one time in every twenty, simply by chance, even if the drug really had no benefit. If you dip two cups into the same jar of white and red beads, every now and then, purely by chance, you will come out with an unusually small number of red beads in one cup, and an unusually large number of red beads in the other. The same is true for any measurement we take in patients: there will be some random variation, and it can sometimes make it look as if one treatment is better than another, on one scoring method, simply through chance. Statistical tests are designed to stop us being misled by that kind of random variation.

So now, let's imagine you're running a trial where you measure ten different, independent outcomes. If we set the cut-off for statistical significance as 'one in twenty', then even if your drug does nothing useful at all, in your single trial you've still got a 50/50 chance of finding a positive benefit on at least one of your outcomes, simply from random variation in your data. If you didn't pre-specify which of the many outcomes is your primary outcome before you started, you could be cheeky, and report any positive finding you get, in any of your ten outcomes, as a positive result from your trial.

Could you get away with doing this openly, and simply saying: 'Hey, we measured ten things, and one of them came up as improved, therefore our new drug is awesome'? Well, you probably could get away with it in some quarters, because the consumers of scientific papers aren't universally switched on to this kind of bait and switch. But generally people would spot it: they would expect to see a 'primary outcome' nominated and reported, because they know that if you measure ten things, one of them is pretty likely to come up as improved simply through chance.

The problem is this: even though people know that you should nominate a primary outcome, these primary outcomes often change between the protocol and the paper, after the people conducting the research have seen the results. Even you – a random punter who's picked up this book on a station platform, and not a professor of either statistics or medicine – can see the madness in this. If the primary outcome reported in the finished paper is different from the primary outcome nominated before the trial started, then that is absurd: the entire point of the primary outcome is that it's the primary outcome nominated *before* the trial started. But people do switch their primary outcomes, and this is not just an occasional problem. In fact, it's almost routine practice.

In 2009, a group of researchers got all the trials they could find on various uses of a drug called gabapentin.[24] They then looked at those for which they could obtain internal documents, which meant they could identify the original, pre-specified primary outcome. Then they looked at the published academic papers that reported these trials. Of course, about half of the trials were never published at all (the scandal of this should not wear off with repetition). Twelve trials were published, and they checked to see if the things reported as primary outcomes in the academic papers really were pre-specified as primary outcomes in the internal documents, before the trial started.

What they found was a mess. Of the twenty-one primary outcomes pre-specified in the protocols, which should all have been reported, only eleven actually appeared. Six weren't reported in any form, and four were reported, but reported as if they were secondary outcomes instead. You can also look at this from the other end of the telescope: twenty-eight primary outcomes were reported in the twelve published trials, but of those, about half were newly introduced, and were never really primary outcomes at all. This is nothing short of ridiculous: there is no excuse, not for the researchers doing the switching, and not for the academic journals failing to check. But that was only one drug. Was it a freak occurrence?

No. In 2004 some researchers published a paper looking at all areas of medicine: they took all the trials approved by the ethics committees of two cities over two years, then chased up the published papers.[25] About half of all the outcomes were incorrectly reported. Of the published papers, almost two thirds had at least one pre-specified primary outcome that had been switched, and this was not being done at random: exactly as you'd expect, positive outcomes were more than twice as likely to be properly reported. Other studies on primary-outcome switching report similar results.

To be clear: if you switch your pre-specified primary outcome between the beginning and the end of your trial, without a very good explanation for why you've done so, then you're simply not doing science properly. Your study is broken by design. It should be a universal requirement that all studies report their pre-specified primary outcome as the primary outcome. This should be enforced by all journals, and things should have been done this way since trials began. It's really not difficult. Yet we have collectively failed to adhere to this simple, obvious core requirement on an epic scale.

For one final illustration of what this means in practice, I shall return to paroxetine, and the studies that were conducted in children. Remember, when an area of medicine is subject to some kind of litigation, documents often become available to researchers that would otherwise be hidden from view, allowing them to identify problems, discrepancies and patterns that would not normally be detectable. For the most part these are documents which should always be in the public domain, but are not. So paroxetine may not be worse than any other drug for this kind of mischief (in fact, as we have seen from the study just described, outcome switching happens across the board): it's simply one of the cases about which we have the most detail.

In 2008 a group of researchers decided to go through the documents opened up by the litigation over paroxetine, and examine how the results of one clinical trial – 'trial 329' – had been published.[26] As late as 2007 systematic reviews were still describing this trial as having a positive result, which is how it was reported in publications of its results. But in reality that was completely untrue: the original protocols specified two primary outcomes and six secondary ones. At the end of the trial there was no difference between paroxetine and placebo for any of these outcomes. At least nineteen more outcomes were also measured, making twenty-seven in total. Of those, only four gave a positive result for paroxetine. These positive findings were reported as if they were the main outcomes.

It would be tempting to regard the reporting of trial 329 as some kind of freak episode, an appalling exception in an otherwise sane medical world. Tragically, as the research above demonstrates, this behaviour is widespread.

So widespread, in fact, that there's room for a small cottage industry, if there are any academics feeling brave enough to pursue the project. Someone somewhere needs to identify all the studies where the main outcomes have been switched, demand access to the raw data, and helpfully, at long last, conduct the correct analyses for the original researchers. If you choose to do this, your published papers will immediately become the definitive reference on these trials, because they will be the only ones to correctly present the pre-specified trial outcomes. The publications from the original researchers will be no more than a tangential and irrelevant distraction.

I'm sure they'll be pleased to help.

Dodgy subgroup analyses

If your drug didn't win overall in your trial, you can chop up the data in lots of different ways, to try and see if it won in a subgroup: maybe it works brilliantly in Chinese men between fifty-six and seventy-one. This is as stupid as playing 'Best of three . . . Best of five . . .' And yet it is commonplace.

Time and again we have come back to the same principle in this chapter: if you give yourself multiple chances at finding a positive result, but use statistical tests that assume you only had one go, then you vastly increase your chances of getting the result you want – if you flip a coin for long enough, you will eventually get four heads in a row.

A new way of doing this is the subgroup analysis. The trick is simple: you've finished your trial, and it had a negative result. There was no difference in outcome – the patients on placebo did just as well as those on your new tablets. Your drug doesn't work. This is bad news. But then you dig a little more, do some analyses, and find that the drug worked great for Hispanic non-smoking men aged fifty-five to seventy.

If it's not immediately obvious why this is a problem, we have to go back and think about the random variation in the data in any trial. Let's say your drug is supposed to prevent death during the duration of the trial. We know that death happens for all kinds of reasons, at often quite arbitrary moments, and is – cruelly – only partly predictable on the basis of what we know about how healthy people are. You're hoping that when you run your trial, your drug will be able to defer some of these random unpredictable deaths (though not all, because no drug prevents all causes of death!), and that you'll be able to pick up that change in death rate, if you have a sufficiently large number of people in your trial.

But if you go to your results after your trial has finished, and draw a line around a group of deaths that you can see, or around a group of people who survived, you can't then pretend that this was an arbitrarily chosen subgroup.

If you're still struggling to understand why this is problematic, think of a Christmas pudding with coins randomly distributed throughout it. You want to work out how many coins there are altogether, so you take a slice, a tenth of the pudding, at random, count the coins, multiply by ten, and you have an estimate of the total number of coins. That's a sensible study, in which you took a sensible sample, blind to where the coins were. If you were to x-ray the pudding, you would see that there were some places in it where, simply through chance clumping, there were more coins than elsewhere. And if you were to carefully follow a very complex path with your knife, you would be able to carve out a piece of pudding with more coins in it than your initial sensible sample. If you multiply the coins in this new sample by ten, you will make it seem as if your pudding has lots more coins in it. But only because you cheated. The coins are still randomly distributed in the pudding. The slice you took after you x-rayed it and saw where the coins were is no longer informative about what's really happening inside there.

And yet this kind of optimistic overanalysis is seen echoing out from business presentations, throughout the country, every day of the week. 'You can see we did pretty poorly overall,' they might say, 'but interestingly our national advertising campaign caused a massive uptick in sales for lower-priced laptops in the Bognor region.' If there was no prior reason to believe that Bognor is different from the rest of your shops, and no reason to believe that laptops are different from the rest of your products, then this is a spurious and unreasonable piece of cherry-picking.

More generally, we would say: if you've already seen your results, you can't then find your hypothesis in them. A hypothesis has to come *before* you see the results which test it. So subgroup analyses are a reasonable thing to do, but unless they're specified before you begin (or unless you've accounted, in your analysis, for the number of subgroup analyses you've done), they are just another way to increase your chances of getting a spurious, false positive result. But such fishing trips are amazingly common, and amazingly seductive, because they feel superficially plausible.

This problem is so deep-rooted that it has been the subject of a whole series of comedy papers by research methodologists, desperate to explain their case to over-optimistic researchers who can't see the flaws in what they're doing. Thirty years ago, Lee and colleagues published the classic cautionary paper on this topic in the journal *Circulation*.[27] They recruited 1,073 patients with coronary artery disease, and randomly allocated them to get either Treatment 1 or Treatment 2. Both treatments were non-existent, because this was a fake trial, a simulation of a trial. But the researchers followed up the real data on these real patients, to see what they could find, in the random noise of their progress.

They were not disappointed. Overall, as you would expect, there was no difference in survival between the two groups, since they were both treated the same way. But in a subgroup of 397 patients (characterised by 'three-vessel disease' and 'abnormal left ventricular contraction') the survival of Treatment 1 patients was significantly different from that of Treatment 2 patients, entirely through chance. So it turns out that you can show significant benefits, using a subgroup analysis, even in a fake trial, where both interventions consist of doing absolutely nothing whatsoever.

You can also find spurious subgroup effects in real trials, if you do a large enough number of spurious analyses.[28] Researchers working on a trial into the efficacy of a surgical procedure called endarterectomy decided to see how far they could push this idea – for a joke – dividing the patients up into every imaginable subgroup, and examining the results. First they found that the benefit of the surgery depended on which day of the week the patient was born on (see following page):[29] if you base your clinical decisions on that, you're an idiot. There was also a beautiful, almost linear relationship between month of birth and clinical outcome: patients born in May and June show a huge benefit, then as you move through the calendar there is less and less effect, until by March the intervention starts to seem almost harmful. If this finding had been for a biologically plausible variable, like age, that subgroup analysis would have been very hard to ignore.

Last, the ISIS-2 trial compared the benefits of giving aspirin or placebo for patients who'd just had a suspected heart attack. Aspirin was found to improve outcomes, but the researchers decided to do a subgroup analysis, just for fun. This revealed that while aspirin is very effective overall, it doesn't work in patients born under the star signs of Libra and Gemini. Those two signs aren't even adjacent to each other. Once again: if you chop up your data in lots of different ways, you can pull out lumpy subgroups with weird findings at will.

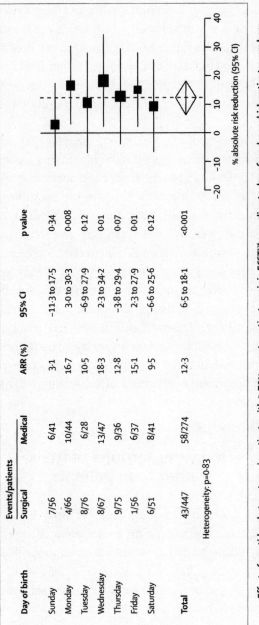

Day of birth	Events/patients		ARR (%)	95% CI	p value
	Surgical	Medical			
Sunday	7/56	6/41	3·1	−11·3 to 17·5	0·34
Monday	4/66	10/44	16·7	3·0 to 30·3	0·008
Tuesday	8/76	6/28	10·5	−6·9 to 27·9	0·12
Wednesday	8/67	13/47	18·3	2·3 to 34·2	0·01
Thursday	9/75	9/36	12·8	−3·8 to 29·4	0·07
Friday	1/56	6/37	15·1	2·3 to 27·9	0·01
Saturday	6/51	8/41	9·5	−6·6 to 25·6	0·12
Total	43/447	58/274	12·3	6·5 to 18·1	<0·001
Heterogeneity: p=0·83					

Effect of carotid endarterectomy in patients with ≥70% symptomatic stenosis in ECST[26] according to day of week on which patients were born

So should patients born under Libra and Gemini be deprived of treatment? You would say no, of course, and that would make you wiser than many in the medical profession: the CCSG trial found that aspirin was effective at preventing stroke and death in men, but not in women;[30] as a result, women were under-treated for a decade, until further trials and overviews showed a benefit.

That is just one of many subgroup analyses that have misled us in medicine, often incorrectly identifying subgroups of people who wouldn't benefit from a treatment that was usually effective. So, for example, we thought the hormone-blocking drug tamoxifen was no good for treating breast cancer in women if they were younger than fifty (we were wrong). We thought clotbusting drugs were ineffective, or even harmful, when treating heart attacks in people who'd already had a heart attack (we were wrong). We thought drugs called 'ACE inhibitors' stopped reducing the death rate in heart failure patients if they were also on aspirin (we were wrong). Unusually, none of these findings was driven by financial avarice: they were driven by ambition, perhaps; excitement at new findings, certainly; ignorance of the risks of subgroup analysis; and, of course, chance.

Dodgy subgroups of trials, rather than patients

You can draw a net around a group of trials, by selectively quoting them, and make a drug seem more effective than it really is. When you do this on one use of one drug, it's obvious what you're doing. But you can also do it within a whole clinical research programme, and create a confusion that nobody yet feels able to contain.

We've already seen that positive trials are more likely to be published and disseminated than negative ones, and that this can be misleading. Essentially, the problem is this: when we systematically review only the published trials, we are only seeing a subset of the results, and a subset that contains more positive results. We've taken a basket out with us to shop for trials, and been given only the nicest trials to put in it. But we'd be foolish to imagine that only nice trials exist.

This same problem – of how you take a sample of trials – can present itself in another, much more interesting way, best illustrated with an example.

Bevacizumab is an expensive cancer drug – its sales in 2010 were $2.7 billion – but it doesn't work very well. If you look on ClinicalTrials.gov, the register of trials (which has its own problems, of course), you will find about 1,000 trials of this drug, in lots of different kinds of cancer: from kidney and lung to breast and bowel, it's being thrown at everything.

Inevitably – sadly – lots of results from these trials are missing. In 2010 two researchers from Greece set about tracking down all the studies they could find.[31] Looking only for the large 'phase 3' trials, where bevacizumab was compared against placebo, they found twenty-six that had finished. Of these, nine were published (representing 7,234 patients' worth of data), and three had results presented at a conference (4,669 patients' worth). But fourteen more trials, with 10,724 participating patients in total, remain unpublished.

That's damnable, but it's not the interesting bit.

They put all the results together, and overall it seems, regardless of which cancer you're talking about, this drug gives a marginal, brief survival benefit, and to roughly the same extent in each kind of cancer (though remember, that's before you take into account the side effects, and other very real costs). That's not the interesting bit either: remember, we're trying to get away from the idea that individual drug results are newsworthy, and focus on the structural issues, since they can affect every drug, and every disease.

This is the interesting bit. From June 2009 to March 2010, six different systematic reviews and meta-analyses on bevacizumab were published, each in a different type of cancer, each containing the few trials specific to that cancer.

Now, if any one of those meta-analyses reports a positive benefit for the drug, in one form of cancer, is that a real effect? Or is it a subgroup analysis, where there is an extra opportunity to get a positive benefit, regardless of the drug's real effects, simply by chance, like rolling a dice over and over until you get a six? This is a close call. I think it's a subgroup analysis, and John Ioannidis and Fotini Karassa, the two researchers who pulled this data together, think so too. None of the individual meta-analyses took into account the fact that they were part of a wider programme of research, with machine guns spraying bullets at a wall: at some stage, a few were bound to hit close together. Ioannidis and Karassa argue that we need to analyse whole clinical trials programmes, rather than individual studies, or clumps of studies, and account for the number of trials done with the drug on any disease. I think they're probably right, but it's a complicated business. As you can now see, there are traps everywhere.

'Seeding Trials'

Sometimes, trials aren't really trials: they're viral marketing projects, designed to get as many doctors prescribing the new drug as possible, with tiny numbers of participants from large numbers of clinics.

Let's say you want to find out if your new pain drug, already shown to be effective in strict trials with ideal patients, also works in routine clinical use. Pain is common, so the obvious and practical approach is to use a small number of community clinics as research centres, and recruit lots of patients from each of these. Running your study this way brings many advantages: you can train a small number of participating doctors easily and cheaply; administrative costs will be lower; and you can monitor data standards properly, which means a better chance of good-quality data, and a reliable answer.

The ADVANTAGE trial on Vioxx was conducted very differently. They set out to recruit over 5,000 patients, but their design specified that each doctor should only treat a handful of patients. This meant that a vast number of participating doctors were required – six hundred by the end of the study. But that was OK for Merck, because the intention of this study wasn't really to find out how good the drug was: it was to advertise Vioxx to as many doctors as possible, to get them familiar with prescribing it, and to get them talking about it to their friends and colleagues.

The basic ideas behind seeding trials have been discussed for many years in the medical literature, but only ever in hushed tones, with the threat of a defamation suit hanging over your head. This is because, even if the number of sites looks odd from the outside, you can't be absolutely sure that any given project really is a seeding trial, unless you catch the company openly discussing that fact.

In 2008, new documents were released during unrelated litigation on Vioxx, and they produced exactly this proof.[32] Although the ADVANTAGE trial was described to patients and doctors as a piece of research, in reality, reading the internal documents, it was intended as a marketing trial from the outset. One internal memo, for example, under the heading 'description and rationale', sets out how the trial was 'designed and executed in the spirit of the Merck marketing principles'. Here these were, in order: to target a select group of critical customers (family doctors); to use the trial to demonstrate the value of the drug to the doctors; to integrate the research and marketing teams; and to track the number of prescriptions for Vioxx written by doctors after the trial finished. The data was handled entirely by Merck's marketing division, and the lead author on the academic paper reporting the trial later told the *New York Times* that he had no role in either collecting or analysing the data.

Seeding trials raise several serious issues. To begin with, the purpose of the trial is hidden from participating patients and doctors, but also from the ethics committees giving permission for access to patients. The editorial accompanying the paper that exposed the ADVANTAGE trial is as damning, on this, as any academic journal article possibly could be.

> [These documents] ... tell us that deception is the key to a successful seeding trial ... Institutional review boards, whose purpose is to protect humans who participate in research, would probably not likely approve an action that places patients in harm's way in order to influence physicians' prescribing habits. If they knew, few established clinical researchers would participate as coinvestigators. Few physicians would knowingly enroll their patients in a study that placed them at risk in order to provide a company with a marketing advantage, and few patients would agree to participate. Seeding trials can occur only because the company does not disclose their true purpose to anyone who could say 'no.'[33]

So seeding trials mislead patients. It's also poignant – for me, as a medic, at any rate – to imagine the hollow boasts from vain, arrogant, hoodwinked doctors. 'We're having some great results with Vioxx, actually,' you can imagine them saying in the pub.

'Did I tell you I'm an investigator on that trial? It's fascinating work we're doing ...'

But there are much more concrete concerns from these trials, because they can also produce poor-quality data, since the design is geared towards marketing, rather than answering a meaningful clinical question. Collecting data from small numbers of patients in multiple different locations risks all kinds of unnecessary problems: lower quality control for the information, for example, or poorer training for research staff, increased risk of misconduct or incompetence, and so on.

This is clear from another seeding trial, called STEPS, which involved giving a drug called Neurontin to epilepsy patients in community neurology clinics. Its true purpose was revealed, again, when internal company documents were released during litigation (once again: this is why drug companies will move hell and high water to settle legal cases confidentially, and out of court).[34] As you would expect, these documents candidly describe the trial as a marketing device. One memorable memo reads: 'STEPS is the best tool we have for Neurontin, and we should be using it wherever we can.' To be absolutely clear, this quote isn't discussing using the results of the trial to market the drug: it was written while the trial was being conducted.

The same ethical concerns as before are raised by this trial, as patients and doctors were once again misled. But equally concerning is the quality of the data: doctors participating as 'investigators' were poorly trained, with little or no experience of trials, and there was no auditing before the trial began. Each doctor recruited only four patients on average, and they were closely supervised, not by academics, but by sales representatives, who were directly involved in collecting the data, filling out study forms, and even handing out gifts as promotional rewards during data collection.

This is all especially concerning, because Neurontin isn't a blemishless drug. Out of 2,759 patients there were 73 serious adverse events, 997 patients with side effects, and 11 deaths (though as you will know, we cannot be sure whether those are attributable to the drug). For Vioxx, the drug in the ADVAN-TAGE seeding trial, the situation is even more grave, as this drug was eventually taken off the market because it increased the risk of heart attacks in patients taking it. We do good-quality research in order to detect benefits, or serious problems, with medicines, and a proper piece of trial research focused on real outcomes might have helped to detect this risk much earlier, and reduced the harm inflicted on patients.

Spotting seeding trials, even today, is fraught with worry. Suspicions are high whenever a new trial is published, on a recently marketed drug, where the number of recruitment sites is suspiciously high, and only a small number of patients were recruited from each one. This is not uncommon.

But in the absence of any documentary proof that these trials were designed with viral marketing in mind, very few academics would dare to call them out in public.

Pretend it's all positive regardless

At the end of your trial, if your result is unimpressive, you can exaggerate it in the way that you present the numbers; and if you haven't got a positive result at all, you can just spin harder.

All of this has been a little complicated, at times. But there is one easy way to fix an unflattering trial result: you can simply talk it up. A good example of this comes from the world of statins. From the evidence currently available on these drugs, it looks as if they roughly halve your risk of having a heart attack in a given period, regardless of how large your pre-existing risk was. So, if your risk of heart attack is pretty big – you've got high cholesterol, you smoke, you're overweight, and so on – then a statin reduces your large yearly risk of a heart attack by a half. But if your risk of a heart attack is tiny, it reduces that tiny risk by half, which is a tiny change in a tiny risk. If you find it easier to visualise with a concrete example, picture this: your chances of dying from a meteor landing on your head are dramatically less if you wear a motorbike crash helmet every day, but meteors don't land on people's heads very often.

It's worth noting that there are several different ways of numerically expressing a reduction in risk, and they each influence our thinking in different ways, even though they accurately describe the same reality. Let's say your chances of a heart attack in the next year are high: forty people out of 1,000 like you will have a heart attack in the next year, or if you prefer, 4 per cent of people like you. Let's say those people are treated with a statin, and their risk is reduced, so only twenty of them will have a heart attack, or 2 per cent. We could say this is 'a 50 per cent reduction in the risk of heart attack', because it's gone from 4 per cent to 2 per cent. That way of expressing the risk is called the 'relative risk reduction': it sounds dramatic, as it has a nice big number in it. But we could also express the same change in risk as the 'absolute risk reduction', the change from 4 per cent to 2 per cent, which makes a change of 2 per cent, or 'a 2 per cent reduction in the risk of heart attack'. That sounds less impressive, but it's still OK.

Now, let's say your chances of having a heart attack in the next year are tiny (you can probably see where I'm going, but I'll do it anyway). Let's say that four people out of 1,000 like you will have a heart attack in the next year, but if they are all on statins, then only two of them will have such a horrible event. Expressed as relative risk reduction, that's still a 50 per cent reduction. Expressed as absolute risk reduction, it's a 0.2 per cent reduction, which sounds much more modest.

There are many people in medicine who are preoccupied with how best to communicate such risks and results, a number of them working in the incredibly exciting field known as 'shared decision-making'.[35] They have created all kinds of numerical tools to help clinicians and patients work out exactly what benefit they would get from each treatment option when presented with, say, different choices for chemotherapy after surgery for a breast tumour. The advantage of these tools is that they take doctors much closer to their future role: a kind of personal shopper for treatments, people who know how to find evidence, and can communicate risk clearly, but who can also understand, in discussion with patients, their interests and priorities, whether those are 'more life at any cost' or 'no side effects'.

Research has shown that if you present benefits as a relative risk reduction, people are more likely to choose an intervention. One study, for example, took 470 patients in a waiting room, gave them details of a hypothetical disease, then explained the benefits of two possible treatment options.[36] In fact, both these treatments were the same, offering the same benefit, but with the risk expressed in two different ways. More than half of the patients chose the medication for which the benefit was expressed as a relative risk reduction, while only one in six chose the one whose benefit was expressed in absolute terms (most of the rest were indifferent).

It would be wrong to imagine that patients are unique in being manipulated by the way figures on risk and benefit are presented. In fact, exactly the same result has been found repeatedly in experiments looking at doctors' prescribing decisions,[37] and even the purchasing decisions of health authorities,[38] where you would expect to find numerate doctors and managers capable of calculating risk and benefit.

That is why it is concerning to see relative risk reduction used so frequently in reporting the modest benefits of new treatments, both in mainstream media and in professional literature. One good recent example comes, again, from the world of statins, in the coverage around the Jupiter trial.

This study looked at the benefits of an existing drug, rosuva-statin, for people with low risk of heart attack. In the UK most newspapers called it a 'wonder drug' (the *Daily Express*, bless it, thought it was an entirely new treatment,[39] when in reality it was a new use, in low-risk patients, of a treatment that had been used in moderate- and high-risk patients for many years). Every paper reported the benefit as a relative risk reduction: 'Heart attacks were cut by 54 per cent, strokes by 48 per cent and the need for angioplasty or bypass by 46 per cent among the group on Crestor compared to those taking a placebo or dummy pill,' said the *Daily Mail*. In the *Guardian*, 'Researchers found that in the group taking the drug, heart attack risk was down by 54 per cent and stroke by 48 per cent.'[40]

The numbers were entirely accurate, but as you now know, presenting them as relative risk reductions overstates the benefit. If you express the exact same results from the same trial as an absolute risk reduction, they look much less exciting. On placebo, your risk of a heart attack in the trial was 0.37 events per one hundred person years. If you were taking rosuvastatin, it fell to 0.17 events per one hundred person years. And you have to take a pill every day. And it might have side effects.

Many researchers think the best way to express a risk is by using the 'numbers needed to treat'. This is a very concrete method, where you calculate how many people would need to take a treatment in order for one person to benefit from it. The results of the Jupiter trial were not presented, in the paper reporting the final findings, as a 'number needed to treat', but in that low-risk population, working it out on the back of an enve-lope, I calculate that a few hundred people would need to take the pill to prevent one heart attack. If you want to take rosuva-statin every day, knowing that this is the likelihood of you receiving any benefit from the drug, then that's entirely a matter for you. I don't know what decision I would make, and everyone is different, as you can see from the fact that some people with low risk choose to take a statin, and some don't. My concern is only whether those results are explained to them clearly, in the newspapers, in the press release, by their doctor, and in the original academic journal article.

Let's consider one final example. If your trial results really were a disaster, you have one more option. You can simply present them as if they were positive, regardless of what you actually found.

A group of researchers in Oxford and Paris set out to examine this problem systematically in 2009.[41] They took every trial published over one month that had a negative result, in the correct sense of the word, meaning trials which had set out in their protocol to detect a benefit on a primary outcome, and then found no benefit. They then went through the academic journal reports of seventy-two of these trials, searching for evidence of 'spin': attempts to present the negative result in a positive light, or to distract the reader from the fact that the main result of the trial was negative.

First they looked in the abstracts. These are the brief summaries of an academic paper, on the first page, and they are widely read, either because people are too busy to read the whole paper, or because they cannot get access to it without a paid subscription (a scandal in itself). Normally, as you scan hurriedly through an abstract, you'd expect to be told the 'effect size' – '0.85 times as many heart attacks in patients on our new super-duper heart drug' – along with an indication of the statistical significance of this result. But in this representative sample of seventy-two trials, all with unambiguously negative results for their main outcome, only nine gave these figures properly in the abstract, and twenty-eight gave no numerical results for the main outcome of the trial at all. The negative results were simply buried.

It gets worse: only sixteen of these negative trials reported the main negative outcome of the trial properly anywhere, even in the main body of the text.

So what was in these trial reports? Spin. Sometimes the researchers found some other positive result in the spreadsheets, and pretended that this was what they had intended to count as a positive result all along (a trick we have already seen: 'switching the primary outcome'). Sometimes they reported a dodgy subgroup analysis – again, a trick we've already seen. Sometimes they claimed to have found that their treatment was 'non-inferior' to the comparison treatment (when in reality a 'non-inferiority' trial requires a bigger sample of people, because you might have missed a true difference simply by chance). Sometimes they just brazenly rambled on about how great the treatment was, despite the evidence.

This paper is not a lone finding. In 2009 another group looked at papers reporting trials on prostaglandin eyedrops as a treatment for glaucoma[42] (as always, the specific condition and treatment are irrelevant; it's the principle that is important). They found thirty-nine trials in total, with the overwhelming majority, twenty-nine of them, funded by industry. The conclusions were chilling: eighteen of the twenty industry-funded trials presented a conclusion in the abstract that misrepresented the main outcome measure. All of the non-industry-funded studies were fine.

All this is shameless, but it is possible because of structural flaws in the information architecture of academic medicine. If you don't make people report the primary outcome in their paper, if you accept that they routinely switch outcomes, knowing full well that this distorts statistics, you are permitting results to be spun. If you don't link protocols clearly to papers, allowing people to check one against the other for 'bait and switch' with the outcomes, you permit results to be spun. If editors and peer reviewers don't demand that pre-trial protocols are submitted alongside papers, and checked, they are permitting outcome switching. If they don't police the contents of abstracts, they are collaborators in this distortion of evidence that distorts clinical practice and makes treatment decisions arbitrary rather than evidence-based, and so they play their part in harming patients.

Perhaps the greatest problem is that many of those who read the medical literature implicitly assume that such precautions are taken by all journal editors. But they are wrong to assume this. There is no enforcement for any of what we have covered, everyone is free to ignore it, and so commonly – as with newspapers, politicians and quacks – uncomfortable facts are cheerfully spun away.

Finally, perhaps most worryingly of all, similar levels of spin have been reported in systematic reviews and meta-analyses, which are correctly regarded as the most reliable form of evidence. One study compared industry-funded reviews with independently funded reviews from the Cochrane Collaboration.[43] In their written conclusions, the industry-funded reviews all recommended the treatment without reservation, while none of the Cochrane meta-analyses did. This disparity is striking, because there was no difference in their numerical conclusions on the treatment effect, only in the narrative spin of the discussion in the conclusions section of the review paper.

The absence of scepticism in the industry-funded reviews was also borne out in the way they discussed methodological shortcomings of the studies they included: often, they simply didn't. Cochrane reviews were much more likely to consider whether trials were at risk of bias; industry-funded studies brushed over these shortcomings. This is a striking reminder that the results of a scientific paper are much more important than the editorialising of the discussion section. It's also a striking reminder that the biases associated with industry funding penetrate very deeply into the world of academia.

5

Bigger, Simpler Trials

So, we have established that there are some very serious problems in medicine. We have badly designed trials, which can suffer from all kinds of fatal flaws: sometimes they're conducted in unrepresentative patients, or are too brief, they can measure the wrong outcomes, they can go missing if the results are unflattering, they can get analysed stupidly, and often they're simply not done at all, simply because of expense, or lack of incentives. These problems are frighteningly common, both for the trials that are used to get a drug on the market, and for the trials that are done later, all of which guide doctors' and patients' treatment decisions. It feels as if some people, perhaps, view research as a game, where the idea is to get away with as much as you can, rather than to conduct fair tests of the treatments we use.

However we view the motives, this unfortunate situation leaves us with a very real problem. For many of the most important diseases that patients present with, we have no idea which of the widely used treatments is best, and, as a consequence, people suffer and die unnecessarily. Patients, the public, and even many doctors live in blissful ignorance of this frightening reality, but in the medical literature, it has been pointed out again and again.

Over a decade ago, a *BMJ* paper on the future of medicine described the staggering scale of our ignorance. We still don't know, it explained, which of the many current treatments is best, for something as simple as treating patients who've just had a stroke. But the paper also made a disarmingly simple observation: strokes are *so* common, that if we took every patient in the world who had one, and entered them into a randomised trial comparing the best treatments, we would recruit enough patients in just twenty-four hours to answer this question. And it gets better: many outcomes from stroke – like death – become clear in a matter of months, sometimes weeks. If we started doing this trial today, and analysed the results as they came in, medical management of stroke could be transformed in less time than it takes to grow a sunflower.

The manifesto implicit in this paper was very straightforward: wherever there is genuine uncertainty about which treatment is best, we should conduct a randomised trial; medicine should be in a constant cycle of revision, gathering follow-up data and improving our interventions, not as an exception, but wherever that is possible.

There are technical and cultural barriers to doing this kind of thing, but they are surmountable, and we can walk through them by considering a project I've been involved in, setting up randomised trials embedded in routine practice, in everyday GP surgeries.[1] These trials are designed to be so cheap and unobtrusive that they can be done whenever there is genuine uncertainty, and all the results are gathered automatically, at almost no cost, from patients' computerised notes.

To make the design of these trials more concrete, let's look at the pilot study, which compares two statins against each other, to see which is best at preventing heart attack and death. This is exactly the kind of trial you might naïvely think has already been done; but as we saw in the previous chapter, the evidence on statins has been left incomplete, even though they are some of the most widely prescribed drugs in the world (which is why, of course, we keep coming back to them in this book). People have done trials comparing each statin against a placebo, a rubbish comparison treatment, and found that statins save lives. People have also done trials comparing one statin with another, which is a sensible comparison treatment; but these trials all use cholesterol as a surrogate outcome, which is hopelessly uninformative. We saw in the ALLHAT trial, for example, that two drugs can be very similar in how well they treat blood pressure, but very different in how well they prevent heart attacks: so different, in fact, that large numbers of patients died unnecessarily over many years before the ALLHAT trial was done, simply because they were being prescribed the less effective drug (which was, coincidentally, the newer and more expensive one).

So we need to do real-world trials, to see which statin is best at saving lives at commonly used doses; and I would also argue that we need to do these trials urgently. The most widely used statins in the UK are atorvastatin and simvastatin, because they are both off patent, and therefore cheap. If one of these turned out to be just 2 per cent better than the other at preventing heart attacks and death, this knowledge would save vast numbers of lives around the world, because heart attacks are so common, and because statins are so widely used. Failing to know the answer to this question could be costing us lives, every day that we continue to be ignorant. Tens of millions of people around the world are taking these drugs right now, today. They are all being exposed to unnecessary risk from drugs that haven't been appropriately compared with each other, but they're also all capable of producing data that could be used to gather new knowledge about which drug is best, if only they were systematically randomised, and their outcomes followed up.

Our large, pragmatic trial is very simple. Everything in GPs' offices today is already computerised, from the appointments to the notes to the prescriptions, as you will probably already know, from going to a doctor yourself. Whenever a GP sees a patient and decides to prescribe a statin, normally they click the 'prescribe' button, and are taken to a page where they choose a drug, and print out a prescription. For GPs in our trial, one extra page is added. 'Wait,' it says (I'm paraphrasing). 'We don't know which of these two statins is the best. Instead of choosing one, press this big red button to randomly assign your patient to one or the other, enter them into our trial, and you'll never have to think about it ever again.'

The last part of that last sentence is critical. At present, trials are a huge and expensive administrative performance. Many struggle to recruit enough patients, and many more struggle to recruit everyday doctors, as they don't want to get involved in the mess of filling out patient report forms, calling patients back for extra appointments, doing extra measurements and so on. In our trial there is none of that. Patients are followed up, their cholesterol levels, their heart attacks, their weird idiosyncratic side effects, their strokes, their seizures, their deaths: all of this data is taken from their computerised health records, automatically, without anybody having to lift a finger.

These simple trials have one disadvantage, which you may already have spotted, in that they aren't 'blinded', so the patients know the name of the drug they've received. This is a problem in some studies: if you believe that you've been given a very effective medicine, or that you've been given a rubbish one, then the power of your beliefs and expectations can affect your health, through a phenomenon known as the placebo effect. If you're comparing a painkiller against a dummy sugar pill, then a patient who knows they've been given a sugar pill for pain is likely to be annoyed and in more pain. But it's harder to believe that patients have firm beliefs about the relative benefits of atorvastatin and simvastatin, and that these beliefs will then impact on cardiovascular mortality five years later. In all research, we make a trade-off between what is ideal and what is practical, giving careful consideration to the impact that any methodological shortcomings will have on a study's results.

So, alongside this shortcoming, it's worth taking a moment to notice how many of the serious problems with trials can be addressed by our study design of simple trials in electronic health records. Setting aside the assumption that they will be analysed properly, without the dubious tricks mentioned in the previous chapter, there are other, more specific benefits. First, as we know, trials are frequently conducted in unrepresentative 'ideal patients', and in odd settings. But the patients in our simple pragmatic trials are exactly like real-world patients, because they are real-world patients. They are all the people that GPs prescribe statins to. Second, because trials are expensive, stand-alone administrative entities, and because they struggle to recruit patients, they are often small. Our pragmatic trial, meanwhile, is vanishingly cheap to run, because almost all of the work is done using existing data – it cost £500,000 to set up this first trial, and that included building the platform that can be used to run any trial you like in the future. This is exceptionally cheap in the world of trials. Third, trials are often brief, and fail to look at real-world outcomes: our simple trial runs forever, and we can collect follow-up data and monitor whether people have had a heart attack, or a stroke, or died, for decades to come, at almost no cost, by following their progress through the computerised health records that are being produced by their doctors anyway.

All this is made possible in Britain because of the GP Research Database, or GPRD, which has been running for many years. This contains anonymised medical records of several million patients from participating family doctors' practices, and is already widely used to do the kinds of side-effects monitoring studies I discussed earlier: in fact, this database is currently owned and run by the MHRA itself. So far, however, it has only been used for observational research, rather than randomised trials: people's prescriptions and medical conditions are monitored, and analysed in bulk, in the hope that we can spot patterns. This can be helpful, and has been used to generate useful information about several medicines, but it can also be very misleading, especially when you try to compare the benefits of different treatment options.

This is because, often, the people given one treatment aren't quite the same as the people given another, even though you think they are. There can be odd, unpredictable reasons why some patients are prescribed one drug, and some another, and it's very hard to work out what these reasons are, or to account for them after the fact, when you're analysing data you've collected from routine medical practice in the real world.

For example, maybe people in a posh area are more likely to be prescribed the more expensive of two similar drugs, because budgets in that clinic are less pressed, and the expensive one is more heavily marketed. If so, then even though the expensive drug is no better than a cheaper alternative, it would appear superior in the observational data, because wealthy people, overall, are healthier. This effect can also make drugs look worse than they really are. Many people have mild kidney problems, for example, which grumble along in the background alongside their other medical problems; they cause them no specific health issues, but their doctor is aware, from blood tests, that their kidneys are no longer clearing things from their bloodstream quite as efficiently as they do for the healthiest people in the population. When these patients are being treated for depression, say, or high blood pressure, maybe they will be put on a drug that is regarded as having a better safety profile, just to be on the safe side, on account of their mild kidney problems. In this case, that drug will look much *less* effective than it really is, when you follow up the patients' outcomes, because many of the people receiving it were sicker to start with: the patients with minor things, like mild kidney problems, were actively channelled onto the drug believed to be safest.

Even when you know these things are happening, it's hard to account for them in your analysis; but often there are gremlins distorting your findings, and you don't even realise they're there. Sometimes this has led to serious problems: hormone replacement therapy is just one memorable case of people being misled, by trusting 'observational' data, instead of doing a trial.

HRT is a reasonably safe and effective short-term treatment to reduce the unpleasant symptoms that some women experience while going through the menopause. But it has also been prescribed much more freely to patients, some of whom have received it for many years on end, for reasons that border on the aesthetic: HRT was regarded as a way to cheat ageing, and it maintained various features of a younger body, in a way that was desirable for many women. But this wasn't the only reason that doctors gave long-term prescriptions for these drugs. By observing the health records of older women, researchers were able to spot what they believed was a reassuring pattern: women who take HRT for many years live longer, healthier lives. This was very exciting news, and it helped to justify the long-term prescription of HRT to an even greater extent. Nobody had ever done a randomised trial – randomly assigning women either to receive HRT or to receive normal management without HRT. Instead, the results of the 'observational' studies were simply taken at face value.

When a randomised trial was finally done, it revealed a terrible surprise. Far from protecting you, HRT in fact *increases* your chances of various heart problems. It had only appeared to be beneficial because overall the women requesting HRT from their doctors were likely to be wealthy, vivacious, active, and many of the other things we know are already associated with living longer. We weren't comparing like with like, and because we accepted observational data uncritically, and failed to do a randomised trial, we continued to prescribe a treatment that exposed women to risks nobody knew about. Even if we accept that some women might have chosen to risk their lives for the other benefits of long-term HRT, all women were deprived of this choice by our failure to conduct fair tests.

This is why we need to do *randomised* trials wherever there is genuine uncertainty as to which drug is best for patients: because if we want to make a fair comparison of two different treatments, we need to be sure that the people getting them are absolutely identical. But randomly assigning real-world patients to receive one of two different treatments, even when you have no idea which is best, attracts all kinds of worried attention.

This is best illustrated by a bizarre paradox which currently exists in the regulation of everyday medical practice. When there is no evidence to guide treatment decisions, out of two available options, a doctor can choose either one arbitrarily, on a whim. When you do this there are no special safeguards, beyond the frankly rather low bar set by the GMC for all medical work. If, however, you decide to randomly assign your patients to one treatment or another, in the same situation, where nobody has any idea which treatment is best, then suddenly a world of red tape emerges. The doctor who tries to generate new knowledge, improve treatments and reduce suffering, at no extra risk to the patient, is subject to an infinitely greater level of regulatory scrutiny and oversight; but above all, that doctor is also subject to a mountain of paperwork, which slows the process to the point that research simply isn't practical, and so patients suffer, through the absence of evidence.

The harm done by these disproportionate delays and obstructions is well illustrated by two trials, both conducted in A&E departments in the UK. For many years it was common to

treat patients who'd had a head injury with a steroid injection. This made perfect sense in principle: after a head injury, your brain swells up, and since the skull is a box with a fixed volume, any swelling in there will crush the brain. Steroids are known to reduce swelling, and this is why we inject them into knees, and so on: so giving them to people with head injuries should, in theory, prevent the brain from being crushed. Some doctors gave steroids on the basis of this belief, and some didn't. Nobody knew who was right. People on both sides were pretty convinced that the people on the other side were dangerously mad.

The CRASH trial was designed to resolve this uncertainty: patients with serious head injury would be randomised, while still unconscious, to receive either steroids or no-steroids, and the researchers would follow them up to see how they got on.[2] This created huge battles with ethics committees, which didn't like the idea of randomising patients who were unconscious, even though they were being randomly assigned to two treatments, both in widespread use throughout the UK, where we had no idea whatsoever which was better. Nobody was losing out by being in the trial, but the patients of the future were being harmed with every day this trial was delayed.

When the trial was finally approved and conducted, it turned out that steroids were harming patients, and in large numbers: a quarter of the people with serious head injuries died whichever treatment they received, but there were two and a half extra deaths for every hundred people treated with steroids. Our delay in discovering this fact led to the unnecessary and avoidable deaths of very large numbers of people, and the authors of the study were absolutely clear who should take responsibility for this: 'The lethal effects we have shown might have been found decades ago had the research ethics community accepted a responsibility to provide robust evidence that its prescriptions are likely to do more good than harm.'

But this wasn't the only harm done. Many trial centres insisted on delaying treatment, in order to get written consent to participation in the trial from a relative of the unconscious patient. This written consent would not have been necessary to receive steroids, if you happened to be treated by a doctor who was a believer in them; nor would you have needed written consent to *not* receive steroids from a doctor who wasn't a believer. It was only an issue because patients were being randomised to one treatment or the other, and ethics committees choose to introduce greater barriers when that happens, even though the treatments patients are randomised to are the exact ones they would have got anyway. In the treatment centres where the local regulators insisted on family consent to randomisation, it delayed treatment with steroids by 1.2 hours on average. This delay, to my mind, is disproportionate and unnecessary: but as it happened, in this case, it did no harm, because steroids don't save lives (in fact, as we now know, they kill people).

In other studies, such a delay would cost lives. For example, the CRASH-2 trial was a follow-up piece of research, conducted in A&E departments by the same team. This study looked at whether trauma patients with severe bleeding are less likely to die if they're given a drug called tranexamic acid, which improves clotting. Since these patients are bleeding to death, there is a degree of urgency about getting them treated. Of course, all patients were given all the usual treatment you would expect them to get; the only extra feature of their management, determined by the trial, was whether they were randomly assigned to get tranexamic acid on top of normal management, or not.

The trial found that tranexamic acid is hugely beneficial, and saves lives. But again, some sites delayed giving it, while they tried to contact relatives and get consent for randomisation. A one-hour delay in giving tranexamic acid reduces the number of patients helped from 63 per cent to 49 per cent, so patients in

the trial were directly harmed by a delay introduced to get consent for randomisation between two options, where nobody knew which was better anyway, and patients throughout the UK are liable to get one or the other on almost entirely arbitrary grounds anyway.

This, once again, is something I would regard as disproportionate – and disproportionate is exactly the correct word. It is vitally important that the rights of patients are protected, and that they are not subjected to dangerous treatments in the name of research. Where trials are examining the effects of new, highly experimental treatments, it's absolutely right that there should be an enormous amount of regulatory oversight, and a wealth of information communicated clearly and compulsorily to the patient: I wouldn't dream of suggesting otherwise. But when someone is in a trial comparing two currently used treatments, both of which are believed to be equally safe and effective, where randomisation adds no extra risk, the situation is very different.

This is the situation for our trial in GPs' practices comparing two statins: in routine everyday practice in the UK, patients are sometimes given atorvastatin, and sometimes simvastatin. No doctor alive knows which is better, because there is absolutely no evidence comparing the two, on real-world outcomes like heart attack and death. When doctors make their arbitrary 'choice' of which to give, on the basis of no evidence, nobody is interested in regulating that, so there is no special process, and no form to complete explaining that there is no evidence for the decision. It seems to me that this everyday doctor, blithely giving one or other treatment in the absence of evidence, who makes *no* attempt to improve our understanding of which treatment is best, is committing something of an ethical crime, for the simple reason that they are perpetuating our ignorance. That doctor is exposing large numbers of future patients around the world to avoidable harm, and misleading their current patient about what we know of the benefits and risks of the treatments they are giving, with their fake certainty, or at best their failure to be honest about our uncertainty, and for no discernible benefit. But there is no special ethics committee oversight of that doctor's activity.

Meanwhile, when a patient is randomly assigned to one or other statin in our trial, suddenly this becomes a major ethical issue: the patient must fill out several pages of paperwork, over the course of twenty minutes, explaining that they understand all of the risks of the treatment they are being given, and that they are in a trial. They have to do this, even though no extra risk is introduced during the course of the trial; even though they were going to get one or other statin anyway; even though the trial imposes no extra burden on their time; and even though their medical records are already in the GP Research Database, and so are monitored for observational research regardless of their participation in the trial. These two statins are already used by millions of people around the world, and have been shown to be safe and effective: the only question for the trial is which is better. If there actually is a difference between them, huge numbers of people will be dying unnecessarily while we don't know.

The twenty-minute delay introduced by the consent form for this trial is interesting, because it's not simply an inconvenience. First, it may not even address the concerns that the ethicists are seeking to address: these committees and experts are keen to tell everyone that their restrictions are necessary, but they have collectively failed to produce research demonstrating the value of the interventions they force researchers to comply with, and in some cases, what little evidence we do have suggests that their interventions may even have the opposite effect to what they intend. One piece of research into what patients remember from consent forms, for example, shows that people remember *more* information, in total, from short forms than they do from long twenty-minute ones.[3]

But more than that, a twenty-minute consent process, to receive a drug you were going to get anyway, threatens the whole purpose of the trial, which is to try to randomise patients as seamlessly and unobtrusively as possible, in routine clinical care. It doesn't just make simple pragmatic trials slower and more expensive; it also makes them less representative of normal practice. When you introduce a twenty-minute consent process to receive a statin the patient was going to get anyway, the doctors and patients being recruited aren't normal doctors and patients, but the unrepresentative ones willing to stop what they're doing and spend twenty minutes going through a form together.

This isn't a problem for the pragmatic statins trial I have described, because the purpose of that trial isn't really to find out which statin is better. In reality, it's about the process, and its aim is to answer a much more fundamental and important question: can we randomise patients in routine care, cheaply and seamlessly? If we cannot, then we need to find out why not, and ask whether the barriers are proportionate, and whether they can be safely overcome. Ethicists appear to argue that the twenty-minute consent process is so valuable that we are better off letting patients die while we continue to practise in ignorance.

I'm not simply saying that I disagree with this; I'm saying that I think the public deserve a say in whether they agree, through an informed, open debate.

But more than that, I worry that these regulations express an implicit fantasy about normal clinical practice, which has never been adequately challenged: that of spurious over-certainty. Perhaps if all doctors were forced to admit to the uncertainties in our day-to-day management of patients, it might make us a little more humble, and more inclined to improve the evidence base on which we base our decisions. Perhaps if we honestly told patients, 'I don't know which of these two treatments is best,' whenever that was the case, patients would start to ask questions themselves. 'Why not?' might be the first, and 'Why don't you try to find out?' might follow shortly afterwards.

Some patients will prefer to avoid randomisation, for the illusion of certainty, and the fantasy that their doctor has been able to make a tailored decision about which statin, or any other drug, is best for them. But I think we should be able to offer everyone the chance to be randomised wherever there is true uncertainty about which is the better of two widely used treatments that are already known to be safe and effective. I think this should be done on the basis of a brief consent form, no more than a hundred words, with the option to access more detailed explanatory material for anyone who wants it. And I think research ethicists should be asked to provide evidence that the harm they are inflicting on patients around the world by imposing inflexible rules, such as a twenty-minute consent process, is proportional to whatever benefit they believe they confer.

More than this, I think we need a cultural shift in the way we all, as patients, view our reciprocal relationship with research in medicine. We only know what works because of trials, and we all profit from the participation of patients before us in these trials; but many of us seem to have forgotten this. By remembering, we could create a social contract whereby everyone expects their health service to be constantly conducting trials, simple A/B tests, comparing treatments against each other to see which is the best, or even the cheapest, if they're both equally effective. A doctor failing to take part in such tests could be regarded as an oddity who is harming future patients. It could be obvious to all patients that participating in these trials is a normal reflection of the need to produce better evidence to improve medical treatments, for themselves in future, and for the others in the community with whom they share their medical system.

In almost every developed country in the world, medicine is provided free at point of access, by the community, funded through taxation. From the perspective of the community, this whole process could be regarded as a simple bargain: we provide medicines free at point of access; in exchange, you need to let us find out what works best for you and others. The NHS could be in a constant cycle of testing and learning, improving its performance, and improving outcomes for everyone in the country, and everyone in the world, by creating better knowledge on what works.

It's a quirk of history that has been largely lost to most doctors and academics, but this was essentially the dynamic around the first ever truly modern randomised trial. In 1946 the antibiotic streptomycin had just been discovered, and after huge effort, 50kg was produced for the UK. It was hoped that this drug could be used to treat tuberculosis, but it was incredibly expensive, and we needed to find out if it actually worked. Patients with TB meningitis, in the brain, weren't a problem: they would die in front of you, and fast, almost every time; so if *any* of these patients survived after being given streptomycin, you knew the drug was probably effective. For pulmonary TB, in the lungs, the story was more complicated: people would often recover by themselves over time, without any medication, so it would be harder to tell if the drug really had improved their chances, or hastened their recovery.

In the US this drug was available on the open market, at huge prices. If you wanted to try it, you simply bought some, took it, and hoped for the best. But the UK Medical Research Council was in sole charge of our 50kg supply, and decided it was going to use this expensive new drug efficiently, in a randomised trial, to find out whether it really did make any difference to survival (and also whether it caused any unpredictable side effects).

Doctors weren't pleased, but in the immediate post-war environment, with rationing still commonplace, the notion of central control for the sake of the greater good was not so unusual. The first proper modern randomised trial went ahead, and the whole world's understanding of streptomycin's effectiveness was generated, essentially because the MRC forced our hands.

If that whole story sounds uncomfortably Stalinist, then I apologise, but you may have misunderstood. I'm not proposing that we coerce every patient to participate in a trial, wherever there is uncertainty about which is the best treatment for them, by exploiting the opportunity that the state has to ration supply: I'm simply suggesting that trials should be routinely embedded in all clinical practice, as the norm, as an everyday act. If people really want to opt out, and take drugs of unknown effectiveness without generating any new knowledge, then of course I accept their desire to be antisocial for no personal gain.

But this is a need that becomes more pressing with every day. Health care is cripplingly expensive, trials are the best tool we have to make our treatment decisions more cost-effective, and they can be run on many of the most important questions in medicine at very little cost, inflicting no harm at all on participants. Irrational prescribing costs lives, and it costs money; while the cost of research to prevent irrational prescribing is trivial in comparison, and large simple routine trials would swamp the bad evidence that has polluted medical practice, in just a few years. Our extreme effort, aiming to run trials at almost no cost in routinely collected electronic health records, is just one example of how this could be done.

Instead, we have occasional, small, brief trials, in unrepresentative populations, testing irrelevant comparisons, measuring irrelevant outcomes, with whole trials that go missing, avoidable design flaws, and endless reporting biases that only per-sist because research is conducted chaotically, for commercial gain, in spuriously expensive trials. The poor-quality evidence created by this system harms patients around the world.

And if we wanted, we could fix it.

6

Marketing

So far, we have established that the evidence gathered to guide treatment choices in medicine suffers from a huge number of avoidable biases and problems. But that is only part of the story: this poorly collected evidence is then disseminated, and implemented, through chaotic and biased systems, which adds a whole extra layer of exaggeration and error.

To understand what is happening here, we need to ask a simple question: how do doctors decide what to prescribe? This is a surprisingly complicated issue, and to feel our way through it we need to think about the four main players exerting pressure: the patient; the funder (which in the UK means the NHS); the doctor; and the drug company.

For patients, things are simple: you want a doctor to prescribe the best treatment for your medical problem. Or rather, you want the treatment that has been shown, overall, in fair tests, to be better than all the others. You will probably trust your doctor to make this decision, and hope that there are systems in place to ensure that it is done properly, because getting involved in every single decision yourself would be enormously time-consuming.

That's not to say that patients are locked out, either by tradition or by design. It's true that it's rare for patients to make decisions about which treatment is best *entirely* for themselves, by reading the primary research literature, and spotting the strengths and flaws in each trial for themselves. I feel bad about that, and wish that this book could teach you everything you need to know, but the reality is that medical decision-making requires a lot of specialist knowledge and skills, which take time and practice to acquire at a safe level of competence, and there is a serious risk of people making very bad decisions when it's not done well.

That said, doctors and patients do make decisions together all the time, when medical practice is at its best, in discussions where doctors act as a kind of personal shopper, eliciting the outcomes a patient is most interested in achieving, and communicating the best existing evidence clearly, to allow an informed decision. Some patients might want a longer life at any cost, for example, while others might hate the hassle of taking a pill twice a day, and prefer to tolerate a greater risk of a bad long-term outcome. We will discuss how this can best be done later, but for now we will settle on the fact that in most cases, patients just want the best treatment.

Our next players are the funders (the NHS, or an insurer), and for them, the answer is also fairly simple: they want the same thing as the patient, unless it's insanely expensive. For common drugs, and common decisions, they might have a set 'pathway' that dictates to GPs (more commonly than to hospital doctors) which drug is to be used, but outside those simple rules for simple situations, they rely on doctors' judgements.

Now we come to our core player in the individual treatment decision: doctors. They need good-quality information, but they need it, crucially, under their noses. The problem of the modern world is not information poverty, after all, but information overload, and even more precisely, what Clay Shirky calls 'filter failure'. As recently as the 1950s, remember, medicine was driven almost entirely by anecdote and eminence; in fact, it's only in the past couple of generations that we have collected good-quality evidence at all, in large amounts, and for all the failures in our current systems, we suddenly now have an over-whelming avalanche of data. The exciting future, for evidence-based medicine, is an information architecture that can get the right evidence to the right doctor at the right time.

Does this happen? The simple answer is no. Although there are many automated systems for disseminating knowledge, for the most part we continue to rely on systems that have evolved over centuries, like the long, meandering essays in academic journals that are still used to report the results of clinical trials. Often, if you ask a doctor whether they know if one particular treatment is best for a particular medical condition, they'll tell you they certainly do, and name it. But if you ask them how they know it is the best, their answer might scare you.

They might say: that's what I learnt at medical school; that's what the person in the office next door told me she uses; that's what I see the local consultant prescribing in his letters on patients I've referred; that's what the local drug rep told me; that's what I picked up on a teaching day two years ago; that's what I think I read in a review article somewhere; that's what I remember from some guidelines I looked up once; that's what the local prescribing guidelines recommend; that's what a trial I read said; that's what I've always used; and so on.

In reality, doctors can't read every scientific article that's relevant to their work, and that's not just my opinion, or even a moan about my own reading pile. There are tens of thousands of academic journals, and millions of academic medical papers in existence, with more produced every day. One recent study tried to estimate how long it would take to keep up with all this information.[1] The researchers collected every academic paper published in a single month that was relevant to general practice. Taking just a few minutes for each one, they estimated it would take a doctor six hundred hours to skim through them all. That's about twenty-nine hours each weekday, which is, of course, not possible.

So doctors will not be going through every trial, about every treatment relevant to their field, meticulously checking each one for the methodological tricks described in this book, diligently keeping their knowledge perfectly current. They will take shortcuts, and these shortcuts can be exploited.

To see how bad doctors are at prescribing efficiently, we can look at national prescribing patterns. The US spends around $300 billion a year on drugs, while the NHS in the UK spends over £10 billion. You know by now that many of the drugs on the market are 'me-too' drugs, which are no better than the drugs they copy, and that often the branded 'me-too' drugs could be replaced with equally effective drugs from the same class which are old enough to have come out of patent.

In 2010 a team of academics analysed the top ten most highly prescribed classes of drugs in the NHS, and calculated that at least £1 billion is wasted, every year, from doctors using branded me-too drugs in a situation where there was an equally effective off-patent drug available.[2]

For example: atorvastatin and simvastatin are both equally effective, as far as we currently know, and simvastatin came off patent six years ago. So you would expect that everyone should be taking simvastatin instead of atorvastatin, unless there's a very good idiosyncratic reason to choose the unnecessarily expensive one in a specific patient. But even in 2009 there were still three million prescriptions a year for atorvastatin, not much down from the six million in 2006: this cost the NHS an unnecessary £165 million a year. And all those prescriptions for atorvastatin were despite major national programmes to try and get doctors switching.

The same pattern can be seen across the board. Losartan is an 'ARB'-type blood-pressure drug: there are lots of me-too drugs in this class, and because high blood pressure is so common, this class of medicines is the fourth most expensive for the NHS. In 2010, losartan came off patent: it is clinically almost indistinguishable from other ARB drugs, so you would expect the NHS to have switched everyone onto it, ready for the big price drop. But even after the price drop came, only 0.3 million of the 1.6 million people taking an ARB were on Losartan, so the NHS lost £200 million a year.

If we can't manage rational prescribing decisions even for these incredibly common medicines, then that is good evidence that prescription is a haphazard affair, where clear information is not efficiently disseminated to the people making the decisions, on either effectiveness or cost-effectiveness. I can honestly say, if I were in charge of the medical research budgets, I would cancel all primary research for a year, and only fund projects devising new ways to optimise our methods for disseminating information, ensuring that the evidence we already have is summarised, targeted and implemented. But I am not in charge, and there are some much more powerful influences out there.

Now let's think through a doctor's prescribing decision from the perspective of a drug company. You want the doctor to prescribe your product, and you will do everything you can to make that happen. You might dress this up as 'raising awareness of our product', or 'helping doctors make decisions', but the reality is, you want sales. So you will advertise your new treatment in medical journals, stating the benefits but downplaying the risks, and leaning away from unflattering comparisons. You will send out 'drug reps' to meet doctors individually, and talk up the merits of your treatment. They will offer gifts, lunches, and forge personal relationships that may be mutually beneficial later.

But it goes deeper than this. Doctors need ongoing education: they practise for decades after they leave medical school, and looking back from today, medicine has changed unrecognisably since, say, the 1970s, which is when many currently practising doctors came out of medical training. This education is expensive, and the state is unwilling to pay, so it is drug companies that pay for talks, tutorials, teaching materials, conference sessions, and whole conferences, featuring experts who they know prefer their drug.

All of this is built on the back of a published academic evidence base that drug companies have carefully nurtured, through selective publication of flattering results and judicious use of design flaws, to give a flattering picture of their product. But those aren't the only tools available to companies for influencing what appears in journals. They pay professional writers to produce academic papers, following their own commercial specifications, and then get academics to put their names to them. This acts as covert advertising, and will get more academic publications on their drug, more rapidly. It also aggrandises the favoured experts' CVs, and helps doctors friendly to the company get the kudos and veneer of independence that comes from a university post.

The company can also give money to patient groups, if those groups' views and values help it sell more drugs, and so give them greater prominence, power and platform. On top of all this, it can then pay academic journals to accept papers, with advertising revenue and 'reprint' orders, and with these academic papers it can foreground the evidence showing that its treatment works, and even expand the market for its drug, by producing work that helpfully shows that the problem it treats is actually much more widespread than people realise.

All of this sounds very expensive, and it is: marketing spending is hard to measure in a vast industry, and the figures are often hotly contested. Some have estimated that the pharmaceutical industry overall spends about twice as much on marketing and promotion as it does on research and development. Regardless of how those two figures compare to each other, the fact that they are in the same ballpark gives one pause, and this is worth mulling over in various contexts. For example, when a drug company refuses to let a developing country have affordable access to a new AIDS drug it's because – the company says – it needs the money from sales to fund research and development on other new AIDS drugs for the future. If R&D is a fraction of the company's outgoings, and it spends a similar amount on promotion, then this moral and practical argument doesn't hold water quite so well.

The scale of this spend is fascinating in itself, when you put it in the context of what we all expect from evidence-based medicine, which is that people will simply use the best treatment for the patient. Because when you pull away from the industry's carefully fostered belief that this marketing activity is all completely normal, and stop thinking of drugs as being a consumer product like clothes or cosmetics, you suddenly realise that medicines marketing only exists for one reason. In medicine, brand identities are irrelevant, and there's a factual, objective answer to whether one drug is the most likely to improve a patient's pain, suffering and longevity. Marketing, therefore, one might argue, exists for no reason other than to pervert evidence-based decision-making in medicine.

This is a very powerful machine: tens of billions of pounds are spent each year, as much as $60 billion in the US alone, on medicines marketing.[3] And, most impressively, this money isn't plucked from the air: it is paid for by patients, funded entirely from the public purse, or patients' payments into medical insurance companies. About a quarter of the money taken by pharmaceutical companies for the drugs they sell is turned around into promotional activity which has, as we will see, a provable impact on doctors' prescribing. So we pay for products, with a huge uplift in price to cover their marketing budget, and that money is then spent on distorting evidence-based practice, which in turn makes our decisions unnecessarily expensive, and less effective.

All of this comes on top of a system for evidence-based medicine that is already gravely wounded, with trials that are often of poor quality, and are poorly communicated to doctors at the best of times.

It's magnificent. Now, on to the details.

Ads to patients

It is doctors who make the final decision about signing a prescription, but in reality the decision on which treatment to choose – and whether to bother with treatment at all – is made between them and their patients. This is entirely how you would want things to be; but it does make patients another lever to be leaned on, by an industry keen to increase sales.

We will see in this chapter that the techniques used by drug companies to do this are many and varied: the invention of whole new diseases and explanatory models; funding patient groups; running star patients who fight (with professional PR assistance) against governments that have refused them expensive drugs; and more. But we will start with advertising, because there is an ongoing battle to bring it to the UK, and compared to the more covert strategies, it seems positively transparent.

Direct-to-consumer drug advertising has been banned in almost all industrialised countries since the 1940s, for the simple reason that it works: ads distort doctors' prescribing behaviour – by design – and increase costs unnecessarily. The USA and New Zealand (along with Pakistan and South Korea) changed their minds in the early 1980s, and permitted a resurgence of this open marketing. That doesn't, however, mean that these ads are someone else's problem. There is a constant battle to reopen new territories, and these ads leak through national borders in the age of the internet; but more than anything, they expose some clear truths about the industry's thinking.

Let's take a look at this mysterious world. When ads were first made legal again in the US, they could only appear in print, because of a requirement to include all the side-effect information from the drug label. Since 1997 the rules have been relaxed, and now the side effects can be abbreviated (they are read out at jabbering speed over the end of the TV ads). After this change the pharmaceutical industry's annual advertising budget rose from $200 million to $3 billion in the space of just a few years. Notable single spends include Vioxx, at $161 million, which was taken off the market because of serious concerns over hidden data; and Celebrex, at $78 million, also later taken off the market because it was harming patients.

Various approaches have been used to try to evaluate the impact of these ads in the real world.[4] One study observed patients visiting their doctors in Canada, where direct-to-consumer drug advertising is still banned, and the US. It found that those in the US were more likely to believe they needed medication, more likely to request specific drugs that were advertised on television, and more likely to receive a prescription for that drug. In other words: the ads worked. The doctors in the US, meanwhile, were more likely to report being worried about whether the drugs requested by their patients were appropriate.

Another study took a more proactive, experimental approach. Trained actors, posing as depressed patients, were sent to visit doctors in three American cities (three hundred visits in all).[5] They all gave the same background story, about the problems they were having with low mood, and then were randomly assigned to act in one of three ways at the end of the consultation: to ask for a specific named drug; to ask for 'medicine that might help'; or to make no specific request. Those who did what ads drive patients to do – ask for a specific drug, or 'medicine' – were twice as likely to receive a prescription for an antidepressant pill. Whether you think that's a good thing might depend partly on whether you think these drugs are actually worth using (the evidence overall shows that they're pretty ineffective for mild and moderate depression). But whatever your view on antidepressants, the evidence very clearly shows that what patients say to their doctor, and what they request, has a significant impact on what is prescribed. Engaged and informed patients are something most doctors would want to see, but here the question is whether the information the patients have received is truly helpful, and whether doctors can resist inappropriate requests for pills.

So the same study sent more actor-patients to see doctors, but this time they gave a clear history of 'adjustment disorder', a term used by some people to describe the simple human phenomenon of feeling bad in the immediate aftermath of a very bad thing happening in your life. This is a normal and appropriate thing for your feelings to do, though it's unpleasant, as any normal feeling person will know, and pills to treat it aren't a good idea. But patients presenting with 'adjustment disorder' who demanded a specific named drug still got it, in 50 per cent of cases, compared with 10 per cent of those who didn't request any medicine. This is the dark side of these ads, and as a doctor, I've always been surprised by people who say that it's doctors who force pills on patients. Doctors are generally nice people, and eager to please. They will get bounced into giving people what they want, and a lot of patients have been persuaded, through whatever social processes are at play in their world, that pills fix things. I'll rephrase that for something that's coming later in this chapter: a lot of *people* have been convinced that they're *patients*.

So the evidence shows that ads change behaviour, and they change it for the worse. This becomes much more worrying when you look at which drugs get advertised. One study gathered data on 169 drugs on the market, and looked for patterns in their promotion.[6] First, drugs are advertised more when the number of potential patients, rather than current patients, is large. This is an interesting finding, because it means that people are turned into patients, which is good news if they're sick, but bad news if they're not. Second, drugs get advertised more when they are new. That may seem inevitable, but it can be very problematic. As we've already seen, new drugs are often not a good idea: they're the drugs we know least about, because they haven't been around long; they've often been shown only to be better than nothing, rather than the best treatments we already have; and last, even if they're equally effective when compared with older drugs, they will be more expensive.

We've already seen how AstraZeneca managed the transition from omeprazole to esomeprazole, the 'me again' drug, and we can also see it in their advertising strategy. The company spent $100 million on omeprazole in 2000, the second-highest ad spend that year. Then in 2001, as it was shortly to come off patent, AstraZeneca dumped the drug, and put $500 million into advertising esomeprazole instead, the 'me again' drug. But as we've already seen, these two drugs are almost identical, and esomeprazole is basically no better than omeprazole, just much more expensive.[7] The advertising campaign was highly effective, so we waste money on drugs that are no better than those that already exist.

As I've previously argued, when you take a step back from pharmaceutical industry marketing, it is simply a process whereby patients pay money to drug companies, in order for them to produce biased information, which then distorts treatment decisions, making them less effective. This is not simply my view, or an interpretation from the basic principles of economics: we can also watch the phenomenon in real time, by tracking the cost of drugs, and their advertising budget. One study looked at clopidogrel, an 'anti-platelet' drug that can help prevent blood clotting, and is given to people at high risk of various heart problems.[8] It's popular, and expensive – in 2005 it was the world's second-biggest-selling drug, at $6 billion. Clopidogrel came to market in 1999 with no advertising, and was used widely, with no advertising, until 2001. Then the drug company introduced television advertising, spending $350 million in total. Oddly, this had no impact on the number of people taking the drug, which continued to increase at exactly the same rate. So nothing changed, except for one thing: the price of clopidrogel went up by 40 cents per tablet. As a result, Medicaid alone paid out an extra $207 million. For me this episode is strong evidence – if it wasn't already obvious – that it's patients and the public who pay for the industry's expensive marketing campaigns.

That would be fine if we were paying for reliable information, cautiously explained, but the reality is that even if the ads are adequately policed (I will review the endless to-and-fro on this later) they are still focusing only on pills and commercial products, which in turn distorts our whole outlook on medical interventions. Any sensible public-health campaign to inform people about reducing the risk or the consequences of a disease would look at prescription drugs, of course. But it might also inform patients with equal vigour about things like exercise, alcohol, smoking, diet, recreational drug use, social engagement, and perhaps even social inequality. A public education and engagement programme costing $350 million – the amount we have just seen spent on clopidrogel alone – would achieve a great deal for all of those goals, but instead, using money from patients and public, it is spent on TV ads for one pill.

This is a theme that we will see recurring: one of priorities being distorted, alongside individual treatments being sold. But first, we must note that ads are not the only way in which drugs are advertised.

Celebrity endorsement

In the 1952 Hollywood movie *Singin' in the Rain*, Debbie Reynolds plays Kathy Selden, a talented singer who hides behind a curtain, and covertly provides the sweet singing voice for an on-stage starlet who merely mimes the words. In a recent interview, Debbie Reynolds suddenly starts explaining that 'Overactive bladder *affects* you because it *defects* you . . . effective treatment is available.'[9] The interview didn't mention that she was working for Pharmacia, a company pushing a new treatment for overactive bladder. In another recent interview, Lauren Bacall encourages readers to get tested for macular degeneration, which she says can be helped with Visudyne. Neither she nor the interviewer mentions that she is being paid by Novartis to promote the drug.[10] The mom from *Serial Mom* (seriously) drops references in interviews to arthritis drugs, and is paid to do so by Wyeth and Amgen.[11]

This is a new phenomenon, but it has become so widespread that American celebrity shows are now having to screen for endorsements before interview segments go ahead. CBS, for example, recently reported that an interview with Rob Lowe – of all people – was dropped by NBC because of concerns about him promoting a drug for people on chemotherapy.[12] So when PR News gushingly reports that a character on *ER* with Alzheimer's disease was treated with the new drug Aricept thanks to the work of a PR firm working for Pfizer, it comes as no surprise.[13]

The TV spots where sponsorship is openly declared are often much more bizarre: although it's only directed at the US market, for a flavour of how this world works, I strongly encourage UK readers to find Barry Manilow's 'Get Back in Rhythm' video online, promoting a drug that treats abnormal heart rhythms ('Hi, this is Barry Manilow, and that's the *rhythm* to my song "Copacabana" . . .'). Jon Bon Jovi's painkiller ad is even slicker. And Antonio Banderas is the voice of a bee for Merck's Nasonex.

But sometimes the hidden hand can be more subtle. You will remember, I'm sure, the media stories around Herceptin, a drug which has a very modest effect on survival for some kinds of breast cancer, at a cost of serious cardiac side effects, and tens of thousands of pounds for each treatment. From 2005, access to this drug became a spontaneous *cause célèbre* for the British press, and the extent of the distortion is best expressed in the words of one doctor who was suffering from breast cancer herself, and later wrote of being caught up in the bombardment of media coverage: 'I began to feel that if I did not receive this drug then I would have very little chance of surviving my cancer.' When she stepped back from the maelstrom and looked at the data as it stood at the time, she was surprised: 'More careful analysis of the "50 per cent benefit" which had been widely quoted in the medical and non-medical press, and fixed in my mind, actually translated into a 4–5 per cent benefit to me, which equally balanced the cardiac risk . . . This story illustrates how even a medically trained and usually rational woman becomes vulnerable when diagnosed as having a potentially life-threatening illness.'[14]

The dominant theme of this media coverage was that NICE should approve Herceptin for use on the NHS. Yet, bizarrely, the campaign was orchestrated before any evidence had even been given to NICE. The health minister, similarly, said the drug should be approved, but again before the data on its effectiveness was available.

What can explain all this? One group of academics tracked down every single newspaper story on Herceptin, to try to understand what had happened.[15] They found 361 articles in total: the overwhelming majority (four out of five) were positive about the drug's efficacy, and the remainder were neutral, with none negative. Side effects were mentioned in less than one in ten articles, and were often played down as minimal. Some articles came right out, in the face of all common sense, and declared that this miracle cancer drug had no side effects at all.

Half of the stories were about the problems in getting a licence for Herceptin's use in early-stage breast cancer, but they almost never mentioned that the manufacturer, Roche, needed to apply for that licence itself, and hadn't even done so yet. Many of them attacked NICE, but they hardly ever mentioned that it couldn't even look at the drug for this use until it was licensed, and until the government asked it to.

Perhaps most remarkable was the use of individual patients, in two thirds of all the articles. Although journalists chose not to mention how they had found these women, in reality they were being provided to the media by lawyers and PR firms. Elaine Barber and Anne Marie Rogers each appeared in dozens of articles: they were handled by Irwin & Mitchell, which was shortlisted for a Chartered Institute of Public Relations award for its work on this project. Lisa Jardine, Professor of Renaissance Studies at Queen Mary, University of London, who was suffering from breast cancer, told the *Guardian* she was contacted by a PR firm working directly for Roche.[16] The charity CancerBackup also appeared in these stories repeatedly, often hawking a survey finding which had been delivered to it by Roche, which also funded the charity's work.[17]

Why would the involvement of a PR firm working for the drug company not be declared up front? Here's an unusually open explanation. In 2010 the British government proposed a scheme allowing pharmacists to substitute all prescriptions for branded medicines with the generic alternative. Generic drugs, as you now know, are identical copies of a molecule, but manufactured more cheaply by another company when the original inventor's exclusive patent has run out. Doctors who have been influenced by advertising from drug companies often write the brand name of the drug on the prescription, rather than the true scientific name. This proposed new law would allow pharmacists to ignore the brand name and substitute an identical generic copy of the drug, made by whichever company sold it most cheaply, potentially saving the NHS huge amounts of money, at no risk to patients. A letter of protest immediately appeared in *The Times*, signed by various patient groups and experts, and received positive coverage elsewhere in the broad-sheets. 'Plan to Switch to Cheaper Medicines will Harm Patients, Say Experts', reported *The Times*. It even had a case study: 'Patient Given Seroxat Substitute Felt Unwell Within Two Days'. But Margaret McCartney, a GP who writes for the *British Medical Journal*, found out that the letter was coordinated and written by PR company Burson-Marsteller, paid for by the drug company Norgine. Peter Martin, chief operating officer of Norgine, was the major influence behind the campaign, but he did not sign the letter himself. Asked why not, he cheerfully replied, 'There was no conspiracy. The frank truth, the honest truth, is that I thought that having a pharmaceutical company in there would sully the message somewhat.'

Despite that story, we shouldn't allow ourselves to become too obsessed with conspiracies. Cancer and health are – as I've written endlessly outside this book – areas where many news-paper journalists distort facts as a matter of casual routine, without any commercial assistance, partly through a lack of knowledge, but also, I suspect, through a desire to be crusaders. We have already seen how figures can be reported misleadingly, in the coverage of the Jupiter trial on statins (pp. 218–19). But the simple tradition of using human stories, even when they do not represent the reality of the evidence, provides an open goal to companies, hoping that their drug will get positive coverage. In 2010, for example, the British media was filled with journal-ists angry about NICE's recommendation not to fund Avastin, a bowel-cancer drug that costs £21,000 per patient. Overall, on average, when added to all the other treatments, the drug had been shown to increase survival by just six weeks, from 19.9 months to 21.3 months. But the newspaper stories featured Barbara Moss, who paid out of her own pocket to have Avastin in 2006, and was still alive four years later.[18] I am pleased for her, but someone who has survived four years is no illustration whatsoever of what happens if you take Avastin for bowel cancer. You could find patients who survived for four years without Avastin too, and neither they nor Barbara Moss tell you anything informative whatsoever about the drug's effectiveness.

Individual stories such as these are the bread and butter of health journalism. But beyond the desire to report on individual 'miracle cures', there is also a murkier problem. Journalists, like all of us, like to explain the world around them. Sometimes, though, a deceptively simple explanation for a complex phenomenon can be very powerful: it can prime the reader to accept a specific treatment, but it can also change our whole cultural understanding of a disease.

More than molecules

The idea that depression is caused by low serotonin levels in the brain is now deeply embedded in popular folklore, and people with no neuroscience background at all will routinely incorporate phrases about it into everyday discussion of their mood, just to keep their serotonin levels up. Many people also 'know' that this is how antidepressant drugs work: depression is caused by low serotonin, so you need drugs which raise the serotonin levels in your brain, like SSRI antidepressants, which are 'selective serotonin reuptake inhibitors'. But this theory is wrong. The 'serotonin hypothesis' for depression, as it is known, was always shaky, and the evidence now is hugely contradictory.[19] I'm not giving that lecture here, but as one brief illustration, there's a drug called tianeptine – it is a selective serotonin reuptake enhancer, not an inhibitor, that should reduce serotonin levels – and yet research shows that it is also a pretty effective treatment for depression.

But in popular culture the depression–serotonin theory is proven and absolute, because it has been marketed so effectively. In drug ads and educational material you can see it recycled, simply and plainly, because it makes absolute sense: depression is caused by too little serotonin, therefore our pill, which raises serotonin levels, will fix it. This uncomplicated notion is attractive, even though it has little support in academia, perhaps because it speaks to us of controllable, external, molecular pressures. As one US newspaper said recently about depression: 'It's not a personal deficit, but something that needs to be looked at as a chemical imbalance.'[20]

This is not a belief that arose spontaneously out of nowhere: it has been carefully fostered and maintained.[21] A recent ad for paroxetine by GSK says: 'If you've experienced some of these symptoms of depression nearly every day, for at least two weeks, a chemical imbalance could be to blame.'[22] Or a patient's guide to Pfizer's SSRI: 'Zoloft may help correct the chemical imbalance of serotonin in the brain.' The same claims are found in ads around the world, directed not only at adult patients, but also at children. One museum exhibition on the human brain sponsored by Pfizer started at the Smithsonian Institution in Washington and then toured the US. Half of us will experience 'brain dysfunction' at some stage in our lives, it happily explains: 'Chemical imbalances in the brain – often involving the neurotransmitter serotonin – are almost certainly involved.'[23]

In 2008, two academics from the US wrote a remarkable paper, describing what happened when they contacted the journalists who were disseminating this idea, to find out if they could justify their claims.[24] In return they were either ignored, fobbed off, or sent irrelevant academic papers that said nothing about serotonin and depression: 'The quote was attributed to a psychiatric nurse practitioner,' said one vignette. 'The author did not respond to emails, and the nurse's email was not available.'

One *New York Times* piece discussed a founder of the chemical theory of depression: 'A groundbreaking paper that he published in 1965 suggested that naturally occurring chemical imbalances in the brain must account for mood swings, which pharmaceuticals could correct, a hypothesis that proved to be right.' When the academics gave chase, 'Emails to [the journalist] requesting a citation to support his statement went unanswered.' An article in *The Times*, titled 'On the Horizon, Personalized Depression Drugs', quoted a professor saying: 'Some depressed patients who have abnormally low levels of serotonin respond to SSRIs, which relieve depression, in part, by flooding the brain with serotonin.' In evidence, the journalist supplied an academic paper on a completely different subject.

The story of the serotonin hypothesis for depression, and its enthusiastic promotion by drug companies, is part of a wider process that has been called 'disease-mongering' or 'medicalisation', where diagnostic categories are widened, whole new diagnoses are invented, and normal variants of human experience are pathologised, so they can be treated with pills. One simple illustration of this is the recent spread of 'checklists' enabling the public to diagnose, or help diagnose, various medical conditions. In 2010, for example, the popular website WebMD launched a new test: 'Rate your risk for depression: could you be depressed?' It was funded by Eli Lilly, manufacturers of the antidepressant duloxetine, and this was duly declared on the page, though that doesn't reduce the absurdity of what followed.

The test consisted of ten questions, such as: 'I feel sad or down most of the time'; 'I feel tired almost every day'; 'I have trouble concentrating'; 'I feel worthless or hopeless'; 'I find myself thinking a lot about dying'; and so on. If you answered 'no' to every single one of these questions – every single one – and then pressed 'Submit', the response was clear: 'You may be at risk for major depression.'[25]

Lower Risk:

You may be at risk for major depression.

* If you have recurring thoughts of death or suicide, call your doctor or any qualified health care provider right away. If you

need immediate assistance or think you may have a medical emergency, call 911.

You replied that you are feeling four or fewer of the common symptoms of depression. In general, people experiencing depression have five or more common symptoms of the condition. But every individual is unique. If you are concerned about depression, talk with your doctor.

This is not a meaningful diagnostic tool in any sense of the word. It is not raising awareness, or helping avoid underdiagnosis: it is marketing material, masquerading as patient information, and in my view it does clear harm, because it encourages people to diagnose themselves with problems they don't have, and ultimately to seek out drugs that they are unlikely to benefit from. But this is common practice: through checklists on depression, social anxiety disorder, premenstrual dysphoric disorder and more, companies can turn people with discomfort into consumers with an intention to obtain their product.[26]

Sometimes, when speaking on this topic to hostile industry audiences, I have been accused of protectionism, and wanting to maintain control of diagnosis for doctors. So let me place my cards firmly on the table (it may already be too late). Medicine works best when doctors and patients work together to improve health: in an ideal world, patients and the public would be well informed and engaged. It's great if they are aware of the true risks in their lives, and informed enough to avoid underdiagnosis. But overdiagnosis is just as much of a worry.

Medicalisation

This whole process has been bundled up under names like 'disease-mongering' and 'medicalisation': social processes, where the pharmaceutical companies widen the boundaries of diagnosis, to increase their market and sell the idea that a complex social or personal problem is a molecular disease, in order to sell their own molecules, in pills, to fix it. Sometimes disease-mongering can feel solid, and outrageous; but sometimes it falls apart in your fingers; because despite the marketing games, these tablets might well do some good. Let me walk you through my changeable thoughts.

There's no doubt that marketing has an impact on uptake of medicines, or that companies try to sell mechanisms that benefit them, and widen markets. We've seen that much already, with the depression checklists and the story of serotonin. Psychiatry, of course, is particularly vulnerable to such marketing devices, but the problems extend way out, into 'unstable bladder' and other syndromes. To my mind, this process reaches its pinnacle in an ad for Clomicalm: 'the first medication approved for the treatment of separation anxiety in dogs'.

Often the diseases used have existed forever, but have been neglected, unused, until reanimated by a pill. Social Anxiety Disorder, for example, is at least a hundred years old, and you could argue that Hippocrates' description of crippling shyness from 400 BC describes it pretty well too: 'Through bashfulness, suspicion, and timorousness, he will not be seen abroad . . . He dare not come in company for fear he should be misused, disgraced, overshoot himself in gesture or speeches, or be sick; he thinks every man observes him.'

Generally, people used to feel this problem was 'rare': in the 1980s the prevalence was stated at 1–2 per cent; but within a decade estimates as high as 13 per cent were being published. In 1999 paroxetine was licensed for social anxiety, and GSK launched a $90 million advertising campaign ('imagine being allergic to other people'). Is it good if stressed students can get better at giving presentations in their classes? I think so. Do I want that to happen with a pill? I guess it depends on how effective that pill is, and the side effects. Is it good if lots of shy people believe they have a disease? Well, that might reinforce negative self-beliefs, or it might improve self-esteem. These are hugely complex issues, with profit and loss on both sides of the equation.

The same quandaries arise with the aggressive 'disease awareness campaigns' for erectile dysfunction, when drugs like Viagra were being developed. This is something that a lot of doctors probably didn't take seriously enough before there was a pill to treat it. I suppose I might prefer it if patients were being offered, maybe, 'sensate focus' therapy before an erection pill. Moreover, I would prefer that this had always happened, for a long time before Viagra was invented, but there you go: jobbing sex therapists – despite having a very telegenic job – aren't glamorous enough to raise awareness of their services to the same degree as a $700 billion industry.

The point, I think, is that we shouldn't walk away, believing that crippling shyness, or disappointment with your sex drive are non-problems, or even that they are unfixable. But, if only for our collective self-worth, we do need to reassert our awareness that there are covert commercial processes beavering away behind the scenes, manipulating these new cultural constructs.

Perhaps the most complete illustration of this phenomenon – the 'hidden hand' in the culture of medicine – is 'Female Sexual Dysfunction'. This was devised in the 1990s as a new way to sell drugs like Viagra to women, and we can trace its rise, and then its subsequent modest fall from grace, over the following two decades.

In the beginning – as is standard – the scale of the problem was overhyped, through a long series of studies and conferences conducted by people who were paid by the pharmaceutical industry. The most widely cited figure for the number of women suffering from Female Sexual Dysfunction comes from 1999: according to this, some 43 per cent of all women have a medical problem around their sex drive.[27] This survey was published in the *Journal of the American Medical Association* (*JAMA*), one of the most influential journals in the world. It looked at questionnaire data asking about things like lack of desire for sex, poor lubrication, anxiety over sexual performance, and so on. If you answered 'yes' to any one of these questions, you were labelled as having Female Sexual Dysfunction. For the avoidance of any doubt about the influence of this paper, it has – as of a sunny evening in March 2012 – been cited 1,691 times. That is a *spectacular* number of citations.

At the time, no financial interest was declared by the study's authors. Six months later, after criticism in the *New York Times*, two of the three authors declared consulting and advisory work for Pfizer.[28] The company was gearing up to launch Viagra for the female market at this time, and had lots to gain from more women being labelled as having a medical sexual problem. Ed Laumann, lead author on the 43 per cent paper, seemed rather haunted by this exposure, and has been much clearer about caveats in his subsequent work. That is a wise move, because it seems to me that if your model says nearly half of all the women in the world have sexual problems, then the problem lies with your model, and not with the women you are describing.

Can we make any sense of such an insanely high figure? Subsequent research has tried to make a better fist of things. A 2007 survey, for example, compares different methods of measuring the prevalence of these problems in four hundred female attenders at a GP practice (already a sicker group of people than the general population, but we'll continue nonetheless). Looking crudely at the symptoms and behaviour that these women reported, and then comparing them against the list of symptoms in the World Health Organization's 'ICD-10' diagnostic manual (which is used only as a guide to diagnosis, not as a bible or checklist) 38 per cent had at least one diagnosis of sexual dysfunction. But if you restrict this – much more sensibly – to

Drug company sponsored meetings to define new disorder

Year	Meeting and host	No of company sponsors*
1997	Sexual function assessment in clinical trials—Cape Cod conference	9
1998	International consensus development conference on female sexual dysfunction: definitions and classifications—American Foundation for Urologic Disease, Boston	8
1999	Perspectives in the management of female sexual dysfunction—Boston University School of Medicine	16
2000	Female sexual function forum—Boston University School of Medicine	22
2001	Annual meeting of the female sexual function forum, Boston	22
2002	The new view "female sexual dysfunction": promises, prescriptions, and profits, San Francisco	0
2002	Annual meeting of the International Society for the Study of Women's Sexual Health,[4] Vancouver	15
2003	International consultation on erectile and sexual dysfunctions—to endorse instruments for assessment of sexual function, Paris	Currently seeking sponsors

*A very small minority of sponsors were not drug companies.

the women who perceive themselves as having a problem, this prevalence falls to 18 per cent. And if you restrict it further – again sensibly – to those who regard the problem as even moderate or severe, the prevalence is just 6 per cent.

These prevalence papers were not the only tool in the industry's armoury. There were also conferences – many conferences – all of them well funded. A researcher called Ray Moynihan unpicked this story as it arose, over a decade ago, and has attracted some interesting attacks for his trouble. His list of conferences from 2003 (above) tells a story all by itself.[29]

Moynihan's paper in the *BMJ* was read by a lot of people, and helped raise awareness of this growing problem. Shortly afterwards, patient organisations found themselves receiving a rather strange email from Michelle Lerner, senior account manager at HCC De Facto PR. Moynihan had questioned whether FSD exists at all, Lerner wrote:

I know many support organisations have been incensed about these claims, and we think it's important to counter them and get another voice on the record. I was wondering whether you or someone from your organisation may be willing to work with us to generate articles … countering the point of view raised in the BMJ. This would involve speaking with select reporters about FSD, its causes and treatments.

Lerner initially denied any involvement with these emails. Then she admitted they came from her, but refused to say who was paying her agency. Finally it was established that she was working for Pfizer, which was – as we know – testing Viagra for women at the time. When it was contacted, Pfizer described this kind of PR activity as 'customary and unremarkable'.[30]

More customary and unremarkable activity was to be seen in the promotion of online teaching materials around this disease.[31] The world of 'continuing medical education' for doctors is, as we shall shortly see, a major focus for covert promotional activity. One clear illustration of how free training for doctors can be used to change the emphasis of medical practice was the online resource femalesexualdysfunctiononline .org. This website contained teaching resources on FSD, helping doctors to find people who would benefit from treatment, and it was sponsored by Procter & Gamble, which at the time was developing testosterone patches, hoping to sell them as a treatment to increase women's libido, and planning a $100 million marketing push to raise awareness of FSD.[32]

The teaching programme at femalesexualdysfunctiononline .org was accredited by the American Medical Association, as they so often are, but my concern is not so much what this website says, but rather what it does not: because now it says nothing at all. P&G failed to get a licence to market testosterone patches for female libido, so this apparently valuable, accredited teaching programme for doctors fell off the internet altogether. If we believe that FSD really is a serious medical problem, affecting large numbers of women, then this freely accessible and lavishly produced educational material is presumably a valuable resource. If we believe that the pharmaceutical industry produces these resources for the better education of doctors, without seeking to influence their practice, as they claim, then surely we would expect it still to be online (since the costs of keeping a website up, once it's been built at great expense, are almost negligible). Instead, it seems that as soon as there's no money to be made, these educational resources have simply disappeared. That tells a story that will recur in the rest of this chapter: information that sells drugs is given a platform; information that does not is on its own.

That's not to say, however, that P&G didn't try hard to get its product licensed, and in the EU it had some success, after a fashion. Medicines regulators know all about 'off-label prescribing'; they know that when they approve a drug only for use in one narrow condition, or in one small group of people, that caveat can be ignored in practice, as doctors prescribe the treatment for use much more widely, and in many other conditions. Sometimes the regulators can see this coming, and try to head it off. So, in the EU, testosterone patches were approved for the treatment of poor libido, but only in women with diagnosed sexual problems that had arisen as a result of surgically induced menopause (that is, having had their ovaries and uterus removed because of cancer, or something similar). Inevitably, these patches are now being used 'off-label', for women without any history of such surgery. The FDA saw this coming a long way off, so it declined to license the product at all, specifically citing concerns about off-label use after the approval committee's unanimous 'no' vote.[33]

This might be a good moment to mention that the evidence for testosterone patches being any use, even after surgery, is extremely weak, from two trials in very unrepresentative 'ideal patients', showing marginal benefits against a massive placebo effect, with common side effects (sometimes apparently irreversible), and no long-term safety data.[34] It's worth noting that almost no treatments for FSD have come to market, and crucially, all of the disease-mongering activity we have seen happened in the lead-up to their approval. This was simply the academic groundwork in the companies' 'publication planning' programme, where they prove that a problem is widespread, and create a desire for a cure.

So, medicalisation is a mixed bag. We may well find new safe and effective drugs for conditions most of us have never thought of as medical problems before, and they may well improve people's quality of life, in all kinds of different ways. There might also be an interesting conversation over where these stand on the continuum between medical treatments and recreational drugs. But these possible benefits come at a cost. It's clear that we can be distracted in medicine by looking where the money tells us to look, and so miss things: the complex personal, psychological and social causes of sexual problems, perhaps, while focusing on the mechanics and the pills, the 'impaired clitoral blood flow' and the twenty-four-hour blood-hormone profiles. We might also suffer a cultural cost when we medicalise everyday life and promote reductionist, molecular, mechanical models of personhood. Similarly, as with size-zero models, when we invent seductive new norms of sexual behaviour, we risk making perfectly normal people feel inadequate.

But the greatest risk is that we fail to notice that our models of personhood, and what is normal, are being quietly engineered by a $700 billion industry.

Patient groups

We now arrive at the final murky and disappointing corner in our whistlestop tour of direct-to-consumer marketing. Patient groups perform a vital and admirable role: they bring patients together, disseminate information and support, and can help to lobby on behalf of people with the condition they represent.

In some respects it's no surprise that many patient groups are funded by the pharmaceutical industry – we'll see how many, and how transparently, in a moment – because on some issues, the desires of these two sectors are neatly aligned. A patient group wants money and resources, to lobby and to support its members effectively, and can benefit from specialist knowledge and business knowhow. A drug company offers this, and then it has its own needs: it wants to disseminate friendly messages for its brand, in a regulatory environment that prevents direct advertising to patients. It also wants to be seen as generous and socially responsible, like any other company, and we should recognise that illness is an emotional experience as well as a physical one: friendly assistance when you are at your lowest ebb can buy a good deal of loyalty.

But some industry interests are not so perfectly aligned with those of patients, as we already know. A company might want to increase sales of a product through the conventional covert marketing we've seen, but also by expanding the diagnostic limits of the disease, to enlarge its market. It has a special interest in selling *new* drugs, even though, as we've seen, these are the very products for which we know the least about the risks and benefits to patients, and for which the cost is – perversely, given the lack of information – highest.

If you read the pharmaceutical industry's own commercial literature, it's easy to see how it views relationships with patient organisations. Here is a PR company's health-care strategist writing in the magazine *Pharmaceutical Executive*. This isn't a smoking gun, but rather, a banal corporate explanation of why drug companies give patient groups money:

> Years before a new drug is launched, pharma companies and advocacy teams should map out how strong ties can advance corporate goals and brand objectives. Product managers see advocacy groups as allies to help advance brand objectives, like

increasing disease awareness, building demand for new treatments, and helping facilitate FDA clearance of their drug . . . But there are a few things to remember: Some advocacy groups, especially the more established ones, will not endorse one product over another. Companies need to determine boundaries early to avoid getting in trouble.[35]

And so on.

How prevalent is industry funding? The health campaigning organisation Health Action International (HAI) looked at the patient groups working with the EMA, the European drugs regulator.[36] Two thirds of them received funding from the pharmaceutical industry, with the average donation rising from €185,500 in 2006 to €321,230 in 2008, which generally represented about half the running costs for each organisation. Most worryingly, it also found that many of them had failed to declare this income clearly. In 2005 the EMA introduced 'transparency guidelines', but even by March 2010 only three patient groups had reported income from as far back as 2006 online. Despite this failure, all the organisations were invited back by the EMA to participate in stakeholder meetings.*

Is there any evidence that this funding changes behaviour? I think so, and I think you will agree, though researching this has not been a priority for funders, despite the influence of these groups on practice. As one example, we can look at the ongoing

---

* Was that a mistake? The 'transparency guidelines' themselves are weak and confused: they give no deadline for reporting, for example. The EMA objected strongly when HAI's report was published in 2010. You are mistaken, it said: patient groups tell *us* about their funding, they don't tell the public, and neither do we. I think it's fair to say that this is indicative of the EMA's approach to transparency more broadly. The one thing that made these organisations declare their funding was HAI calling up and asking questions about it. That goes to show the power of embarrassment as a public-policy tool.

cat-and-mouse game between industry and regulators over whether companies should be allowed to advertise, or give information directly to patients. This is recognised by most companies as an effective way to increase use of their pills, so they are keen to see the laws liberalised. Can we see any trace of this agenda in the patient groups funded by pharma? Another report by HAI, from 2011, looks at patient groups lobbying the European Commission and their lobbying patterns.[37] All those that did not receive any funding from a pharmaceutical company wanted to keep the current regulations, preventing companies from promoting their drugs to patients. Groups receiving money from drug companies were significantly more likely to think that the industry should have a larger role in providing information about drugs to patients.

This is a worrying finding in itself, but it also undermines the very purpose of having independent patient charities involved on 'multi-stakeholder' policy-making forums: these already contain industry voices, formally representing themselves; the patient groups are supposed to represent patients.

But this correlation between voting patterns and industry income may not be simple evidence of foul play. While there will obviously be episodes of bad behaviour – people altering their views simply to attract industry money – I think there's something much more interesting happening here. Patient groups' interests and industry interests can legitimately overlap, as we've seen. So there is no need for people to explicitly change their views for the overall voice of patient groups to be distorted: the industry can simply give funding, and therefore a more prominent platform, to people who spontaneously express the views that it prefers. In this way, everyone can feel OK about themselves, while still participating in a broader system that produces a biased and distorted picture of patient opinions. This helps to explain why patient charities receiving industry funding are so angry and baffled if you suggest that their output is biased; even though, overall, it's clear that their sector's output is biased.

But that moral sop doesn't change the reality we see on a regular basis, some of which is frankly ugly. As an example, the *Independent* newspaper recently examined some major media outcries around patient groups attacking NICE, and married up the attacks with the funding.[38] When NICE advised against expensive arthritis drugs, the Arthritis and Musculoskeletal Alliance (ArMA) organised a letter criticising it in *The Times*, signed by ten rheumatology professors. Half of the charity's income is from drug companies, and it didn't raise a single word to criticise its industry funders over the cost of these drugs, despite this being a major policy issue, and plainly relevant to NICE's decision. The National Rheumatoid Arthritis Society (NRAS) launched an appeal against the same decision, standing alongside ArMA and three drug companies, calling it 'another nail in the coffin' for its patients. NRAS receives over £100,000 a year from the industry, and it didn't criticise the industry over pricing policy either. The National Kidney Federation attacked NICE over its rejection of new, hugely expensive, marginally beneficial treatments.[39] Its press release was vicious, describing the decision as 'barbaric', 'damaging' and 'unacceptable'. Half of the National Kidney Federation's annual £300,000 budget comes from the pharmaceutical industry. Nowhere does the press release criticise the company charging tens of thousands of pounds for each person receiving its drug.

The head of NICE, Professor Sir Michael Rawlins, points out that the cost of manufacturing these drugs is often a tenth of the price for which they are sold, that we pay high prices in part for marketing (some of which goes directly to the patient groups), and that when we spend money on one thing, we can't spend it on something else.[40] This last gritty reality, which would present itself in any medical system without infinite money, is not often welcomed by patients and the public.

Repeatedly, we come back to the same circle: we pay high prices for drugs; a quarter of what we pay goes on marketing; our money is then spent on things like patient groups; who in turn insist that we should pay very high prices for these drugs, undermining the very groups, like NICE, that try to determine the best choices for patients overall.

What can you do?

1. Drug advertisements do not serve to inform the public, and should be banned. Their expansion into Europe should be resisted.
2. If drug companies really want to help inform patients about health, they could pay into a central, independent repository, which can give grants to people with a good track record of giving evidence-based information to the public.
3. Patients, journalists and the public should be wary of people selling new diseases if they are also selling the cure.
4. Any firm running a disease-awareness campaign should declare in its advertisements that it is doing so because it is developing or marketing a product to treat it.
5. All educational materials should bear the same declaration.

Ads to doctors

Addressing doctors directly is the most tangible way that drug companies try to influence prescribing practice, and this usually happens through print ads, in academic journals. As with most marketing activity in medicine, we can be fairly sure that if the companies spend money on it, they know it has some value. The published academic evidence supports that view, as far as it goes, but again, this has not been a funding priority.[41] So, drugs are used more after their advertising programmes start, and less when they stop. Doctors who recognise the ad for a drug are more likely to prescribe it. Econometric models – as far as any mortal can follow them through – suggest that marketing has more influence on drug-usage patterns than the publication of new evidence, and so on.[42]

As you might imagine, drug ads are supposed to be regulated for things like truthfulness and accuracy, but there are good reasons to worry that this is not done well. In the UK, the Prescription Medicines Code of Practice Authority administers the Association of the British Pharmaceutical Industry's code of self-regulation. To find out more about the general tone of ads, a Health Select Committee looking at the influence of the pharmaceutical industry in 2005 got the Institute of Social Marketing to examine a sample, and it found that the goal-posts were mobile. While ads for drugs are supposed to contain 'objective and unambiguous' information, in reality they associated them with attributes you'd associate with any other product: 'energetic', 'passionate', 'desirable', 'sexy', 'romantic', 'intimate' and 'relaxed'. The PMCPA clarified that 'emotional messages' were OK, if the material was 'factual [and] balanced'.

But this is a side issue: what we care about most is whether ads make assertions that are factually correct, and supported by good-quality evidence. That's a simple thing to assess: you just gather together the claims from a representative sample of ads, then check them against the available evidence, and one good example of this kind of study was published in 2010.[43] Some researchers from Holland went through all the biggest medical journals in the world – the *JAMA*, the *Lancet*, the *New England Journal of Medicine*, and so on – between 2003 and 2005. Every ad that appeared in that time was included, once each, if it made a claim about the effect of a drug. Then they checked the references for all the claims in the ads, found the trials they referred to, and passed them to an easily exploited workforce of assessors (250 medical students who'd just finished their evidence-based-medicine training).

Each student independently checked the methods of two trials, and the associated ad, using objective criteria, like a well-established scoring system for assessing the quality of trials. Medical students are cheap, but they might not be reliable raters, so each trial was scored by between two and six students, and if there was a discrepancy in scores, the trial was reviewed by a panel of four academics. The results were abysmal. Only half of the claims in the ads were supported by the trials the ads themselves referenced as evidence; only half of the trials got a score of 'high-quality'; and less than half of the ads – in the leading medical journals in the world – referenced a high-quality trial which supported their claim.

This is just one study, but it's fully representative of what has been found before. Another study from the *Lancet* in 2003 looked at all the claims from cardiac medication ads in six Spanish medical journals: of the 102 references the researchers could trace, 44 per cent simply didn't support the promotional statement.[44] Another from 2008 on psychiatric drug ads found similar results.[45] The story is the same for rheumatology drugs.[46] Am I cherry-picking? The best current systematic review is free to read, well worth the time, and found twenty-four similar studies.[47] Overall, it found that only 67 per cent of the claims in ads are supported by a systematic review, a meta-analysis or a randomised-control trial.

Despite this overwhelming evidence, the British Department of Health has rejected calls for drug companies to be forced to publish corrective statements when they've been found to have made incorrect claims in their ads.[48] So doctors will never know when they have been misled.

As far back as 1995, around half of all medical journal editors responding to a survey agreed that they should check the content of the ads they accepted for accuracy, and even submit them for peer review.[49] In reality, this almost never happens.[50] If the factual claims in these ads aren't reliably backed up by evidence, then you already know all you need to know about whether the regulations governing these claims work: around the world, they don't.

'Drug reps'

Drug reps are the people who visit doctors in their offices, and try to convince them in person that their company's drugs are the best (though it's not just drugs: 'device reps' for surgeons present a very similar set of issues). These people are often young and attractive; they also bear gifts, and the promise of a long, mutually beneficial relationship with a drug company. It's hard to know how well these interactions are policed: as with all relationships, they are built incrementally, on mutual trust, so the most egregious behaviour will happen between friends. It's also fair to say that things have tightened up on these interactions over the past few years in the UK (albeit from a position where behaviour tantamount to outright bribery was the subject of routine jokes among doctors). Here, because this world is harder to penetrate, I've stepped away from the drier world of evidence, and spoken to some drug reps in confidence: if you're feeling melodramatic, we could call them whistleblowers, though I don't think they've said anything to me that you wouldn't hear from them in a pub.[51]

First, though, before we look at how they operate, there is already a wealth of published evidence on their activities. This is a huge business: the overwhelming majority of the industry's promotional budget goes on influencing doctors, rather than patients, and about half of that gets spent on drug reps. They are not cheap, and though their numbers fluctuate, they have doubled in the past two decades,[52] with one rep for every three to six doctors, depending on how you measure it.[53] A systematic review found that the majority of medical students have contact with drug reps before they even qualify.[54] Because the industry spends so much money on drug reps, you can be sure they influence prescribing.

Doctors repeatedly assert, in both qualitative and quantitative research – not to mention when you chat with them socially – that drug reps have no impact on their prescribing (many claim that they improve it).[55] Cheeringly, they also report that their own behaviour won't be changed by interactions with drug reps, but that other doctors' behaviour probably will be.[56] And the more drug reps you meet, the more likely you are to think they're not having an effect on you at all.[57]

This is naïve arrogance. From the most current systematic review, there have been twenty-nine studies looking at the impact of drug rep visits.[58] Seventeen of those twenty-nine studies found that doctors who see drug reps are more likely to prescribe the promoted drug (six had mixed results, the rest show no difference, and none show a drop in prescribing). Doctors who see drug reps also tend to have higher prescribing costs, and are less likely to follow best-practice prescribing guidelines.

To give a flavour of this research, one classic study took forty doctors who had requested that a drug should be added to their hospital formulary – the list of locally approved drugs – in the preceding two years.[59] Eighty doctors from the same places who hadn't applied to put a drug onto the formulary were then randomly selected, and the contact these two groups had had with the industry was compared. The doctors asking for new drugs to be made locally available were thirteen times more likely to have met drug reps, and nineteen times more likely to have directly accepted money from drug companies.

These visits – repeatedly shown to distort prescribing practice – take place on time that patients have paid for, and generally without the blessing of the people commissioning local services, who know that such activities increase costs, through foolish prescribing. They're also spreading: since the new 'nurse practitioners' are now able to prescribe drugs in many places (a development I welcome, although it annoys many doctors), they too have become a target for promotional activity. The most recent US study in this new area found that 96 per cent of nurse prescribers reported regular contact with drug reps, and the overwhelming majority thought such contact was 'helpful'.[60]

Individual visits aren't the only way drug reps can get time to persuade doctors. One of the most prevalent exposures – and one of the hardest to avoid – is at meetings. 'Grand Round', for example, is a tradition in most hospitals, where one medical team presents a complex or interesting patient for discussion in front of the rest of the hospital. This is a big deal – especially for the quaking junior doctor who presents the basic history of the patient being discussed – and it's attended by everyone, from medical students to professors, as an educational event. Grand Round generally happens at lunchtime, with sandwiches by the door, and is sponsored by a drug company: it either presents for a minute or two at the beginning from the stage, or runs a stall, with reps on hand to engage doctors in discussion.

I wouldn't say that working hospital doctors are either particularly rich or particularly poor, compared to other graduates with similar abilities and qualifications. The UK scales are all publicly accessible: junior doctors are paid between £25,000 and £40,000 per annum for the first five or ten years, and then consultants go on to around £70,000. It's a gritty world, without the kind of glossy corporate perks you might see in the City; but then, it's a different kind of world. However you cut it, doctors can afford either to buy or to make their own sandwiches, and don't need these paid commercial ad breaks embedded in normal hospital work. In the many surveys that have been published, junior doctors attend between 1.5 and eight industry-sponsored lunches or rounds each month.[61]

The problem isn't just that this kind of sponsorship looks

bad. Junior doctors are more likely to choose a sponsor's drug, even if it is inappropriate, after seeing a drug rep present on it at a Grand Round meeting.[62] These interactions generally start at medical school, and doctors can be very naïve about the interest being shown in their career and their well-being.[63] To really understand the human impact drug reps have in the workplace, we have to veer – against my better judgement – into personal stories.

When I was doing a junior job in the middle of nowhere, I went for a team meal that was paid for by a drug rep. The evidence, going back many years, shows that people who go on drug-company dinners are more likely to prescribe that company's drugs.[64] But every other junior doctor was going, and since we all lived in hospital accommodation, if I hadn't gone I'd have been sitting in an undecorated institutional bedroom on my own with nobody to talk to. This is not a sob story, but rather a description of how objections are eroded. At the end of the meal, the friendly rep asked where everyone was going next, because we were all soon moving on to our new training jobs. Applying for these was all we'd thought about for weeks, and everyone was bubbling with information.

It was only years later, talking to other drug reps, that I realised this wasn't a friendly chat: she wanted to know where we were going next so she could pass on her notes about us to the rep covering our new area. You might think we were naïve, but in the many years I've been lecturing students and doctors on how to deal with industry marketing, every single time, the doctors in the room are surprised by this creepy realisation: the drug reps who you thought were impressed by your new job are actually keeping notes on what you think and say.

It goes much further. Once you start chatting to reps, you rapidly learn that they break doctors down into types, and these have even been documented in academic papers.[65] If they think you're a crack, evidence-based medicine geek, they'll only come to you when they have a strong case, and won't bother promoting their weaker drugs at you. As a result, in the minds of bookish, sceptical evidence geeks, that rep will be remembered as a faithful witness to strong evidence; so when their friends ask about something the rep has said, they're more likely to reply, 'Well, to be fair, the reps from that company have always seemed pretty sound whenever they've brought new evidence to me …' If, on the other hand, they think you're a soft touch, then this too will be noted.

One classic paper written by a drug rep in collaboration with an academic describes these techniques in detail, and if you're a doctor, I highly recommend reading it, because you might see your own discussions reflected back to you in an unexpected light.[66] They go through various situations, and the training and methods used: how to manage the acquiescent doc who says yes to everything, just to get you out of the room? How to set boundaries on the mercenary doctor who wants more expensive dinners at restaurants like Nobu? What about the lonely GP who wants a friend? This kind of social strategic information may well appear in the notes your local drug rep keeps on you. In fact, since we have a Data Protection Act that gives you the right to this information, using a 'Subject Access Request', some mischievous fun could be had by any informal group of doctors who gathered and then published this information.

For myself, I stopped seeing drug reps about two years after qualifying. But that doesn't stop me running into them. I can't block my ears when they're presenting at the beginning of a meeting in the place where I work, and often, in an outpatients department corridor, in a part of a building where only staff are supposed to be, you will find one waiting for you. Generally they're let in by admin staff, often by temps. Sometimes the person who let them in has a fresh bunch of flowers on their desk when you go downstairs to ask – in your nicest voice, treading on eggshells – why an intruder who has nothing at all to do with patient care has been let through to stand in a corridor surrounded by confidential patient notes.

To NHS admin staff, in the cosmetic shambles of the public sector, a competent-looking person in a smart suit has the air of someone who is supposed to be allowed into doctors' offices. In fact, more than many people around the NHS, reps look as if they come from a real workplace. They're charming, well-presented, engaged, attentive; they remember details about your children (from their notes), and they've got expensive biscuits and free memory sticks. Good sales people are good schmoozers, and I have watched them work their magic.

But they can also be insidiously divisive. Drug reps will bring food and treats for a whole team, but the people they want to influence are the key doctors. If those don't go on a team outing, the drug rep won't pay next time. I've watched a new consultant create resentment and dislike in his first week at a new community outpatients clinic by saying he doesn't want free drug-company treats at the weekly lunchtime team meeting. As you can probably imagine, the changeover after the departure of a longstanding consultant is a fragile and anxious time, when a service might be in transition between two very different approaches. Resentment over free food from people advertising products is just another new pressure to introduce.

So, what do drug reps do? First, their presentations are as partisan as you might expect. This isn't an area where quantitative research is well-funded – a recurring theme, in this part of the book, you'll notice – but in general they will hand out copies ('reprints') of academic papers describing trials that support their drug, for example, though they won't hand out reprints of those which show it in a bad light, for obvious reasons. This plants erroneous, distorted pictures of the research literature in doctors' memories, and if you're like me, you often can't remember where you learnt something, or how you know it: you just know it.

They'll also have lines ready to respond to objections from doctors. One rep told me he never saw a doctor pull out an academic paper in objection to his claims, unless it had been handed out by a competing company's rep. Once reps know what objections and what papers are being rolled out by the competition, they can discuss them with the marketing department, and develop rebuttals, ready to go, elsewhere on their patch. If the issue comes up more than once, it can be passed up the chain, and all reps on that drug are trained in how to combat these new objections to prescribing their drug that are regularly coming from doctors, primed by the competition.

Since most drug reps cover a number of doctors, and aim to see each one every three months or so, this level of monitoring and refutation is fairly easy to arrange. They also have flash-cards or iPad shows, with the company branding, key words about their drug, and misleading graphs. Sometimes these graphs will play the same games that newspapers and political pamphlets do: a vertical axis that doesn't start at zero, for example, exaggerating a modest difference. But sometimes they will be smarter: a graph that shows a huge difference on a bar chart between people having the rep's drug, for example, and people on another treatment, but where the 'other treatment', on close examination, is something rubbish.

They also hand out gifts, though the regulations on this are always shifting, and vary from country to country. Since May 2011, in the UK, under a change in the ABPI code, promotional pens, mugs and trinkets have been voluntarily banned. As these regulations haven't been heavily resisted, my hunch is that the gifts don't achieve much, and they also have the disadvantage of being obviously seedy: a doctor can end up with an office covered in drug-company logos – on biros, calendars, memory sticks – and that's not a good look.

In any case, from my own experience, any regulations are applied elastically: a couple of years back, when gifts were supposed to have a value of less than £6, and to have some medical use, the justifications were often tenuous ('A doctor might need some tea from a nice posh flask on a home visit'). And I still don't understand how the laptops I've seen handed out for 'working on a project together', to doctors I know who will read this book (I chose not to name you), fell within the £6 rule.

The question of why these gifts work is an interesting one, since their value is often fairly modest, once we set extreme cases of overt bribery aside. Social scientists writing on the culture of drug reps suggest that by giving gifts, they become part of the social landscape; and also that doctors develop an unconscious sense of obligation, a debt to be repaid, especially when stronger relationships are built through social events.[67] In some respects these are obvious observations that apply to sales techniques in many fields: how easy is it to boot someone out and disregard their opinion when you've laughed together drunkenly over dinner? But in any case, as with most anaesthetic drugs, we don't know exactly *how* they work, but we know that they *do* work.

Even where the gifts are regulated, there is still hospitality; and it's clear that meals, travel and accommodation will continue to be available as before. A quick browse through the PMCPA website shows that the self-regulation guidelines on sensible limits are regularly broken. There's the odd visit to a strip club, business-class flights around the world, golf hotels and so on.[68] One recent case concerns an unwise conference feedback document from Cephalon, describing how the company paid for doctors to go to an educational medical conference in Lisbon. Alongside the £50-a-head meals and early-morning bar bills for spirits and cocktails are comments from doctors like: 'dinner was fantastic', 'great night again', 'we then went to a few bars and to a club until 3 a.m. . . . good photos to prove it!!!'[69] 'All the customers were really looked after and spoke positively about Effentora – let's make sure they start Rxing [prescribing] now!'

The cases which reach the public domain are only the tip of the iceberg, since there is little or no investigative work, so their exposure relies on competitors discovering and reporting a transgression, or doctors who are personally engaged in unethical behaviour reporting themselves to the authorities, which doesn't happen often. Trips like the one described in the previous paragraph are used to influence the prescribing behaviour of doctors seeing patients like you, and spending NHS money: some with the booze, and some without, but in any case, the evidence shows that they are effective at changing behaviour.

One classic study followed a group of doctors before and after an all-expenses-paid trip to a symposium in 'a popular sunbelt vacation site'.[70] Before they left, as you'd expect, the majority said they didn't think this kind of thing would change their prescribing behaviour. After they got back, their prescription of the company's products increased threefold. In fact, this behaviour has become so widespread that the Serious Fraud Office announced in 2011 that it would be using new powers from the Bribery Act 2010 specifically to investigate corporate entertainment for NHS doctors, nurses and managers which goes beyond 'sensible proportionate promotional expenditure'. When it's even imaginable that doctors, nurses and NHS managers should be specifically targeted for fraud and bribery investigations, over their hospitality and other contacts with drug reps, you know there's a problem.

Finally, alongside the gifts, the travel and the hospitality, drug reps are the conduit through which other benefits flow. They are the eyes and ears of the company on the ground, gathering information about local 'key opinion leaders', senior or charismatic doctors who influence their colleagues. These people are identified for special attention, but also – if they already like your drug – they are taken up, given extra staff, and put to good use in ways that I shall shortly discuss.

There is one final twist of data in our story. People working in drug sales teams are frequently paid by results. How can they know what drugs a doctor is prescribing, when that information is only in patients' and doctors' records? In the US, data on individual patients' prescriptions are sold freely, and have become one of the most lucrative health-information markets around. Although it might come as a surprise to patients, American pharmacies sell their prescribing records to companies like Verispan, Wolters-Kluwer (an academic publisher) and IMS Health:[71] that last company alone holds the data on two thirds of all prescriptions ever filed at community pharmacies.

Patient names are removed (though if you're the only person in your town with multiple sclerosis, everyone can see what you're taking), but more important for drug reps, the doctors involved are identifiable. Using this information, a company can see exactly what drugs people are prescribing, hone its sales pitch, and get proof of whether doctors have kept the promises that they made to drug reps.

These promises are very important in the world of drug reps: they will sit down, explain the benefits of their pill, and try to get the doctor to commit to a concrete plan, for example to start the next five patients he or she sees with diagnosis X on the new drug. With a little peer pressure and a persuasive argument, a commitment might be made, and this can then be monitored using IMS data. As a result, favours to a doctor can be adjusted, and tailored pressure can be planned before the next visit. For the easily persuaded, the rep might ask: 'Why are you prescribing that cheaper drug, doctor, when ours has fewer side effects? Look at this graph, comparing the two, which proves it.' For the 'spreader', in drug rep parlance, the rep might ask: 'Why are you prescribing such a random mix of antidepressants from the same drug class?'

Since the prescription data also includes doctors' medical registration numbers, it can be married up with demographic and career information on them from other databases. So drug companies can browse through a region's statistics, looking for young starters or influential seniors. One company called Medical Marketing Service will 'enhance' the prescriptions data with 'behavioural and psychogeographic selections that help you better target your perfect prospects'.

Inevitably, this has become another area of cat and mouse. The American Medical Association has tried to implement a Physician Data Restriction Program, whereby individual doctors who dislike this kind of spying can opt out,[72] and individual states occasionally try to restrict the sale of this data. But these restrictions result in lobbying, vast lawsuits, and the usual appeals. Vermont, for example, banned the sale of prescription data in 2007; the issue went to the Appeals Court, and then the US Supreme Court, whose judges finally overturned the decision, after great legal expense.[73]

What about the UK? The day may yet come when your prescription record is sold to any casual purchaser, but for the moment, drug reps have told me they rely on more human systems. Sometimes they ask the doctor if they can see their prescribing records – plenty say yes – but otherwise they go to the source: 'The main way was going to the nearby chemists and asking them. Chemists will see you, and let you see the computer screen for a doctor's prescriptions, so you're able to see exactly how many prescriptions they're writing.' Which is nice. 'And patient names, of course.'

What can you do?

1. *Don't see drug reps!* If you're a doctor, or a prescribing nurse, or a medical student, don't see drug reps. The evidence shows that they will influence your practice, and that you are wrong to believe that they won't.

2. *Ban drug reps from your clinic or hospital.* Drug reps increase costs and work against evidence-based medicine. All staff, whether medical or non-medical, can legitimately raise concerns about this in their workplace, and patient representatives can too. Hospital managers could consult on banning them (though many are on the gravy train too). Individual consultants may have more influence. In a smaller clinic setting, you could address the objections from colleagues, and explain why drug reps worry you. If reps can only be banned from some meetings, for local political reasons, then make good use of those meetings. You could make a poster display explaining why it is better not to have drug reps, and how commercial pressures distort evidence-based medicine. The six-foot-high pull-up displays that drug reps use to advertise their products and provide information are called 'banner stands': these can be ordered online, with whatever poster you choose to design, setting out the evidence on how drug reps harm medical practice, for as little as £50. If you make a good 'no drug rep' banner stand, send it to me, so I can share it.

3. *Encourage people to declare all gifts and hospitality to their patients.* If doctors, nurses and managers won't stop accepting these benefits and visits, then ask them to publicly declare what they've taken, online and in waiting rooms, in a place that's easily visible to patients and the public. Since they believe that these gifts and visits have no impact on their prescribing behaviour, they should be happy for the information to be shared with the NHS patients who pay their salaries.

4. *Ban drug reps from your medical school.* If you're a medical student, and you believe, as I think I've shown, that drug reps are harmful, you could move to ban them from educational activities. If this proves difficult, you could audit industry promotional activity, and report back on it publicly, to shame

your institution. This is important, as guidelines are often very different from reality. In one medical school where I have taught, drug reps have been banned from the hospital by the lead clinical pharmacologist, but the students say this is ignored by individual consultants. Collaborating between universities, you could also help produce data showing which medical schools are the worst for industry influence. Remember, the industry spends around a quarter of its revenue trying to influence doctors, and half of that on drug reps. This is a vast spend, amounting to billions of pounds, that you can influence.

5. *Report breaches of the drug rep behaviour code to the PMCPA.* By reporting what you see and hear, you can help to make self-regulation a little less farcical.

6. *Train medical students and doctors about the dangerous influence drug reps can have on medical practice.* To my mind, this is not a political act, but rather a legitimate part of training in evidence-based medicine. Doctors will be subjected to marketing activity throughout their whole working lives, for four decades of practising medicine after they leave medical school, and the majority report that they were not adequately trained to deal with this marketing activity.[74] The references throughout this book will help you, and I'd be delighted if you were able to use this book as a starting point: if you produce good teaching materials, then do please share them.

7. *Regulations should change to prevent pharmacists from sharing confidential doctor and patient information with drug reps.* This is obvious, and it should be policed. You could ask your local pharmacist and your doctor if they share your confidential prescription data with local drug reps, and if they do, ask them not to.

8. *Purge your drug-company junk.* If you work in medicine, and your office is filled with branded promotional material from drug companies, then gather up all those pens, mugs, calendars, memory sticks and trinkets, and put them in a bin. Or possibly a museum.

Ghostwriters

If I tell you that Katie Price did not – necessarily – write every word of her best-selling autobiography, then this is probably not a revelation to you. But then, nor is it a problem: you want something entertaining, and everyone knows that celebrities don't write their own books. That is the culture and tradition of this kind of publication, and it's an open secret.

From doctors and academics we have higher expectations. The reader of an academic journal reasonably assumes that what they are reading is an independent academic's study, or review article, or opinion in an editorial. But this is quite wrong. In reality, academic articles are often written by a commercial writer employed by a pharmaceutical company – covertly, to a greater or lesser extent – with an academic's name placed at the top to give the imprimatur of independence and scientific rigour. Often these academics have had little or no involvement in collecting the data or drafting the paper.

What's more, while you might assume that academic publications are spontaneous projects from independent academics, they may also be part of a carefully choreographed timetable of publications, all running to the marketing schedule for one company's product. So, papers will appear before a new drug's launch describing cross-sectional survey data which reveals that the prevalence of a medical condition is much higher than previously thought; others will review the field, and say that current treatments are widely regarded as ineffective and dangerous; and so on. By this mechanism, the entire academic literature, used by doctors to guide decisions – the only tool we have – can be ghost managed, behind the scenes, to an undeclared agenda.

People who work in the industry are keen to point out that sometimes commercial writing assistance is quietly declared with a small note in the acknowledgements section at the bottom of the paper (often in a small font, and often just naming a company rather than individuals). This can only be found by those who know to look for it – did you know? – and very often the declaration is simply not made. Furthermore, even if some readers are alerted to the presence of writing assistance by such notes, the many other negative impacts of ghostwriting on the literature and the professions still stand, as we shall see.

So how common is it for academic papers to be written by someone other than those listed as 'author'? As with most shady activity, it's hard to gather clear data: the very point of ghostwriting is that it's hidden from the audience, and academics and the industry are generally too ashamed to discuss it openly. But through a combination of leaked documentation, and cautious surveys promising anonymity, various estimates have been gathered.

One study from 2011 took a representative sample of all the papers in six leading medical journals – including the *Lancet*, *JAMA*, and so on – then contacted the corresponding authors (the 'lead' author on the paper, whose contact details are always published).[75] They included every kind of publication, from original research to review articles and editorial opinion pieces. Reviews and editorials are particularly important for the companies that employ ghostwriters, because they represent an opportunity to summarise the evidence for a whole field, in a readable format, and so set the frame for subsequent discussion and research.

Knowing that the response rate was likely to be low, the researchers made a point of promising in their initial contact email that all responses would be treated with the utmost confidentiality. Remarkably they managed to get replies from over two thirds of the people they contacted, and 630 articles are included in this study. The responses they got indicated that 8 per cent of all articles had ghost authorship: 12 per cent of research articles, 6 per cent of reviews, and 5 per cent of editorials.

If you're wondering how hidden this activity was, that was also analysed: there were forty-nine papers with ghost authors, and only seven of those – 14 per cent – made any attempt to declare that there had been extra writing assistance from unnamed commercial writers, even in an acknowledgements section hidden away at the bottom of the article, where only those forewarned of this practice would even know to look. The vast majority of articles with ghost authors in this 2011 paper gave no such hint.

Whether or not an 8 percent prevalence for ghostwriting sounds high or low to you is a matter of personal taste.

But there is every reason to believe that this paper represents an underestimate. For one thing, it's looking at some of the most eminent journals in the world, so standards may be higher. For another, it relies on self-report data. If a stranger contacted you out of the blue, and asked you to admit career-torpedoing anti-social and unethical behaviour from your work email address, you might be a little reluctant. A guarantee of anonymity from someone you've never heard of and never met is small reassurance against the real and foreseeable consequences of confessing all. So respondents may have dishonestly denied involvement in ghostwriting. And there is every reason to believe that the 30 per cent of study authors who declined to participate at all were more likely to have been involved in ghostwriting.

Another study from 2007 took all the industry-funded trials approved by ethics committees in two Danish cities, and compared the people documented as prominently involved in the trials against those listed as authors on the academic papers reporting their results, finding evidence of ghost authorship in 75 per cent of cases:[76] the company statisticians, the company staff who designed and wrote the protocol for the trial, and the commercial writers drafting the manuscript would somehow disappear when the final paper was published, to be replaced by neat, independent academics.

Since this activity is so hard to trace, it is, I think, legitimate simply to ask people who work with academic authors about their experiences. One editor-in-chief of a specialist journal told US Senate Finance Committee staff recently that in his estimation at least a third of all papers submitted to his journal were produced by commercial medical writers, working on behalf of a drug company.[77] The editor of the *Lancet* described the practice as 'standard operating procedure'.[78]

It's worth pausing for a moment here to think about why ghostwriting is used. If you saw that an academic paper describing a new scientific study was designed, conducted and written by drug-company employees, you would be very likely to discount its findings. At the very least, alarm bells would be ringing, and you'd worry more than usual about whether there was data missing, or a flattering emphasis placed on the results. If you were reading an opinion piece explaining why a new drug is better than an old one, and you saw that it was written by someone from the company making the new drug, then in all likelihood you'd laugh at it. This, you would mutter, is a promotional piece. What's it even doing in an academic journal?

In this context, it's not hard to understand the language, strategies and intentions of people producing these papers. Generally, a 'Medical Education and Communication Company' will get involved early on in the research process, to help plan a whole programme of apparent academic publications around the marketing of a drug. This is to lay the groundwork, as we've already seen – to produce papers arguing that the condition being treated is more common than previously thought, and so on. It will also go through the quantitative data available from each study, to look at how it might be sliced up. A good publication planner will help identify ways that more than one paper could be produced from each piece of research, so creating a broader palette of promotional opportunities.

This is not to say that the commercial writer will see all the data: in fact, one of the extra benefits of this way of working – from a company's perspective – is that the writer will often only see tables and results that have already been prepared by a company statistician, tailored to tell a specific story. This is, of course, just one way in which the production of a paper by company personnel differs from the normal run of academic work.

At this stage of proceedings, the discussion may only take place between the company and the commercial writing agency. After a plan has been made, and articles outlined, both will set about identifying academics whose names can go on their papers. For an experiment, or a piece of research, there may already be some academics involved. For an editorial, a review article or an opinion piece, the articles can be devised and even drafted autonomously. Then they will be delivered to the waiting academic for some comments, and most important of all, for their name and independent status to be used as author and guarantor of the work.

Writing an academic paper is a long and arduous business for the lecturers and professors who actually do it themselves. First, you have to review the literature for a whole field, avoid embarrassing omissions, and write a coherent introduction to your paper describing the work that has gone before. This is of course a golden opportunity to frame the field. Then, if it's a report of a piece of research, you have to conduct the work – which can take forever – get through all the bureaucratic hurdles, the ethics committees, coordinate the data collection, and more. Finally you have to get the data into a state where it can be analysed, find and clean the errors and duplicates, run the analysis, and build tables (my God, the days I have wasted making tables work). Before that, you need to decide what tables to make, which findings to highlight, and more. After all that, you need a discussion section that makes sense of the findings, discusses the strengths and weaknesses of the methods, and so on. Even for a straight opinion piece or a casual review article, you need to have the idea, and the time, in the first place.

After the paper is written, the horror begins. Several colleagues who have their name on it will have small comments, suggestions, tweaks. These will all come in at random times on email, and everyone's suggestion has to be approved by everyone else. All papers go through multiple drafts, maddeningly similar, and you can never be sure if someone has introduced an insane sentence that you might miss on a casual reading, so it all has to be checked – and rechecked – competently, and repeatedly.

Finally, the submission process is a pig. Every academic journal has different pernickety requirements; every one wants the references to be formatted in a different way – the tables as a separate document, at the bottom, there's a word limit, some bizarre house style about never using the word 'this' to back-reference the immediately preceding clause, even though that's a completely normal way of using the English language, and so on.

Because of all this, academics don't produce a huge number of papers every year, even though their performance is measured specifically on how many they get out, and how good they are. For that very reason, you should perhaps be a little suspicious of any medical academic who publishes a lot of papers, at the same time as holding down a clinical job seeing patients.

So professional assistance, for this arduous process, is a huge advantage; and this is why the selection of the academic or doctor who receives 'guest authorship', benefiting from the commercial writer's work, is a complex and interesting business. They are not picked at random: as we shall see, drug companies keep a close rota of 'key opinion leaders', academics and doctors who are influential in their field, or their local area, and are friendly to either the company or the drug.

The relationship between the key opinion leader and the drug company is mutually beneficial, in ways that are hard to see at first. Of course the drug company is able to give a false impression of independence for a paper that it effectively conceived and wrote, and of course money can change hands. (Did I not mention that? Some academics are paid an 'honorarium' by the drug company for putting their name on a paper.) But there are other, more hidden benefits.

The academic gets a publication on their CV, for very little work, so they look like a better academic. They are more likely to be recognised as a key opinion leader in the future – which is great for the company, since they're friends – and they are also more likely to be promoted in their university. A junior lecturer with an impressive record of academic journal publications will be much more likely to get on and become a senior lecturer, and then a 'reader', and then a professor. By this means the ambitious academics receive a benefit in kind, as good as any cheque, for which they are grateful to the drug company. Most important of all, the key opinion leader, who has the opinions that benefit the company, becomes much more senior and influential, a rising star. This is why, even when commercial writing assistance is declared, in a small font, in an acknowledgements section at the end of an academic paper (and that's still not done consistently, as we saw above), that's not enough. First, it still misleads readers, who expect the author of an academic paper to be the person who wrote it, and may not know to look for this tiny caveat. And second, this tiny disclosure does nothing to address the distortion of the academic career pathway.

I wouldn't want you to take any of this on faith. Most of the activity in this field takes place behind closed doors, but occasionally there are court cases, and from these come leaked documents, and sometimes, if we are lucky, emails and memos describing the ghostwriting process. As I've said before, nothing should be read into the specific drugs or companies involved in the stories here, because the activity they describe is, as we can see from the data above, widespread throughout the industry, for all companies and in all fields of medicine. These are simply the individual drugs for which we happen to be able to see, in black and white, the internal memos and discussions that lie behind the ghostwritten papers. Keeping documents on these matters out of the hands of people like me is, incidentally, one reason why companies tend to settle out of court, and avoid a public hearing where such documents are likely to be exposed.*

One interesting example comes from the antipsychotic drug olanzapine (brand name Zyprexa), used to treat conditions like schizophrenia.[80] Lots of documents on the ghostwriting strategy of Lilly, the drug's manufacturer, became public during a court case concerned with whether it had overstated the benefits of the drug, and marketed it for conditions where it wasn't licensed.

Lilly set a goal of making Zyprexa 'the number one selling psychotropic in history', and its emails discuss using ghostwrit-

---

* The other reason is that court cases set precedents, and make future cases against a company much easier to fight. Because of this, companies settle before cases come to court when they think the result might go against them – meaning that they control the public legal discourse as well as the public academic discourse. A paper on this subject, called 'Why the Haves Come Out Ahead', is one of the most widely cited in legal academia.[79]

ers to present it in a positive light: 'The paper for the Progress in Neurology and Psychiatry supplement has been completed and sent to the journal for peer review,' says one from its marketing person. 'We "ghost" wrote this article and then worked with author Dr Haddad to work up the final copy.'

The paper being talked about did appear in a supplement to the journal *Progress in Neurology and Psychiatry*. Peter Haddad, whose name appears as its author, is a consultant psychiatrist in Manchester, seeing patients and training junior doctors.[81] He's neither very senior, nor very junior, and I'm not telling you his job because I think it's an outrage, or because I think it makes him sound impressive: I'm telling you because this is the banal, everyday reality of how this process works, with everyday doctors playing their part, around the country. The emails explain how Lilly's global team approved the draft of Peter's paper, but final approval had to come from Lilly UK, since the journal was based in the UK. Nice work, Peter Haddad.

From the same case, there is also an entire briefing document discussing how Lilly will place an article saying that the injectable form of olanzapine might be useful in containing challenging behaviour when people with schizophrenia are very agitated and disturbed.[82] I recommend downloading this online if you have any doubts about the truth of what I've been saying, or about the casual way in which these projects are pursued. Remember that the subject of this document is an article to be written by an independent academic. The company lists its aims, and they don't sound very academic:

Rationale/ Aims

- Prepare the market for the launch of intramuscular olanzapine.
- Create a need for intramuscular olanzapine through the promotion of awareness of safety issues surrounding current typical IM treatments for acute agitation associated with schizophrenia.
- Build awareness of the need for atypical intramuscular management of acute phase.

It talks about how to work around the fact that the drug isn't yet licensed:

- Consider the content carefully - never make it seem that olanzapine already has a license for intramuscular use (although it is acceptable to allude to the fact that this may be a future outcome). One way to do this is to avoid definitive language.

It attaches an outline that can be used as a guide to write the pieces (because, remember, there may be many similar articles in this aspect of the publication plan, written by different people in different countries):

- The outline may be used when briefing a freelance writer. A third party spokesperson/authoring Key Opinion Leader (KOL) should also be identified to the freelance writer in conjunction with such discussions.

Then it discusses choosing an appropriate 'author':

- Utilise a local KOL to author the feature, either by drafting the full feature for KOL review or by giving the outline to a KOL to develop the feature.
- The KOL has full editorial control over the feature; if they wish to make changes, this should be encouraged and accommodated.
- The KOL could be:
    - An investigator in an olanzapine intramuscular trial.
    - Lilly-friendly key opinion leader, who has previously been involved in other areas of olanzapine.
    - A participant of a Lilly advisory board.
- KOLs can be paid honorarium for their time/input in reviewing/authoring of the article. It should be made clear that this payment is for their time and not to sway the content of the article. If using a member of a relevant advisory board, KOL authoring could be covered by a retainer given for participation in advising the board. This is the best option to use, if possible.

What happens once a paper is in progress? For this, let's switch to another study, on an antidepressant called paroxetine. You can read all of these documents and more at the Drug Industry Document Archive, built by the University of California, San Francisco, to house materials released during legal cases involving the pharmaceutical industry.[83] Professor Martin Keller of Brown University is discussing the content of 'his' paper with a PR person working for the drug company GSK: 'You did a superb job with this, thank you very much. It is excellent. Enclosed are some rather minor changes from me.'[84]

The ghostwriter gets back to him with everything nicely organised and ready to go, because, of course, the academic must be the one who sends the paper to the journal.[85] You may remember the earlier description of what a time-consuming hassle it is for academics to pull a paper together and submit it to a journal themselves. When you're working with GSK, this is all rather more straightforward: 'Please retype on your letter-head and revise as you like.'

Dear Dr Keller:

We are pleased to enclose all of the necessary materials for you to submit your manuscript, "Efficacy of Paroxetine but Not Imipramine in the Treatment of Adolescent Major Depression: A Randomized, Controlled Trial," to the *Journal of the American Academy of Child and Adolescent Psychiatry.*

Please find enclosed the following items:

* Five copies of the manuscript (submit four to the journal; keep one for your files)

* One set of glossy prints of the figures (submit to the journal)

* A draft cover letter to Dr Dulcan, editor of JAACAP (please retype on your letterhead and revise as you like)

And so it goes. To some academics – to those in the know, and on the key opinion leader circuit – this has all become so commonplace, so obvious, that they have even used it to try to dodge responsibility for the content of the papers on which their name appears. After a crucial study on the painkiller drug Vioxx was found to have failed adequately to describe the deaths of patients receiving it,[86] the first author told the *New York Times*: 'Merck designed the trial, paid for the trial, ran the trial . . . Merck came to me after the study was completed and said, "We want your help to work on the paper." The initial paper was written at Merck, and then it was sent to me for editing.' Well that's OK, then.

It doesn't stop at journal articles. Medical writing company STI, for example, wrote a whole physician textbook which appeared with the names of two senior doctors on it.[87] If you follow through the documentation, now in the public domain, a draft of the textbook says it was paid for by GSK, and written by two staff members at the medical writing company they paid. But in the preface to the final published textbook, the doctors whose names appear on the cover merely thank STI for 'editorial assistance', and GSK for 'an unrestricted educational grant'.

Dr Charles Nemeroff, one of the 'authors' of this textbook, responded to these allegations in the *New York Times* in 2010. He said he conceptualised the book, wrote the original outline, reviewed every page, and the company had 'no involvement in content'.[88] So you can draw your own conclusions, I've included, below, a scan of the letter sent to Nemeroff at the outset of the project by the writing company.[89] I promise this is the last time I'll print such documents, but they are so refreshingly explicit. To my eyes, this letter features STI, the commercial writers employed by GSK, saying things like 'We have begun development of the text', and 'A complete content outline is enclosed for your comment.' There's also a timeline, according to which the manuscript is repeatedly sent to the sponsor for 'sign-off' and 'approval'.

So, the person conducting a study, analysing the data, writing a paper, steering it into the hands of a journal, and even writing your medical textbook, may not be quite who you imagine.

As a result of all this, as we've seen, key opinion leaders who favour the industry's drugs are given shining CVs and rise to ever higher academic status, conferring even more independent kudos on the treatments they prefer. The academic literature is overwhelmed with repetitive and unsystematic discussion papers, acting as covert promotional literature rather than genuine academic contributions. It is also distorted, with publications that repeatedly reframe treatments in ways that the industry prefers. Even if they're not promoting just one drug,

RE: **PRIMARY CARE HANDBOOK OF PSYCHOPHARMACOLOGY**

Dear Charlie:

I am pleased to provide an update on the status of this project. We
have begun development of the text, and Diane Coniglio, PharmD is the
primary technical writer and project manager. I will be working closely
with Diane at all times and will serve as technical editor. You and
Alan are in good hands with Diane; she has many years of experience and
is a creative and accomplished technical writer.

We have developed a timeline for completion of work as follows:

• Sample text for preliminary comment	Feb 21
• Draft I to co-authors/APPI/sponsor	May 2
• Comments to STI	May 30
• Draft II to co-authors/sponsor	June 20
• Comments to STI	July 11
• Draft III to co-authors/sponsor for sign-off	July 25
• Production begins	August 1
• Page proofs to co-authors/APPI/ sponsor for final approval	August 15
• Disk to publisher for printing	September 1

A complete content outline is enclosed for your comment. We have made
several key content assumptions as listed below. Please comment on
these issues.

this means that academics working on commercial areas of
medicine, those involving new pills, have an increased prominence compared to those who don't have professional writers to
do their donkeywork for them. So people studying social
factors, or lifestyle changes, or side effects, or medicines that are
out of patent, are edged out.

It goes without saying that when we're dealing with medical
treatments, which can be hugely harmful as well as helpful, it's
vitally important that all our information is reliable, and transparent. But there is another ethical dimension, which often
seems to be neglected.

These days, in most universities, we send a long and threaten-
ing document to every undergraduate student, explaining how
every paragraph of every essay and dissertation they submit will
be put through a piece of software called TurnItIn, expensively
developed to detect plagiarism. This software is ubiquitous, and
every year its body of knowledge grows larger, as it adds every
student project, every Wikipedia page, every academic article, and

everything else it can find online, in order to catch people cheating. Every year, in every university, students are caught receiving undeclared outside help; every year, students are disciplined, with points docked and courses marked as 'failed'. Sometimes they are thrown off their degree course completely, leaving a black mark of intellectual dishonesty on their CV forever.

And yet, to the best of my knowledge, no academic anywhere in the world has ever been punished for putting their name on a ghostwritten academic paper. This is despite everything we know about the enormous prevalence of this unethical activity, and despite endless specific scandals around the world involving named professors and lecturers, with immaculate legal documentation, and despite the fact that it amounts, in many cases, to something that is certainly comparable to the crime of simple plagiarism by a student.

Not one has ever been disciplined. Instead, they have senior teaching positions.

So, what do the regulations say about ghostwriting? For the most part, very little. A survey in 2010 of the top fifty medical schools in the United States found that all but thirteen had no policy at all prohibiting their academics putting their name to ghostwritten articles.[90] The International Committee of Medical Journal Editors, meanwhile, has issued guidelines on authorship, describing who should appear as a named author on a paper, in the hope that ghostwriters will have to be fully declared as a result. These are widely celebrated, and everyone now speaks of ghostwriting as if it has been fixed by the ICMJE. But in reality, as we have seen so many times before, this is a fake fix: the guidelines are hopelessly vague, and are exploited in ways that are so obvious and predictable that it takes only a paragraph to describe.

The ICMJE criteria require that someone is listed as an author if they fulfil three criteria: they contributed to the conception and design of the study (or data acquisition, or analysis and interpretation); they contributed to drafting or revising the manuscript; and they had final approval on the contents of the paper. This sounds great, but because you have to fulfil *all three* criteria to be listed as an author, it is very easy for a drug company's commercial medical writer to do almost all the work, but still avoid being listed as an author. For example, a paper could legitimately have the name of an independent academic on it, even if they only contributed 10 per cent of the design, 10 per cent of the analysis, a brief revision of the draft, and agreed on the final contents. Meanwhile, a team of commercial medical writers employed by a drug company on the same paper would not appear in the author list, anywhere at all, even though they conceived the study in its entirety, did 90 per cent of the design, 90 per cent of the analysis, 90 per cent of the data acquisition, and wrote the entire draft.[91]

In fact, often the industry authors' names do not appear at all, and there is just an acknowledgement of editorial assistance to a company at the end of the paper. And often, of course, even this doesn't happen. A junior academic making the same contribution as many commercial medical writers – structuring the write-up, reviewing the literature, making the first draft, deciding how best to present the data, writing the words – would get their name on the paper, sometimes as first author. What we are seeing here is an obvious double standard. Someone reading an academic paper expects the authors to be the people who conducted the research and wrote the paper: that is the cultural norm, and that is why medical writers and drug companies will move heaven and earth to keep their employees' names off the author list. It's not an accident, and there is no room for special pleading. They don't want commercial writers in the author list, because they know it looks bad.

Is there a solution? Yes: it's a system called 'film credits', where everyone's contribution is simply described at the end of the paper: 'X designed the study, Y wrote the first draft, Z did the statistical analysis,' and so on. Apart from anything else, these kinds of credits can help to ameliorate the dismal political disputes within teams about the order in which everyone's name should appear. Film credits are uncommon. They should be universal.

If I sound impatient about any of this, it's because I am. I like to speak with people who disagree with me, to try to change their behaviour, and to understand their position better: so I talk to rooms full of science journalists about problems in science journalism, rooms full of homeopaths about how homeopathy doesn't work, and rooms full of people from big pharma about the bad things they do. I have spoken to the members of the International Society of Medical Publications Professionals three times now. Each time, as I've set out my concerns, they've become angry (I'm used to this, which is why I'm meticulously polite, unless it's funnier not to be).

Publicly, they insist that everything has changed, and outright ghostwriting – with no hidden hints at all for the reader – is a thing of the past. They repeat that their professional code has changed in the past two years. But my concern is this. Having seen so many codes openly ignored and broken, it's hard to take any set of voluntary ideals seriously. What matters is what happens, and undermining their claim that everything will now change is the fact that nobody from this community has ever engaged in whistleblowing (though privately many tell me they're aware of dark practices continuing even today). And for all the shouting, as we've seen, this new code isn't even very useful: a medical writer could still produce the outline, the first draft, the intermediate drafts and the final draft, for example, with no problem at all; and the language used to describe the whole process is oddly disturbing, assuming – unthinkingly – that the data is the possession of the company, and that it will 'share' it with the academic.

They gloss over the fact that even if this was true, it would do nothing to address the impact of their work on the overall emphasis in the literature, giving greater weight and volume to messages that companies will pay to see written. They gloss over the impact that writing assistance has on the academic career pathway, where the very people most accepting of industry assistance paradoxically climb the academic ladder, gaining the appearance of independence.

But more than that, even if we did believe that everything has suddenly changed just now, as they claim, as everyone criticized in this book always claims – and it will be half a decade, at least, as ever, before we can tell if they're right – not one of the longstanding members of the commercial medical writing community has ever given a clear account of why they did the things described above with a clear conscience. There are people around who paid guest authors to put their names on papers they had little or nothing to do with; who ghostwrote papers covertly, knowing exactly what they were doing, and why, and what effect it would have on the doctors reading their work. These are the banal, widespread, bread-and-butter activities of the industry. So, a weak new voluntary code with no teeth from people who have not engaged in full disclosure – nor, frankly, offered an apology – is not, to my mind, any evidence that things have changed.

What can you do?

1. Lobby for your university to develop a strong and unambiguous code forbidding academic staff from being involved in ghostwriting. If you are a student, draw parallels with the plagiarism checks that are deployed on your own work.
2. Lobby for the following changes in all academic journals you are involved in:
- A full description of 'film credit' contributions at the end of every paper, including details of who initiated the idea for the publication.

- A full declaration of the amount paid to any commercial medical writing firm for each paper, in the paper, and of who paid it.
- Every person making a significant contribution should appear as a proper author, not 'editorial assistance'.

3. Raise awareness of the issue of ghostwriting, and ensure that everyone you know realises that the people who appear as authors on an academic paper may have had little to do with writing it.
4. If you teach medical students, ensure that they are aware of this widespread dishonesty among senior figures in the academic medical literature.
5. If you are aware of colleagues who have accepted guest authorship, discuss the ethics of this with them.
6. If you are a doctor or an academic, lobby for your Royal College or academic society to have a strong code forbidding involvement in ghostwriting.

Academic journals

We put a lot of trust in academic journals, because they are the conduit through which we find out about new scientific research. We assume that they take scientific articles based on merit. We assume that they make basic checks on accuracy (though we've seen that they don't prevent misleading analyses of trial data being published). And we assume that the biggest, most famous journals – which are read regularly by many more people – take the better articles.

This is naïve. In reality, the systems used by journals to select articles are brittle, and vulnerable to exploitation.

First, of course, there are the inherent frailties in the system. There is a huge amount of confusion for the public – and for many doctors – around what 'peer reviewed' publication actually means. Put very simply, when a paper is submitted to a journal, the editor sends it out to a few academics who they know have an interest in a particular field. These reviewers are unpaid, and do this work for the good of the academic community. They read the paper, and come to a judgement about whether it's a newsworthy piece of research, a well-conducted study, fairly described, and whether its conclusions broadly match its findings.

This is an imperfect and subjective set of judgement calls, standards vary hugely between journals, and there's also room to stick the knife into competitors and enemies, since most reviewers' comments are anonymised. That being said, the reviewers are often not very anonymous, because a comment like 'This paper is unacceptable because it doesn't cite the work of *Chancer et al.* in the introduction' is a pretty good sign that Professor Chancer himself has just peer reviewed your paper. In any case, good journals often take papers that aren't perfect, on the grounds that they have something, of some small scientific interest, in their results. So the academic literature is a 'buyer beware' environment, where judgement must be deployed by expert readers, and you cannot simply say, 'I saw it in a peer reviewed paper, therefore it is true.'

Then there is the clear conflict of interest. This problem is now openly discussed for academics – their industry grants, their drug-company stock portfolio – and every scientist is compelled by journal editors to declare their financial interests when publishing a paper. But the very editors who impose this rule on their contributors have, for the most part, exempted themselves from the same process. That is odd. The pharmaceutical industry has global revenues of $700 billion, and it buys a lot of advertising space in academic journals, often representing the greatest single component of a journal's income stream, as editors very well know. In some respects, taking a step back, it's odd that journals should only take ads for drugs (and the occasional body scanner): the rates in *JAMA* are cheaper than those in *Vogue*, taking circulation into account (300,000 against a million), and doctors buy cars and smartphones like everyone else. But journals do like to look scholarly; and it was only recently that they were trying to persuade the government that drug ads are educational content, and should therefore be tax exempt. You will remember, I hope, just how educational these ads are, from the discussion earlier in this chapter of how often they make claims that are not supported by the evidence.

To reduce the risk that this income strand will pervert decisions on whether to publish an article, journals often claim that they introduce 'firewalls' between editorial and advertising staff. Sadly, such firewalls are easily burnt through.

In 2004, for example, an editorial was submitted to the respected journal *Transplantation and Dialysis*, questioning the value of erythropoietin, or 'EPO'.[92] Although this molecule is made by the body, it can also be manufactured and given medically, and in this form it is one of the biggest-selling pharmaceutical products of all time. It is also, unfortunately, extremely expensive, and the editorial was submitted in response to a call from Medicare, which had asked for help in reviewing its policy on giving the treatment to people in end-stage renal disease, since there were fears that it might not be effective. The editorial agreed with this pessimistic stance, and was accepted by three 'peer reviewers' at the journal. Then the editor sent the following unwise letter to the author:

> I have now heard back from a third reviewer of your EPO editorial, who also recommended that it be published . . . Unfortunately, I have been overruled by our marketing department with regard to publishing your editorial.
>
> As you accurately surmised, the publication of your editorial would, in fact, not be accepted in some quarters . . . and apparently went beyond what our marketing department was willing to accommodate. Please know that I gave it my best shot, as I firmly believe that opposing points of view should be provided a forum, especially in a medical environment, and especially after those points of view survive the peer review process. I truly am sorry.

The letter was made public, and the journal reversed its decision. As ever, it is impossible to know how often decisions like this are made, and how often they are hidden. All we can do is document the scale of the financial incentive for journals, and the quantitative evidence showing a possible impact on their content.

Overall, the pharmaceutical industry spends around half a billion dollars a year on advertising in academic journals.[93] The biggest – *NEJM, JAMA* – take $10 or $20 million each, and there is a few million each for the next rank down. Strikingly, while many journals are run by professional bodies, their income from advertising is still far larger than anything they get from membership fees. In addition to the large general journals, and the small specialist ones, some journals are delivered to doctors for free, and subsidised entirely by advertising revenue. To see whether this income has an impact on content, a 2011 paper looked at all the issues of eleven journals read by GPs in Germany – a mix of free and subscription publications – and found 412 articles where drug recommendations were made. The results were stark: free journals, subsidised by advertising, 'almost exclusively recommended the use of the specified drugs'. Journals financed entirely through subscription fees, meanwhile, 'tended to recommend against the use of the same drugs'.[94]

Advertising is not the only source of drug company revenue for academic journals; there are several other strands of income, some of which are not immediately apparent. Journals often produce 'supplements', whole extra editions outside of its normal work. These are often sponsored by a drug company, based on the presentations at one of its sponsored conferences or events, and have much lower scientific standards than are found in the journal itself.

Then there are 'reprints'. These are special extra copies of individual academic papers that are printed off and sold by academic journals. These are then handed out to doctors by drug reps to promote their drugs, and are bought in huge quantities, with spends of up to $1 million to buy crates of copies of just one paper. Those are the sorts of figures that haunt editors' imaginations when they try to choose which of two trials they should publish. Richard Smith, a former editor of the *British Medical Journal*, framed the dilemma: 'Publish a trial that will bring in $100,000 of profit, or meet the end-of-year budget by firing an editor.'[95]

Sometimes the implicit reasoning behind these choices can find its way into the public domain. A recent investigation from the UK Prescriptions Medicine Code of Practice Authority, for example, ruled that the company Boehringer Ingelheim was responsible for the content of an article making unacceptable claims for its diabetes drug linagliptin, even though it was written by two academics, and appeared in the Wiley Publishing academic journal *Future Prescriber*, because 'although Boehringer Ingelheim did not pay for the article per se, it in effect commissioned it through an agreement to pay for 2,000 reprints'.[96]

For the most part, however, even the most basic numbers on this huge source of income are hard to obtain. A research project I was involved in found that the biggest and most lucrative reprint orders come overwhelmingly from the pharmaceutical industry (this was a lot of work, and we've just had it published in the *BMJ*,[97] though it might have happened faster with a commercial medical writing firm handling the legwork for us). This simple finding is exactly what you might expect, but there was something else that happened during this study, which many people found much more concerning. We asked all the leading journals in the world for information about their income from reprints, but only the *BMJ* and the *Lancet* were willing to give us any data at all: the *Journal of the American Medical Association* said this information was proprietary; the vice president of publishing for *Annals of Internal Medicine* said they did not have the resources to provide the information; and the managing director of publishing for the *New England Journal of Medicine* said it would conflict with their business practices to tell us. So this huge source of pharmaceutical industry income, paid to the gatekeepers of medical knowledge, remains secret.

Is there any evidence to show that journals are more likely, on a fair comparison, to take industry-funded studies?

This has been studied only rarely – because, as we need to keep reminding ourselves, this whole area has hardly been a research funding priority – but the answer appears to be yes. A paper published in 2009 analysed every study ever published on the influenza vaccine[98] (although it's reasonable to assume that its results might hold for other subject areas). It looked at whether funding source affected the quality of a study, the accuracy of its summary, and the eminence of the journal in which it was published.

Academics measure the eminence of a journal, rightly or wrongly, by its 'impact factor': an indicator of how commonly, on average, research papers in that journal go on to be 'cited' or 'referenced' by other research papers elsewhere. The average journal impact factor for the ninety-two government-funded studies was 3.74; for the fifty-two studies wholly or partly funded by industry, the average impact factor was much higher, at 8.78. This means that studies funded by the pharmaceutical industry were hugely more likely to get into the bigger, more respected journals.

That's an interesting finding, because there is no other explanation for it. There was no difference in methodological rigour, or quality, between the government-funded research and the industry-funded research. There was no difference in the size of the samples used in the studies. And there is no difference in where people submit their articles: everybody wants to get into a big, famous journal, and everybody tries them first. If they get rejected they will try lesser and lesser journals, until someone takes the paper. It's possible that the industry-funded researchers were simply more dogged, or more shameless; and that maybe, when they were rejected by one major journal, they hawked their paper to other equally large ones. It's possible that they could do this more rapidly than those without industry funding, because they had administrative assistance from professional writers to deal with the tedious bureaucracy of each journal's submission system, and tolerated the long delay in publication that this strategy would cause. Or perhaps lucrative industry-funded studies are simply favoured by editors.

Either way, getting published in a higher-impact journal is a huge advantage, for a number of reasons. First, it is prestigious, and implies that your research is regarded as higher-quality. But second, papers in bigger journals are simply more likely to be read. As we've already seen, our systems for disseminating knowledge are ad hoc and antiquated, built on centuries-old platforms where science is presented in essay form, and printed on paper, with no clear mechanism for getting the right information to the right doctor at the right time. In a world where the information architecture of medicine is so massively flawed, simply getting under someone's nose counts for a lot.

This brings us on to one final, dismal tale. In medicine, appearances are important: the appearance of an independent study, the appearance of lots of individual papers all saying the same thing, can help build a case in the minds of busy prescribing doctors. We have seen how individual academic papers can be ghostwritten. But in 2009 a court case in Australia involving Merck revealed a much stranger new game.

Elsevier, the respected international academic publisher, was producing, on behalf of Merck, a whole range of journals, entirely as advertising projects for that one company. These publications looked like academic journals, and they were presented as academic journals, published by the academic journal publisher Elsevier and containing academic journal articles. But they only contained reprinted articles, or summaries of other articles, almost all of which were about Merck's drugs. In issue 2 of the *Australasian Journal of Bone and Joint Medicine*, for example, nine of the twenty-nine articles were about Merck's Vioxx, and twelve of the remainder were about Fosamax, another Merck drug. All of these articles presented positive conclusions, and some were bizarre, including a review article containing just two references.

As well as specialist 'journals', Elsevier also produced a journal aimed at family doctors, which was sent to every single GP in Australia. Again, it looked like an academic journal, but was actually promotional material for one company's products.

In a statement to the *Scientist* magazine after only one of these journals had been uncovered, Elsevier tried to defend itself by arguing that it 'does not . . . consider a compilation of reprinted articles a "Journal"'. This defence was optimistic at best. We are talking about a collection of academic journal articles, published by the academic journal publisher Elsevier, in an academic-journal-shaped package, laid out like an academic journal, with an academic journal name: the *Australasian Journal of Bone and Joint Medicine*. It has since been discovered that Elsevier put out six journals like this, all sponsored by industry.[99] Chief Executive Michael Hansen finally issued a statement admitting that these were made to look like journals, and lacked proper disclosure.[100]

As we've seen, it has been estimated that it would take six hundred hours a month to read the thousands of academic articles relevant to being a GP alone. So doctors skim, they take shortcuts, they rely on summaries, or worse. The simple and predictable consequence of these journals sent out by Merck – and all the other distortions we have seen, from ads to drug reps, ghostwriting and so on – is that a misleading picture of the research on these drugs will lodge in doctors' memories.

Up to a quarter of the pharmaceutical industry's revenue is

spent on marketing, comparable to the amount it spends on research and development, and this all comes from your money, for your drugs. We pay an enormous extra mark-up in price, so that tens of billions of pounds can be spent every year producing material that actively confuses doctors, and undermines evidence-based medicine. This is a very odd state of affairs.

What can be done?

1. Journals should publish all advertising revenue from each individual drug company annually, and for each individual issue.
2. Journals should publish all reprint orders retrospectively for all papers at the end of each year, disclosing income for each; and for each new industry paper, they should declare how much they have previously made in reprint orders from that company (and any *pre*publication reprint orders should be disclosed in the article, although, if such an audit was implemented, one might imagine prepublication reprint orders could become less frequent).
3. Editors should anonymously disclose all cases where pressure is applied for commercial reasons.
4. Editors and senior journal staff should declare their own conflicts of interest, funding sources if they are working academics, stocks, and so on.
5. More research should be done looking at whether projected advertising and reprint income has an impact on journals accepting papers.

Pharma's medical school

At the beginning of this chapter, I presented you with what I hope is a chilling thought: the most senior doctors working today qualified in the 1960s. Today's medical students will qualify at the age of twenty-four, and will then work for five decades. When you're at medical school you're told which treatments work best, in lectures and textbooks, and then you're tested on it. A few years later you're still doing specialist exams, and training in a safe and constrained world, with smart people actively teaching you. Then, suddenly, you're out on your own, seeing patients and getting on with it. Medicine changes around you, unrecognisably over the course of decades: whole new classes of drugs are invented, whole new ways of diagnosing people, and even whole new diseases. But nobody sets you an exam, nobody gives you a reading list, Prof. MacAllister doesn't tell you what works and how. You're alone.

Doctors need to learn about new drugs all the time, but we leave them to get on with it by themselves. Privately organised professional education is extremely expensive – hundreds or thousands of pounds for every course – so individuals tend not to pay for it themselves. The state doesn't want to pay for it either. So the pharmaceutical industry pays instead.

The Department of Health spends a few million pounds a year providing independent medicines information to doctors. The industry spends tens of billions on providing biased information. This presents a bizarre situation: doctors' continuing education is paid for, almost exclusively, by the industry whose products they buy with public money, and by the industry that has been shown routinely to mislead them.

In fact, in the UK, doctors are now actively forced to collect Continuing Medical Education (CME) points, which are counted up each year. This has been tightened up since the changes at the GMC on account of a GP called Harold Shipman, who turned out to be a serial killer, murdering older women with overdose injections of opiates. In a rather odd game of consequences, this means that a set of new regulations, brought in to prevent doctors from murdering people, in reality has simply shepherded them even more into the hands of expensive industry-sponsored promotional activity, where they are misled about the benefits of expensive medicines, and so harm patients.

The basic design of the pharmaceutical industry's medical school for qualified doctors is simple: doctors who already like a company's drug are identified by local drug reps, and then given a platform. In detail, this process can take on many different shapes. Sometimes a company will pay for its favoured doctor to give a talk to other local doctors. If they're good, it will pay for them to give a talk to other doctors further away. If they're reasonably senior, or influential, or have some kind of academic track record, it will pay for them to go to conferences, or give lectures around the world. Sometimes these lectures will be part of the conference sessions, but sometimes there is a whole separate industry strand, with an eerie edge to it.

In fact, it's worth noting that the look of a 'medical conference' is what most people in other industries would recognise as a 'trade fair', and in some respects it's odd that we don't call them that now in medicine. The hall outside the lecture theatre is filled with promotional stands in which nice stuff is given away, brightly coloured floor-to-ceiling banners advertising various products, and attractive drug reps stepping imperceptibly into your path to engage you in conversation about their wares. This is what a trade fair looks like, although sometimes the cues are easy to miss.

I recently found myself eating some salmon at a boring doctors' conference in Cardiff, out near the academic poster displays. It was pretty good salmon, but gradually I noticed that I was standing, eating, in a kind of temporary autonomous zone, denoted by a change in the colour of the carpet, and some brightly coloured promotional displays. I was approached by an attractive, smiling woman in a suit. She asked me where I worked, and whether I saw patients who might use her company's drug; it was only then that I realised whose fork was in my mouth. The food was for people going to some special lectures, paid for by a drug company, in a parallel session, featuring its own chosen speakers. There was no drama, nothing rude, the rep was happy to chat, and the food was really good. She just wanted my contact details.

The paid speakers at these events are the 'key opinion leaders' (KOLs) we met earlier, and it's an odd scene, not just for the audiences, but also for the KOLs themselves. Nobody is obliged to change their views in return for money, in an overt act of corruption, though that may well happen: for the most part, these people are simply saying what they already thought about the drug anyway. But views favouring industry are given a platform, a microphone, and a nice projector for their slides; while those less favourable to industry are left to fend for themselves. In this way, as with negative results being buried, a biased picture is created of the overall swathe of viewpoints and evidence; but no individual doctor or academic has done anything they would regard as unethical.

I have good friends, around the same age as me, just finishing research work and entering their first medical consultant jobs, who give paid talks as KOLs. For them, it isn't about the money, which is often no better than working for an extra day on locum rates. It's not even about the other benefits, like top-flight training in nice venues on how to write and deliver a smooth presentation. I'm going to quote an ex here, and I'm sorry if that's weird, but I'm a doctor, so this kind of KOL activity is happening all around me all the time. Here's what she told me: 'None of those benefits matter. I do it because, in the speakers' room at a conference, or at a country hotel to learn about presentations, I'm spending time with the giants of my field. I'm thirty-six, and I'm getting drunk with the people who write the clinical guidelines! There's no *way* I could do that, unless I was a KOL.'

This is not unusual: often at conferences there will be a posh evening party hosted by one company. Only the people that company knows and likes are invited. If you go to its party, you get to meet top, influential people; if you don't, you don't, and this can have a corrosive effect on a conference generally, by sucking out the top ranks. I have a friend who complains that since vaccines have been invented for the disease he works on (which is mostly found in the developing world) the conferences are suddenly held in much more expensive hotels, and most of the senior figures disappear in the evenings to expensive restaurants, paid for by drug companies. Previously – am I being too utopian here? – they'd be falling over drunk with the junior researchers.

Is the content of this industry-funded teaching systematically biased? One study took a 'mystery shopper' approach, sending attenders along to some industry-sponsored CME teaching about calcium channel blockers, a class of blood-pressure drug.[101] Usefully, lots of companies make their own version of this kind of drug, and there were two pieces of sponsored teaching on them within a year of each other: one on one company's drug, the other on another's. The researchers went to both, and recorded every mention of every drug, noting whether it was positive, negative or equivocal. On each course, the sponsor's drug was mentioned more frequently, and much more positively: scoring three times as many positive mentions as negative ones. When the rival company's drug was mentioned, it got a much tougher ride: on the first course, these mentions were more likely to be negative, and on the second they were more likely to be equivocal. In the few statements directly comparing the sponsor's drug with the competitor's, the teacher usually said that the sponsor's drug was best.

The university supervising this teaching had a clear policy on how bias should be excluded. It obviously didn't work very well. These policies don't, which is why I don't ever take them at face value, unless there is very good evidence that they're followed.

In the second study, researchers followed up doctors' prescribing patterns after they attended some industry-sponsored CME teaching, again on blood-pressure medication.[102] This research found that after doctors attended the sponsored course, their prescription of the sponsor's drug increased.

Are these two studies perfect? No, but only because they have one simple flaw: they were both conducted twenty-five years ago, and nobody has done anything similar since.

This is extraordinary to me. It was established that the most senior doctors in the profession were receiving money to give talks that were, in effect, promotional, under the guise of educational activity; it was established that this distorted content changed prescribing behaviour; and then we just left it alone. The industry says, with no evidence, that everything has changed. I see absolutely no reason at all to believe that. Is a drug company really going to pay for a KOL to be shipped around the country, at great cost, to tell audiences of doctors that a cheap off-patent drug is the most effective first-line treatment for hypertension? The industry regards this activity as promotional; that's why it pays for it. In almost any medical circle you'll find stories of biased local consultants who give these talks, and who always seem to prefer one company's drugs. Leaving them unmonitored, without even the most basic 'mystery shopper' research to monitor their content, is a collective scandal.

And these teaching sessions aren't even just about the benefits of specific drugs: in recent years the manufacturers of the antipsychotic drug olanzapine, for example, have had lawyers running special teaching sessions for doctors.[103] They aren't about the medicine: they are to reassure doctors that they are unlikely to be successfully pursued in the courts over side effects from the drug.

So how big is this scene? Amazingly, although there is extensive work documenting KOL and CME activity in the US, there are very few openly available figures for Britain, because of our secretive regulations. As with drug reps, it's often a matter of two cultures: some doctors and clinics engage in industry-sponsored education all the time, as a matter of routine; while some never do, and think the whole idea is laughable. I can tell you that it is absolutely routine, at a conference run by a Royal College, to find a sponsored section, with posh food and parallel sessions of lectures, given by doctors and academics paid by the sponsoring companies. I can tell you that friends' and colleagues' travel and hotels and registration fees are routinely paid by drug companies. I can tell you that low-key local events, sponsored by a drug company, where a 'key opinion leader' gives a talk about a subject area and a drug, are commonplace (and those speakers do always seem to love the sponsor's drug). But in Europe we have very incomplete data.[104]

In the US, governments are more interested in transparency. As a result we can see much more, and there is little reason to believe that industry marketing activity there is any different to how it is in the UK. So we know that tens of billions of dollars are spent by the industry on drug marketing in America, of which only 15 per cent goes on marketing to patients, even though TV drug ads are permitted in the US. We know their spending priorities are likely to reflect their own research on what marketing activities bear the best fruit, so it's clear that marketing to doctors is effective. In 2008 the US industry body, the Accreditation Council for Continuing Medical Education (ACCME), reported that CME companies – the private firms acting as intermediaries between industry and some teaching – offered 100,000 teaching activities, amounting to more than 760,000 hours in total.[105] More than half of this was paid for directly by industry.

In case you think the US is a very different country to Britain, we can talk about Europe. In France, as of 2008, three quarters of all CME activity is paid for by the pharmaceutical industry, and of the 159 accredited providers, two thirds receive industry money.[106] In Germany, a researcher conducted an anonymous survey of members of a major medical society attending an international conference, and got 78 per cent of them to respond.[107] Two thirds said they'd got an allowance to attend from a drug company, most of them said they couldn't have travelled to attend it without that money, and two thirds said they had no ethical concerns about taking the cash. Similarly, they were sure it would have no effect on their prescribing behaviour. They were, as we have seen, wrong: doctors attending a conference paid for by a pharmaceutical company are significantly more likely to prescribe and request that company's drugs in future.

When drug companies were asked about their attitudes and concerns in the same German study, only one expressed ethical worries, but only 20 per cent responded at all. They may not want to speak openly, but from a brief scan of the industry literature we can get a slightly clearer picture of how they view these educational sponsorship opportunities. This is from *Pharmaceutical Market Europe*, an industry publication, talking about how CME companies can get business. Again, this is not a smoking gun: it's just the everyday corporate reality of how the industry sees this teaching.

> The overwhelming majority of support for European CME still comes from the pharmaceutical industry. Theoretically, anyone can be a supporter; but, as with any sponsorship however 'hands-off', it is the company which has the most to gain that will be supporting CME. An insistence that pharma companies support education in areas not of interest to them ... will receive no backing.[108]

Of course there are regulations, to try to prevent malpractice, but these vary around the world, and commonly, as we see time and again, they are simply ignored. Norway is a wealthy country, with an efficient public sector, and industry is banned from funding any CME at all, whether directly or indirectly, without any problems. In the UK the pharmaceutical industry is allowed to support all CME. In the US there are various regulations and guidelines, which have the usual holes. In 2007 the Senate Finance Committee noted, for example, that drug companies are not *compelled* to follow these guidelines, and that no agency ever puts monitors in the audience to see what's being taught, or engages in any kind of proactive assessment of content.[109] Even if a CME provider is reported, found out, and has its accreditation revoked, this can take nine years.

The committee was witheringly clear on why the industry funds this activity: 'It seems unlikely that this sophisticated industry would spend such large sums on an enterprise but for the expectation that the expenditures will be recouped by increased sales. Press reports and documents exposed in litigation and enforcement actions confirm these suspicions in some instances.'

They are referring, here, to an almost endless stream of leaked internal documents from legal cases which have revealed how the industry thinks, plans and acts. Many of these cases involve companies using CME to promote 'off-label' uses of drugs, expanding their prescription beyond the marketing authorisation, to other diseases where its use is not licensed. Warner-Lambert was accused of using 'independent educational grants' to fund CME programmes that taught doctors to use its drug Neurontin – licensed only for prescription in epilepsy – for completely different conditions, for which the drug has no licence at all. It paid $430 million to settle. Serono paid over $700 million to settle claims that it promoted its drug Serostim for uses for which it had no licence, through various methods including educational grants that funded 'independent educational programmes'. Merck explicitly carried out an internal study to establish the 'return on investment' from discussion groups led by doctors, which was leaked in a court case.[110] It estimated that for every dollar it spent on teaching, it got back almost $2 in revenue from doctors prescribing more of its drugs.

When ACCME reviewed the CME providers it accredits, it found that one in four were openly breaching its guidelines – not in clever, covert ways, but blatantly, not even bothering to hide it. These companies were fully accredited to teach doctors, and they allowed sponsors to influence decisions about content; allowed sponsors to choose who spoke; failed to check for conflicts of interest; repeatedly used the sponsor's drug's brand name to the exclusion of all others; and so on. This is hardly surprising – in fact, to imagine such transgressions wouldn't take place is absurd. In a large, competitive and expensive market, which CME company is going to get the repeat business: the one that respects the rules? Or the one that gives the drug company what it wants?

Perhaps even more remarkably, doctors themselves recognise that the content of this teaching is biased. This includes – specifically – the ones who accept it. In a survey from 2011, 88 per cent of attenders of sponsored educational activities thought that commercial support introduces bias, but only 15 per cent thought such free teaching should be banned, and the majority were unwilling to pay properly for CME themselves.[111] Reports from the American Medical Association, the Institute of Medicine, the Senate Finance Committee, the American Association of Medical Colleges and more have all called for commercial support of CME to be ended. They have all been ignored.

So there you have it. Doctors around the world – except in Norway – are taught which drugs are best by the drug companies themselves. The content is biased, and that's why companies pay for it. For decades people have stood up, shown that the content is biased, written reports against it, demonstrated that weak guidelines fail to police it; and still it continues.

What can you do?

1. Ask your doctor if they accept industry-funded teaching.
2. If you're a doctor, you could refuse to accept any industry-funded teaching, and refuse to give it, too.
3. Don't be ashamed to ask colleagues whether they think it's OK to attend or give industry teaching, and tell them about the research and legal cases demonstrating bias.
4. Receiving free teaching is a benefit in kind, of an expensive professional service. Doctors should be forced to declare publicly, to their colleagues and their patients, whether they accept free teaching from the pharmaceutical industry. Doctors who see patients should place a prominent notice in the waiting room and on their desk, stating exactly what companies they've accepted money or services from, and exactly what drugs those companies make.
5. It's sad that individual doctors don't volunteer to pay thousands of pounds a year, of their own money, for the continuing medical education that governments have made compulsory throughout the decades of their working lives, but they don't. Governments need to think about forcing doctors to pay, or paying from state funds. There is no need for teaching to take place in expensive conference centres.

What does it mean to take money?

So now we enter the closing pages, and there are just a few loose ends to mop up. Many of the concerns we've seen in the preceding pages revolve around one idea: people who receive money from a company might have different views to those who don't.

This might seem like an obvious truth to you, but there are plenty of people who will angrily deny it, as they write another cheque for the school fees. Before we close, I will pick at this last open sore.

First, we should be clear about what a conflict of interest means. The broadest definition says that you have a conflict of interest when you have some kind of financial, personal or ideological involvement that an outsider might reasonably think could affect your reasoning. It is not a behaviour, therefore, but rather a situation: to say that you had a conflict of interest doesn't mean that you acted on it, but simply that you had one, and almost everybody has one, in one regard or another, depending on how you draw the lines.

For example: I don't accept medical training sponsored by the pharmaceutical industry, I don't do research or promotional work for industry, I don't see drug reps, I've never been a KOL, or been flown anywhere nice with a pharmaceutical company. For the easy things, of medicine and academia, it's a simple story. But if we broaden out, to the entirely unpoliced world of conflict of interest for popular-science writers, then the pharmaceutical industry could claim I've got an ideological position – that they are dodgy – and that I make money from selling this. Of course, I think I make fair arguments, giving a clear unbiased view of the evidence from systematic reviews, and I also don't think I'd sell more books by exaggerating. But it is a conflict of interest: a situation, not a behaviour.

You could also argue the other way. For example, I have received two cheques that were partially related to the pharmaceutical industry. A decade ago, in my twenties, the *Guardian* entered me for the 2003 Association of British Science Writers prize. I arrived on the night and won: wandering drunkenly to the stage, I saw the prize was partly sponsored by GSK, alongside some august scientific bodies. I took the cheque, with some muttering. Then, in 2011, I gave two unpaid talks to ghostwriters' associations, explaining how their work harms patients. I give a lot of these 'lion's den' talks, to groups whose work I criticise – angry quacks, journalists, academics, medics, and so on – explaining the harm done by their industries, and often picking up good stories from worried insiders. When the ghostwriters asked me to give the same talk a third time, a day's travel away from London, I apologised and said I was busy. They offered money, I took it, and I gave the same talk again. Am I a stooge for the ghostwriters? I hardly think so, but you can disagree.

So it is important, in my view, to be clear about the importance of conflict of interest, but also, not to be unrealistic and shrill. To understand how much a conflict of interest matters, we need one basic, simple piece of evidence: overall, do academics and doctors with some kind of major interest have more pro-industry opinions than those without? We have already seen, from the very earliest pages of this book, that *trials* with industry funding are more likely to report positive results. We're now talking about the next level, when people discuss the findings of other people's trials, weigh up their strengths and weaknesses, or write opinion pieces, editorials and so on. In these kinds of discussion papers, do authors' conclusions correlate with the extent of their industry funding? The answer, as you might expect, is yes.

As we have seen, the diabetes drug rosiglitazone has had an interesting and chequered history, with the FDA and the manufacturer, GSK, both failing to draw attention to the fact that it was associated with an increased risk of grave cardiac side effects. The drug was recently taken off the market, after billions of dollars of sales, because of problems spotted by academics which the regulators had failed to act on. One group of researchers recently pulled out all the academic papers discussing whether rosiglitazone is associated with an increased risk of heart attack.[112] More specifically, they identified all 202 pieces citing and commenting on either of the two key publications examining this question: a meta-analysis by Steve Nissen, which showed that rosiglitazone does increase heart attacks; and the RECORD trial, which suggested that the drug was fine (although, you may now be concerned to hear, this trial was stopped rather early). The papers discussing these findings were from every category you can think of – review essays, letters, commentaries, editorials, guidelines and so on. As long as they discussed the link between rosiglitazone and heart attacks, and cited one of the two papers, they were in.

Around half of the authors had a financial conflict of interest, and analysing the findings by who said what gave a dismal but predictable result: people who thought rosiglitazone was safe (or, to be absolutely clear, who had a favourable view on the risk of heart attack after taking it) were 3.38 times more likely to have a financial conflict of interest with manufacturers of diabetes drugs generally, and with GSK in particular, when compared with people who took a dim view of the drug's safety. Authors who made favourable recommendations about using the drug were similarly three and a half times more likely to have a financial interest. When the analysis was restricted to opinion articles, the link was even stronger: people recommending the drug were six times more likely to have a financial interest.

It's important to be clear about the limitations of an 'observational' paper like this, and to think through alternative explanations for the observed correlation, just as we would with a research paper showing, for example, that people who eat lots of fruit and vegetables live longer. People who eat lots of vegetables tend to be wealthier, and are more likely to live healthier lives in all kinds of different ways, many of which have nothing to do with eating vegetables, so maybe that's why they take longer to die. Likewise, in the case of favouring rosiglitazone and having a financial interest, maybe you buy stocks in a company, or go and work for it, or take a grant from it, *after* you've developed a favourable view about whether its treatment is good or bad. That may be the case for some people; but in the broader picture of all that we know about how financial interests impact on behaviour, it's hard to believe that the findings are entirely innocent: and it certainly reiterates the fact that we need to be told, in detail, about people's financial dealings with these companies.

How do we deal with this problem? In the most extreme view, anyone who has a conflict of interest in a particular area should be barred from expressing any view on it. Radio DJs, after all, were supposedly forbidden from accepting 'payola' from record companies, and their world didn't collapse (though I'm sure there are other jollies for radio DJs).

A straight ban raises interesting problems, however. First, in some areas of medicine you might struggle to find any experts at all who've never done any work with industry. Here we should pause for a moment to remind ourselves what we really think about the drug industry, and the people who work in it. Although this book is about problems, my goal is that pharma should be adequately regulated and transparent, to the extent that academics can feel positive and enthusiastic about collaboration with it. There is no medicine without medicines; companies can produce great products; and working with people who are focused on completing a project for profit, however distasteful you might find some aspects of that world, can be very exciting.

It's also odd to take our frustrations out on individual doctors and academics, when they're simply doing what governments have told them to do over the past three decades: get out there and work with industry. From the 1980 Bayh-Dole Act in the USA, which helped academics register patents on their ideas, to the Thatcher drive for 'university entrepreneurs', academics have consistently been told they must engage with industry, and find commercial applications for their output. Discounting all of these academics, having pushed them to engage with industry, and successfully convinced some of our finest minds to do so, would be bizarre.

There are other problems with a straight ban. Even if you can find experts with no conflicts, sometimes the people you most want to hear from are those from industry: they might have special inside knowledge of the processes that have shaped new medicines, for example. And once you start to go down the path of listening to their commercial insights, you run up against a new problem. Sometimes, though this is a *very* tricky area, it might be useful to allow industry people with huge conflicts of interest to speak discreetly, without attribution, on something like a medicines regulatory committee.

Journalists know that deep background, off the record, from a source inside a story they are trying to understand can be extremely valuable. Sometimes an industry person will speak more candidly, but unattributably, to a drug approval committee that doesn't publish its minutes. I was told one story of an honorary professor of medicine, now working full time in drug development, standing up on an approvals committee to say, 'Honestly? Everyone knows that drug is rubbish, it won't last two years, and you'd save me some fuss if you killed it now.' I'm not telling you this to persuade you that we should permit secrecy in regulation: I don't think we should. I'm telling you so you can know that you've thought it through fairly.

On occasion, some journals have taken the view that industry is simply not to be trusted, even with declarations, and have then made rules accordingly. *JAMA*, for example, decided a few years ago that it would no longer accept industry-funded studies unless they had an independent statistician analysing the results rather than an industry one. It's an interesting stipulation – it implies that the analysis is where the black magic happens – and it caused an interesting fuss. Stephen

Evans is an eminent statistician who works in the same building as me: he is upright, an expert on fraud detection, and a movingly compassionate Christian (truly) in the way he talks about the dishonest academics he has exposed. He argues that we cannot simply discount the work of individual professionals out of hand, on the grounds that there is an observed association between working for industry and producing biased results:

> Suppose that a biomedical journal invoked a new policy requiring that all authors based in western Europe or North America would receive ordinary peer review, but authors from other countries would receive a peer review with additional hurdles. This policy may seem unfair, but suppose the journal claimed that research has shown that there is a greater prevalence of fraud, bias, and sloppy work among papers coming from these other countries.[113]

I think he's probably right, and that we should judge each paper on its merits, although I do slightly wish he was wrong. It's also interesting to note that after *JAMA* brought in its 'independent statistician' rule, the number of industry-funded trials published in its pages dropped significantly.[114]

In general, the most common approach to conflict of interest is that it should be declared, rather than outlawed, and there are two reasons for pursuing this policy. First, we hope it will allow the reader to decide whether someone is biased; and second, it is hoped that it might change behaviour. When I suggest that doctors should be forced to tell their patients, with prominent notices in their waiting rooms and on their desks, exactly what companies they've accepted money or services from, and exactly what drugs those companies manufacture, it's partly because I think it might elicit a small amount of shame. Sunlight is a very powerful disinfectant, and has been proven to be so in many different areas. In Los Angeles, the simple act of posting every restaurant's kitchen hygiene score in its window improved standards, and statistics on car safety led consumers to demand safer cars.

With medicine, however, declaration is more complicated than a single hygiene rating, or safety score, because it's not always clear what should be declared. Conflict of interest, after all, goes well beyond simple drug industry payments to individual doctors. In the US – this will sound strange to readers in the UK – oncologists get more money if they treat their patients with intravenous drugs rather than simple pills: over half a community oncologist's income comes from giving chemotherapy, so there is room for a conflict of interest. Similar problems may arise in the UK, with GPs managing the budget for their area, and taking a profit from some of the services they provide. And similar problems can arise when people are writing about a treatment they provide, even where there is no corporate involvement, simply through a vague sense of professional allegiance.

One study, for example, looked at whether academic papers said radiotherapy was a good idea for patients who'd had a particular kind of tumour removed, but where the stage of the cancer was not known: twenty-one out of twenty-nine radiotherapists thought it should be given, compared with five out of thirty-four clinicians from other specialities.[115] A similar bias has been shown for surgeons and coronary bypass operations, surgeons and surgery for a bleeding ulcer, and so on. There has been startlingly bad behaviour from advocates of breast cancer screening, who have overstated the benefits and underplayed the harms (such as the medical risks from unnecessary procedures in wrongly diagnosed women) simply because they were passionately wedded to the procedure.

Conspiracy theorists – who are naturally attracted to the problems in medicine – go further, and build vast castles in the clouds, with huge interlocking tales of conflicts of interest. For them, someone is biased for all time, on every topic, because she has a sister who works for the government; or because somewhere in the university where she works, a person she may never have met has a view on a topic that the industry might find favourable. The conspiracy theorists will then announce that these are secrets which have been *deliberately* withheld, when in reality nobody could possibly have anticipated such elaborate and tenuous fantasies.

So for the most part, if only because it's practical, academics and doctors tend to concentrate on getting declarations of major financial interests, often just within the past three years, and to leave these more exotic and intangible elements alone. Some do go further. The *BMJ* staff often declare their membership of political parties and other organisations – which is great, but when you step away from money, you drift into territory that starts to feel like an intrusion into someone's personal life; more than that, as things become more tangential, the decisions about what to declare become more arbitrary, and so perhaps even more misleading, in the selection of what is declared and what is not. As younger people worry less and less about their Facebook security settings, perhaps the future will bring radical transparency for everyone.

But we have other fish to fry. Do people take a declared conflict of interest into account when they read someone's claims? The evidence suggests that they do. In a trial from 2002, three hundred readers were randomly selected from an academic journal's database and divided into two groups.[116] Both groups were sent a copy of a short report which described how the pain from herpes zoster, or shingles, could have a substantial impact on patients' daily functioning; but each group got a slightly different version. The readers in group 1 saw a paper with different named authors from the actual ones, and with a declaration of competing interests, stating that they were employees of a fictitious company treating the condition, and potentially held stock options in it. Readers in group 2 were sent the same paper, but instead of the information about its authors' employment and stock options, it had a statement that the authors had *no* competing interests. The people in each group were then asked to rate the study, on scales of one to five, for interest, importance, relevance, validity and believability. Fifty-nine per cent of the questionnaires came back (which is remarkably high), and the results were clear: people who were told the authors had competing interests thought the study was significantly less interesting, less important, less relevant, less valid and less believable.

So it is clear that people care about conflicts of interest. And for that reason, specific financial relationships with drug companies are usually declared on academic papers. This system seems to operate reasonably well, but even when conflicts are clearly declared, it may only be on the academic paper, and not in the subsequent work derived from it, such as guidelines or review papers. One study from 2011 took a representative sample of meta-analyses – systematic summaries of all the trials in a field – and looked to see if they described the conflicts of interest of the individual trials they summarised. Of twenty-nine meta-analyses reviewed, only two reported the funders of the trials they included.[117] This is clear evidence that information on conflict of interest is not passed up the chain, and that meta-analyses – widely read and influential documents – simply gloss over this important issue.

We should be clear that declaring conflicts is not a final fix, and that like any intervention it can have side effects which should at least be considered alongside the headline benefit. For example, some have argued that forced disclosure of conflict of interest leads doctors to engage in 'strategic exaggeration',[118] knowing that their utterances will be discounted if it is believed that they are acting as shills, and there is some evidence for this in the behavioural economics literature, although only from psychology experiments conducted under laboratory conditions.[119] They may also be affected by a sense of 'moral licensing': once you've declared your interest, you feel free to let rip with biased advice, because you know the recipient has been warned. These are interesting ideas: overall, I would rather have disclosure.

But these are details. I have a strong suspicion that when you see the scale of this problem you might be slightly amazed. A recent survey in the US looked at senior doctors. Sixty per cent of department heads were receiving money from industry to act as consultants, speakers, members of advisory boards, directors and so on.[120]

ProPublica, the US non-profit investigative journalism foundation, has done an astonishing piece of work with its Dollars for Docs campaign, creating a huge, publicly accessible database of payments made to doctors.[121] Individual drug companies have been forced to post this pooled information on their websites, mostly after losing various legal cases. ProPublica has now aggregated data on over $750 million in payments from AstraZeneca, Pfizer, GSK, Merck and many more. The latest slice of data includes details of dinners: so I can tell you that a Dr Emert in West Hollywood ate $3,065 worth of food paid for by Pfizer in 2010, to take just one random example.[122] But while for me this is a curiosity, for patients and others in the US this database has produced a remarkable series of insights, showing the power of putting a lot of information together in one place, where it can be searched and indexed. An individual can look up their own doctor, and see how much they pocketed, to the horror and anger of medics across the country. And anyone can look up whole groups of doctors, to see what horrors lie beneath: overall, 17,700 doctors received money, and 384 got more than $100,000.

What's more, universities around the country seemed to have

little idea what was happening on their own premises, until the data was presented to them clearly. When the University of Colorado, Denver, saw that over a dozen of its senior academics were giving paid promotional talks for pharma, it launched a complete overhaul of its conflict-of-interest policies.[123] The Vice Chancellor was unambiguous: 'We're going to just have to say we're not going to be involved with these [CME] speakers' bureaus, because they're primarily marketing.' In some places, university policy was being routinely ignored. Five faculty members at Stanford were shown to be taking money to give industry-sponsored lectures, and had disciplinary cases launched against them.[124]

The database also made it possible to see what kinds of characters were being paid by industry.[125] By cross-checking the doctors who had taken the most money against records of disciplinary proceedings, in just the fifteen biggest states, ProPublica found 250 doctors with sanctions against them for issues such as inappropriate prescribing, having sex with patients, or providing poor care; twenty doctors with two or more malpractice judgements or settlements; FDA warnings for research misconduct; criminal convictions, and more. Three different drug companies paid one rheumatologist $224,163 over just eighteen months to deliver talks to other doctors, even though the FDA had earlier ordered him to stop 'false or misleading' promotion of a painkiller called Celebrex, saying he had minimised its risks and promoted it for unlicensed uses. Eli Lilly paid a pain doctor $84,450 over a year, although he was censured by his medical board for performing unnecessary and invasive nerve procedures and tests on his patients. Eli Lilly and AstraZeneca paid $110,928 to a doctor who admitted unethical and unprofessional conduct over allegations of improper prescribing of addictive painkillers, receiving several years' probation from his medical board. And so on. Most companies admitted that they never check for this kind of thing. It's a pretty damning judgement on the doctors and companies operating in this dark corner of medicine.

Remarkably, this transparency seems to be changing behaviour, and there is already some evidence that industry payments to doctors have begun to fall since they have become more visible to patients and the public through ProPublica's site.[126] It's disappointing, in some respects, to think that doctors' behaviour should be affected simply by whether their patients can find out what they're doing, but for many that seems to be the reality, and we should at least applaud their change of heart. So Veena Antony, a professor of medicine, received at least $88,000 from GSK during 2009 to give promotional talks.[127] Now she says she has given them up, wary of what patients might think: 'You don't even want the appearance that [you] might be influenced by anything that a company gave.'

Her anxiety tells a wider story: many doctors are worried about how the public might react to this kind of information, especially in a health-care market like the US, where patients can exert a lot of choice. When you take a drug, you want to know that it's the safest and most effective treatment, chosen for you on the basis of the best possible evidence. Informed consumers might avoid doctors who accept industry teaching and hospitality, because these have been shown – as you've now seen – to change the decisions that doctors make for their patients. In the US, a new law called the Sunshine Act will shortly come into force, and it will make lots more information available, so that patients can find out about their own doctor's involvement with industry.

You could be forgiven for believing that we are about to enter the same era of radical transparency in the UK, with patients able to make informed choices about whether their own doctor is independent and trustworthy. From 2013, after all, a new UK ABPI code of practice says that all drug companies must publicly declare how much money they have paid to doctors for their services: this figure includes speakers' fees, consultancies, advisory board memberships and sponsorship for attending meetings. It's a move that has been greeted with huge fanfare, and claims that it heralds a new era in transparency.[128] Celebratory headlines have exclaimed: 'Drug Companies to Declare All Payments Made to Doctors from 2012'.

But even if we excuse the way the starting date for this new era has already slipped back in time, inexplicably, from 2012 to 2013 since it was first announced, the new code faces a much bigger problem. Because it is yet another fake fix, and although it's the last we will see in this book, it follows the same familiar pattern of everything we've seen already, from the International Committee of Medical Journal Editors promising it would only publish pre-registered trials (they didn't stick to it, though everybody acted as if the problem was fixed, p. 51), through the FDA's new rules demanding publication within a year (not enforced, though everybody acts as if the problem is fixed, pp. 52–3), to the European Union's bizarre clinical trials register (a transparency tool whose content has been kept secret for almost a decade, p. 52), and so many more.

To understand why this code is so flawed, you have to dig deeper than the news coverage, because in reality the ABPI has defined 'Declare All Payments Made to Doctors' with such elaborate sophistry and wiliness that it's genuinely difficult to explain its plan briefly, in plain English: the reality is too far from what any sensible person would expect. The code simply requires that companies declare the total amount they've paid to all doctors. Is that clear enough? No: it makes it sound as if drug companies will be saying how much they paid to each doctor, because that would be the obvious thing to do. But I said 'all doctors'?

I'll try again: each company must simply declare two numbers, on a single piece of paper, and that is all. One number is the total amount of money it has paid to all doctors in the UK over that year, all rolled up into one big figure, of however many tens of millions of pounds; the other number is how many separate payments have been made. Is that clear yet? It might be easier with an example. Imagine one drug company paid £10,000 to a Dr Shill, £20,000 to a Dr Stooge, and 998 other similar-sized payments to another 998 different doctors. All it will tell you at the end of the year is: 'We paid out £12 million, split between 1,000 doctors'.

This is meaninglessly uninformative, and tells us nothing at all.

Could we build a database ourselves, from scratch? In reality, no, because we lack a culture of transparency and litigation around drug companies, so there is no legal framework for obtaining the kind of information that ProPublica has curated. It is possible to try to work out which academic doctors have taken money, very crudely, from the declarations that individual doctors and academics make at the end of each academic paper, but these are only made at all if they are relevant to the specific research area of that single study. As a result, sourcing information from here would produce an incomplete patchwork of declarations; and what's more, these declarations rarely give any figures. Since some people simply work for every company, giving an impression of universal obligation without favouritism, this can be very misleading (but has the added advantage of making you look like a very popular expert).

Sourcing information from the declarations on academic papers would also tell you nothing about the huge number of doctors who do no academic work, but who see patients, and are senior key opinion leaders in their local or professional area, and who are being paid large sums by drug companies to teach other doctors; it would tell you nothing about clinicians who accept hospitality; and it would tell you nothing about whether your GP sees drug reps, or accepts money to attend conferences. Essentially, we know nothing about which doctors take what.

What we need, ideally, is a centrally held register of personal or financial interests in the pharmaceutical industry: it could be voluntary, or it could be compulsory, and people have recommended for years that one should be created, but it has never happened. You might note that the most senior figures in medical politics – the people with medals, on the Royal College committees – are the people who would have to drive this through, and they are often the very people receiving the greatest income from industry work.

Doctors reading this would do well to note a lesson that has been learnt in recent years by journalists over phone tapping, and by MPs over their expenses: just because you think something is normal – just because everyone you know is doing it – that doesn't mean outsiders will agree, when they find out. In Germany, following an investigation by *Stern* magazine, the police searched four hundred drug reps' flats and 2,000 medical premises, finding that doctors were routinely accepting money and gifts (as we know). In 2010 two German doctors were convicted and sentenced to a year in prison for accepting bribes to prescribe one company's drugs, on the grounds that this defrauded the insurance company that was ultimately paying for the treatment.[129] Over 13 per cent of fraud cases in America involve the pharmaceutical industry, covering either marketing or pricing issues.[130] Pfizer has agreed to pay over $60 million to settle a foreign bribery case in the US courts, and several other drug companies are in the spotlight for similar charges. The things that doctors have always regarded as normal are gradually giving rise to serious prosecutions.

But, of course, it's not just doctors and academics who can have conflicts of interest: and this is the final part of our long, sorry story.

First, these issues extend beyond medicine. In October 2011 the *Australian* newspaper began a 'Health of the Nation' series, sponsored by the Australian Medicines Industry.[131]

Money is given to newspapers like this to buy goodwill, to build closer relations, and to make it harder for them to be awkward in the future. Since newspapers have no culture of declaring such donations, there's no standard slot for doing so at the bottom of the article, as you'd find in an academic journal, so they often go unmentioned, much like the free holidays for travel writers. On top of that, journalists are frequently paid by drug companies to attend academic medical conferences, with hotel and flights provided, and asked to attend promotional events while they're there, in exchange. I have names that I will give you socially, but not in print (just know that I have a list).

But more than that, this problem extends to the heart of the most powerful institutions in medicine, which can often become dependent on industry for core support and funding. This is well illustrated by a small recent case from the PMCPA, where the hospital representative from Lilly became frustrated with a diabetes consultant, who kept prescribing another company's drugs. 'We are basically paying you to use Novo Nordisk's insulin,' he complained, before explaining that funding for an educational post in the doctor's institution was soon to be 'reviewed' by the Lilly Grants and Awards Committee,[132] and probably cut, since the managers had noted that the doctor was failing to prescribe their drug.

Such funded positions are extremely widespread. Of course they are: they are the bread and butter of medical academia, because the vast majority of trials research is funded by industry, and much of it is situated in universities. Do all of these posts come with menaces? Of course not. At the extremes there are terrifying scandals – famous cases of people such as David Healy, Nancy Oliveri and others – where doctors have been pushed out of a university job because of criticisms they made of companies. As a young doctor, at the early and unglamorous phase of a clinical academic career, I should probably be more afraid than I am. But the fate of occasional individuals who speak out is only part of the problem. The real story here is hidden from view: the doctors and academics who read stories of overt bullying and decide never to pressure their head of department, never to disappoint a funder, never to raise a concern about the appropriateness of a particular engagement with industry, in their academic unit. In every case, you can be sure it was dressed up in the individual's mind as a small concession, necessary to keep a broader project on track, for the good of the department, of the patients, of everyone.

Beyond universities, there are other important institutions in medicine, such as membership organisations and the professional bodies, which all have their own engagements with industry. Here are some fairly random pickings from around the world. In America in 2009, the Heart Rhythm Society received $7 million, half of its revenue, from industry.[133] The American Academy of Allergy, Asthma and Immunology took 40 per cent of its income from industry.[134] The American Academy of Pediatrics officially supports breastfeeding, but receives about half a million dollars from Ross, manufacturers of Similac infant formula[135] (Ross's logo even appears on the cover of the AAP's 'New Mother's Guide to Breast Feeding'). The *British Journal of Midwifery* runs ads from powdered-baby-milk manufacturers, and baby-milk companies run 'training days' for midwives in hospitals around the UK, which are well attended, because they are free. The American Academy of Nutrition and Dietetics is sponsored by Coca-Cola.[136] In 2002 the American College of Cardiology thanked Pfizer for $750,000, Merck for $500,000, and so on.[137]

Payments to these groups will not be covered by the Sunshine Act in 2013, and for their equivalents in the UK, there is no law to help us see what lies on the balance sheets. It's a very troubling state of affairs, and this is not simply an aesthetic concern: these organisations run conferences that are internationally attended, and set ethical norms for their members. More than that, they make the guideline documents which are followed around the world, and creating these often requires subjective judgement calls, especially where the evidence is thin. One study asked 192 authors on forty-four guidelines documents if they received money from industry, and on average four out of five said yes.[138]

This problem is vast and complex, and it won't go away. We need to think very carefully about how to manage it.

What can be done?

1. All doctors should declare all payments, gifts, hospitality, free teaching and so on, to their patients, to colleagues, and to a central register. The conventional cut-off is for everything within the past three years, but we could consider making it longer. We should display the contents in our clinics, to our patients, and let them decide if such activities are acceptable.

2. Drug companies should declare all payments to doctors to a central database, naming each doctor, and giving the amount paid to them, and what it was for. This will permit cross-checking and make declarations easier.

3. Governments should create a publicly accessible national database of payments by companies to doctors, and make it compulsory for doctors and companies to declare everything to it. Until they do, someone else could make a voluntary one.

4. The US Sunshine Act is a good starting model for legislation: companies will be compelled to declare who they gave money to, how much, and on what date; but also what drug the payment related to. This information, displayed by doctors in waiting rooms, would be just fine.

5. Conflict-of-interest policies vary hugely between institutions, and have never been reviewed in the UK. In the US this work has been done by the American Medical Students Association. Its website, www.amsascorecard .org, is a model to us all: it grades over a hundred institutions on their conflict-of-interest policies for gifts, consulting, speaking, disclosure, samples, drug reps, industry support for education and so on, using a transparent methodology and giving a summary grade for each institution, from A to F. Honestly, I feel weepy when I look at it.

CONCLUSION: BETTER DATA

You may be feeling overwhelmed; I couldn't blame you. We should take a moment to recap, to think about how an industry executive would defend themselves, and then work out how to fix things.

For me, missing data is the key to this whole story. Bad behaviour in marketing departments is unpleasant, but it's the one thing that has already received public condemnation, because the issues are tangible, with covert payments, misleading messages, and practices that are obviously dishonest, even to the untrained eye. But for all that they may be disappointing, these distortions can be overcome by any good doctor. If you go straight to the real evidence, and read systematic reviews of good-quality trials, then all the distortions and spin of drug reps and 'key opinion leaders' are nothing more than wasteful, irrelevant noise.

Missing data is different, because it poisons the well for everybody. If proper trials are never done, if trials with negative results are withheld, then we simply cannot know the true effects of the treatments that we use. Nobody can work around this, and there is no expert doctor with special access to a secret stash of evidence. With missing data, we are all in it together, and we are all misled. I will say this only once more, but I think it bears repeating: evidence in medicine is not an abstract academic preoccupation. Evidence is used to make real-world decisions, and when we are fed bad data, we make the wrong decision, inflicting unnecessary pain and suffering, and death, on people just like us.

In a moment, we will look at what can be done: because there are some simple fixes that would put all this behind us, and hugely improve patient care, globally, at almost no cost, if patients and politicians were willing to fight for them. But before that, I'd like to look at what the pharmaceutical industry will say in response to this book.

First, I'm sure – perhaps after some dismissive personal smears – there will be accusations of cherry-picking. People will claim, incorrectly, that I have focused on rare and exceptional cases. On this, I would encourage you to remember how much of this book is based on systematic reviews, which summarise all of the evidence ever collected on a given question. Go back and check, if you like. Our best estimate was that half of all clinical trials go unpublished, and that doesn't come from a story, or an anecdote: it comes from the most current systematic review, containing the results of every study ever conducted on this issue. Where we have walked through individual, shameful cases – like paroxetine, or Tamiflu, or Orlistat – it was only to put narrative meat on these very ugly bones.

So I am confident that you will agree, from the evidence set out in this book, that these are systemic problems; and that it would be shameful, or even dishonest, simply to dismiss them. What's more, where the evidence is lacking – and this isn't often – I have been clear, and I have set out what work is needed to fill those gaps. For example: lectures from key opinion leaders paid by industry are one of the most significant ways in which qualified doctors are educated today, and two decades ago 'mystery shopper' research found that these lectures are systematically biased. The fact that this work hasn't been repeated in the past five years should be a source of shame for the industry and for my profession. It's not a cause for celebration, and it certainly doesn't exonerate anyone.

As their next tactic, we can be sure – because we've watched them do this already – that people from industry will point to their guidelines. Look at all these miles and kilograms of rules, these vast offices filled with regulators: this is one of the most closely monitored industries in the world, they will say, drowning in red tape. But we have proved, I think, that these regulations simply do not do their job. The rules on registering trials were ignored; the FDA rules on posting results within a year have only been obeyed for a fifth of trials; the ICMJE regulations on ghostwriting – absurdly – permit ghostwriting; and so on. These regulations have been tested, and they have been shown to fail.

But the most dangerous tactic of all is the industry's enduring claim that these problems are all in the past. This is deeply harmful because it repeats the insult of all the fake fixes we have seen throughout this book: and it is this recurring pattern of flat denial that allows the problems to persist.

The clearest window onto this strategy comes from the industry's response to its most recent public scandal. In July 2012, GSK received a $3 billion fine for civil and criminal fraud, after pleading guilty to a vast range of charges around unlawful promotion of prescription drugs, and failure to report safety data. The full list of charges and evidence is vast – you can browse it all at the Department of Justice website – but the methods they used will be very familiar to you by now.

GSK bribed doctors with gifts and hospitality; it paid doctors millions of dollars to attend meetings, and to speak at them, in lavish resorts; it used, in the justice department's own words, 'sales representatives, sham advisory boards, and supposedly independent Continuing Medical Education (CME) programs'. It withheld data on the antidepressant paroxetine. It engaged in off-label promotion and kickbacks for the asthma drug Advair, the epilepsy drug Lamictal, the nausea drug Zofran, Wellbutrin, and many more. On top of all this, it made false and misleading claims about the safety profile of its diabetes drug rosiglitazone; it sponsored educational programmes suggesting there were cardiovascular benefits from the drug, when in reality even the FDA label said there were cardiovascular risks; and most damningly of all, between 2001 and 2007, it withheld safety data on rosiglitazone from the FDA.[1]

The industry claimed that these crimes were all in the past. GSK's own press release said they were from a 'different era'. Stephen Whitehead, the head of the ABPI (who previously worked in policy and PR for drug companies, Barclays and the alcohol industry), said: 'The global pharmaceutical community has fundamentally changed during recent years; where we have made mistakes in the past, we have tried to rectify them.'

To examine this claim – even if we set aside the vast amount of evidence in this book – it's useful to trace the current positions of the people who were in senior roles at GSK, during the period of proven fraudulent behaviour, and then look at where they are now. Chris Viehbacher of GSK was singled out in the court ruling: he is now CEO of Sanofi, the third-biggest drug company in Europe. Jean Pierre Garnier was CEO of GSK from 2000 until 2008, only five years ago: he is now chairman of Actelion, a Swiss pharmaceutical company.[2] There is no suggestion that these companies have been involved in improper behaviour. The court also specifically mentioned Lafmin Morgan, who worked at GSK in marketing and sales for twenty years: Morgan was still working for GSK in 2010, just two years ago.[3]

So while GSK and the ABPI may claim that these problems are in the past, in reality: one of the charges involves withholding safety data as recently as 2007, on a drug only suspended from the market in 2010; two of the most senior figures pulled out in the court case are at the helm of pharmaceutical companies in Europe right now; and another senior figure in marketing at GSK continued to work for the company until just two years ago.

It doesn't end there. Richard Sykes was the head of GlaxoWellcome from 1995 to 2000, and then chair of GSK from 2000 to 2002, when many of these fraudulent acts took place. He is now the chair of Imperial College Healthcare NHS Trust, running health care for a large part of West London. He is also the chair of the Royal Institution of Great Britain, the UK's oldest and most eminent science communication establishment. That, more than anything, is a clear illustration of the extent to which this world penetrates British academia and medicine to its absolute core.

And contrary to the spin, this GSK fine was no isolated incident either. Eli Lilly was fined $1.4 billion in 2009 over its off-label promotion of the schizophrenia drug olanzapine (the US government says the company 'trained their sales force to disregard the law'). Pfizer was fined $2.3 billion for promoting the painkiller Bextra, later taken off the market over safety concerns, at dangerously high doses (misbranding it with 'the intent to defraud or mislead'). Abbott was fined $1.5 billion in May 2012, over the illegal promotion of Depakote to manage aggression in elderly people. Merck was fined $1 billion in 2011. AstraZeneca was fined $520 million in 2010.

These are vast sums of money: Pfizer's in 2009 was the largest criminal fine ever imposed in the US, until it was beaten by GSK. But when you consider these figures alongside the revenue for the same companies, it becomes clear that they are nothing more than parking tickets. For the period of time covered by the $3 billion GSK settlement, sales of rosiglitazone were $10 billion, paroxetine brought in $12 billion, Wellbutrin $6 billion, and so on.[4] A graph of GSK's share price over the past year appears on page 349: decide for yourself if you can see any impact from a $3 billion fine and criminal fraud case in July 2012.

So these are not isolated problems, and they are most definitely not a thing of the past, because many of them happened recently, and the people concerned are all still around, in very senior positions of power.

This is also a sobering context in which to view GSK's most recent promises, from October 2012, of greater transparency. The company has announced that it will consider sharing individual patient data from clinical trials with external researchers after they pass through an application process. This is an admirable promise, the sort of thing that was called for in Chapter 1. It is, of course, only a promise: no data has yet been shared, and GSK has a proven track record of making promises for greater transparency and then breaking those promises, as we have seen in this book: they made a trials register, then took it down, and so on. But it is a promise, and a nice one, from one company. What is that worth?

Let me tell you something from my personal life. I know people who work in various drug companies, because I am a nerd, and nerds work in biotechnology. I talk to these friends, and people I trust tell me, when they're earnestly drunk at parties, that Andrew Witty, the current head of GSK, who took over in 2008, is a lovely and honest man. He wants to do the right thing, they say. He bangs his fist on the table and talks of integrity. And I am entirely prepared to believe that this is true.

But it's also completely irrelevant: because this is the serious global business of health, affecting every single one of us. We cannot allow the behaviour of the pharmaceutical industry to swing on a pendulum, one moment dismal, one moment acceptable, oscillating wildly in different companies at different times, with our chances of getting proper data forever at the whim of whether the person at the top is *nice*.

We need clear regulations, with clear public auditing, to ensure that compliance is tested, and documented. And we need it to be applied muscularly, to everyone, without exception. We should remember, after all, that drug companies compete

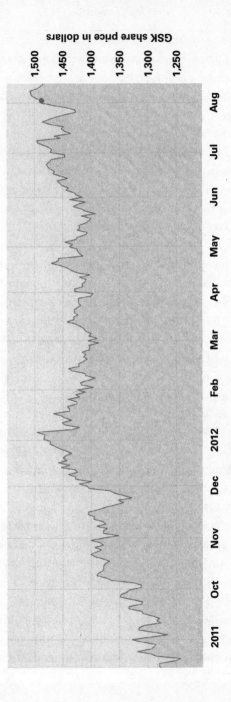
GSK share price in dollars

against each other, and they play by the rules we set as a society. If the rules permit dodgy practices, then companies are practically compelled to play dirty, even if the personnel know that their actions are morally wrong, and even if they want to do the right thing.

This is particularly well illustrated by a recent episode from Australia. The government commissioned a lengthy review on how to regulate bad pharmaceutical marketing. This review made clear policy recommendations to prevent misleading and dangerous practices, and those new regulations would have brought all companies in line with a best practice code that was already followed by members of Medicines Australia, the major industry body. But in December 2011 the government rejected this review. It was going to leave the industry free to engage in dodgy dealings, and the clearest criticism came, not from campaign groups, but from the drug companies themselves. Why would anyone follow the best practice in their voluntary code now? The press release put out by Medicines Australia was brutally honest: 'our member companies are [going to be] disadvantaged by doing the right thing'.[5]

In a moment we will look at what good regulations would look like in practice (it's not a difficult problem to solve), and what individuals can do to make them happen. We will also, more excitingly, think through the future of medicine, in an age of 'big data', when evidence is cheaper and easier to produce than ever before.

But before we get there, we need to remember that this isn't simply about fixing the problem, starting from now. Because even if we set aside the ongoing failure of industry and regulators to address these problems, patients are still being harmed, every day, by the actions of the pharmaceutical industry over the past few decades. It's not enough for companies simply to promise change in the future (and they've not even delivered on that). If the industry wishes to make amends for past crimes, then it needs to take serious action, today, to reverse the ongoing harm that still results from their previous behaviour.

Clearing the decks

First, we need full disclosure, and I don't say this out of some waffly notion of truth and reconciliation. Medicine today is practised using drugs that have come onto the market over several decades, supported by evidence that has been gathered since at least the 1970s. We now know that this entire evidence base has been systematically distorted by the pharmaceutical industry, which has deliberately and selectively withheld the results of trials whose results it didn't like, while publishing the ones with good results.

A vague admission that this happened is no more than a weak gesture: it's the very beginning of a journey back to being an ethical industry. For the sake of patients, we need every single hidden trial to be made available, now, today. There is no way that we can practise medicine safely, as long as the industry continues to withhold this data. It's not enough for companies to say that they won't withhold data from new trials *starting from now*: we need the data from older trials, which they are still holding back, about the drugs we are using every day.

This material is held in old salt mines, in secure dry storage archives, on ageing disks, in huge clunky 2002 laptops, and in cardboard boxes. Every moment that the pharmaceutical industry continues to hide it from us, more patients are harmed: this is an ongoing crime, against all of humanity, and it is happening under all of our noses.

What's more, there is no safe alternative to full disclosure. Doing more trials won't help: trials are expensive, and small, and when their results become available, we combine them with the existing results of all the trials ever conducted, to get the most robust answer, to iron out random error and fluke results. If we do more trials, we simply add these to a pool of data that already exists, and has already been polluted.

In fact, there is only one way we could work around the industry still withholding trials: we would have to throw out everything, every single trial that predates this imaginary moment when companies stop hiding results (which still hasn't yet come, in any case), and then start again. This is an absurd suggestion, but its absurdity is eclipsed by the men and women who sit in offices in the UK and around the world, knowing full well that their companies have trial results that they are still deliberately withholding. Their choice to continue withholding that data, right now, distorts prescribing decisions and harms patients every day. They sleep in their beds, just like you and me.

But the need for an amnesty does not end with trial data.

What are we supposed to do, for example, with the ghost-written papers of the past? Commercial medical writers now admit publicly that this practice was rife (when I ask them, in conversation, 'How can it not have seemed wrong, when you were paying academics to put their names on papers?', they smile sheepishly and shrug). The pharmaceutical companies, too, have been forced to admit that they did this, after endless embarrassing revelations from leaked documents and humiliating court cases on individual drugs. But these are islands: we have no idea of the full scale across the whole of medicine, and, crucially, we have no idea *which* academic papers were corrupted, because so much of this behaviour went undeclared.

These industries now accept that they rigged the academic literature, and that this practice was widespread. That's only partially useful: now, please, we need a list of the papers that were rigged. Some may need to be formally retracted, but at the very least, let us go back, annotate the literature, and see which academic papers were covertly written by paid industry staff.

Let us see which were the product of undeclared publication plans. At the very least, tell us which academics were 'guest authors', who contributed nothing but their name, the illusion of independence, and the reputation of their university, in exchange for a cheque. Tell us how much money they were paid; but most important, tell us their names, so we know how to judge their other work.

Because the medical academic literature isn't like a newspaper: it's not a transient first draft of history, or tomorrow's chip wrapper. Academic papers endure. Many of the studies affected by ghostwriting will still be regarded as canonical, they will be widely cited, and their contents will be used to inform practice five, ten and twenty years from now. This is how evidence-based medicine works, and it's how it's supposed to work: we rely on the research that has been published in order to write textbooks and make decisions. It's not enough to say that you won't use dishonest ghostwriting practices in the future: we all need to know which papers you rigged in the past, right now, to prevent your actions from causing more harm. Patients deserve to know too.

So, if we're to make any sense of the mess that the pharmaceutical industry – and my profession – has made of the academic literature, then we need an amnesty: we need a full and clear declaration of all the distortions, on missing data, ghostwriting, and all the other activity described in this book, to prevent the ongoing harm that they still cause. There are no two ways about this, and there is no honour in dodging the issue.

But going forward, separately, we need to be sure that these practices will not persist. The details of how to do this are set out at the end of each section throughout this book, but the basics of the wish list are clear.

We need to prevent badly designed trials from ever being run in the first place. We need to ensure that all trials report their results within a year at the very latest; we need to measure compliance with that; we need extremely stiff penalties for companies who transgress; and we need doctors and academics who collaborate in withholding trial data to be held personally responsible, and struck off. When it comes to disseminating evidence, we need to make sure this is done cleanly, so that doctors, patients and commissioners of health services have easy access to unbiased summaries of information. It is clear from the evidence presented in this book that the pharmaceutical industry does a biased job of disseminating evidence – to be surprised by this would be absurd – whether it is through advertising, drug reps, ghostwriting, hiding data, bribing people, or running education programmes for doctors. There is much to fix.

This mess has sat, hidden in plain sight, for lack of a clear explanation. It has persisted because it's complex, and because the people we would normally trust to manage such technical problems have failed us. The government, the great and the good of medicine – the silverbacks of the Royal Colleges, the faculties and the learned societies – know everything that you have just read. They know full well, and they have decided, for their own reasons, that they are unconcerned. In some cases, like the regulators, they have actively conspired in the secrecy.

It's hard to imagine a betrayal more elaborate, or more complete, across so many institutions and professions. This is a story of pay-offs, of course, but more than that, it's a story of complacency, laziness, banal self-interest and people feeling impotent. You have been failed by the people at the very top of my profession, for decades now, on matters of life and death, and as with the banks, we're suddenly discovering a terrifying reality. Nobody took responsibility, nobody was in control, but everybody knew something was wrong.

We only have one hope, though it is a small one: you.

Things you can do

If you are concerned by what you have read in this book, then here are some suggestions of things you can do.

There are detailed points on what needs to change at the end of each chapter, which I hope you'll revisit. Here I have pulled out big-picture points, with something for everyone. Creating change is a complex process, especially when problems are diffuse and deeply embedded in the culture of powerful industries and professions: the people who need to be lobbied are doctors and patient groups, as much as politicians, and this is reflected below.

Everyone

The first thing you might do is write to your doctor, or briefly mention your concerns when you see them. To be clear: I don't think it is helpful to waste valuable individual clinical time on political conflict with your doctor. However, if doctors know that their patients are concerned about these issues, they will be more inclined to take them seriously, and it is enough to do this in passing. Many are already very ethical on these issues anyway, and so they may find your concerns heartening. Here are some things you may wish to do:

You may wish to ask a question:
- For example, you might want to know if your doctor accepts drug-company hospitality, or sponsored teaching.

You may wish to make your wishes clear:
- For example, if you do not think it is acceptable for your doctor to go through your medical history with a sales representative from a drug company, as discussed above, then you could ensure that they know that, just in case.

You may wish to make a request:

- For example, you might suggest that they post a list of interactions with industry in their waiting room, as suggested earlier in this book.
- Or you might ask them to disclose whether a drug company has played any role in helping to develop their career pathway.

There are also the usual outlets that anyone can pursue for general political activism, and for lobbying politicians. It would be good to raise what you regard as the key concerns with your representative, but there is no clear legislation in the pipeline (in fact, quite the opposite, as you've seen).

If you have time, there is a clear need for organisers and public campaign groups on the issues raised in this book. You may wish to join existing campaigns, such as alltrials.net or HAI, or find out if your patient group is active on these issues, as discussed below.

Last, since changing laws is complex business, I would like to see work from *policy people*, wonks who know how government works, with suggestions of how some of the problems raised here could be fixed in legislation, or otherwise.

Patients

Patients are at the centre of this story, and you are in a strong position. First, we need to be clear that you cannot fix this problem in your own clinic room, for your own personal treatment decisions. These are systemic flaws, and you can only fix the system. Opting out of medical care entirely is self-destructive, and you should remember that drugs are unlikely to actively do you harm, since they must at least be better than *nothing* to get on the market. The question is whether you are getting the best treatment for you.

I hope that you will be asked to participate in a trial at some stage in your disease. Please say yes: trials are the only way we have of finding out what works, they are generally safe, and they save lives. There are four simple questions you should ask about any trial you're invited to participate in, and if you're refused on any count, I would like to hear about it:

1. Ask for a written guarantee that the trial has been publicly registered before patients are recruited, and ask where you can see it.
2. Ask for a written guarantee that the main outcome of the trial will be published within a maximum of one year after completion.
3. Ask for the name of the person who will be responsible for that.
4. Ask whether, as a participant in the trial, you will be offered a copy of the report describing its results.

If you have an ongoing medical problem, there will be a patient group that covers it, run by people with patients' interests at heart. There are problems with some of these groups, as described above, which you could address, but I would strongly encourage a different path: you can join them, and then encourage these groups to lobby the companies with which they have relationships.

For example, there is one important letter that every special-interest patient group should be sending to every drug company in the world, containing a simple query: 'We are living with this disease: is there anything at all that you're withholding? If so, tell us today.' This letter serves two purposes. Thinking optimistically, it might prompt a declaration: someone might disclose trial data that they were previously withholding, and this will improve patient care. But if they don't, and they have something they should be sharing, then you have still done something valuable: you have created anxiety; you have forced someone to attach their name to the responsibility of misleading you; and you have put a clear date-stamp on a company's ongoing dishonesty. If a company denies withholding trial data on drugs for your disease today, in 2012, but is then caught out in 2014, and puts out a press release saying 'Everything has changed now,' you will know, for certain, that it was still willing to mislead and harm patients in 2012.

Quacks

Alternative therapists who sell vitamins and homeopathy sugar pills, which perform no better than placebo in fair tests, have no role to play in fixing these problems. These business people often like to pretend, with an affectation of outsider swagger, that their trade somehow challenges the pharmaceutical industry. In reality, they are cut from the same cloth, and simply use cruder versions of the same tricks, as I have written many times elsewhere. Problems in medicine do not mean that homeopathic sugar pills work: just because there are problems with aircraft design, that doesn't mean magic carpets really fly. Equally, it's absurd to believe that ineffective sugar pills are a meaningful policy response to widespread regulatory failure in the pharmaceutical industry. If quacks profit at all from the justified anger people feel about the problems you have read here, then it comes at the expense of genuinely constructive activity.

Patient groups

There is much more that patient groups can do, in their role as collective organisers, and I would strongly encourage these groups to meet, and consider what they can do to address the issues in this book, using the unique resources they have. At present, for example, it is nobody's job to monitor missing trial results: so even though we have huge registries filled with details of ongoing studies, nobody is flagging up the trials which have completed, but failed to publish. We should remember that it was independent academics, investigating on a whim, who discovered that only one in five trials had met the reporting requirements of the FDA's new law from 2007. The absence of proper, centralised compliance auditing for withheld trial results is a catastrophic failure in the information architecture of evidence-based medicine, but since it hasn't been fixed, patient groups are in a very strong position to do the job.

They can act as observers for their own area, monitor the registers, look at the completion dates of trials, and then hunt for the publications. If researchers fail to produce results within a year, patient groups should first name them – since this carries a valuable public sting, which can change behaviour in future – then contact them, and ask for the data that will improve their members' treatment. Patient groups are also in a strong position, with their extensive grass roots membership, to find out about trials which are being conducted, but which have not been placed on a trials register. If there are patient groups who are willing to address the problems in this book, I would be happy to work with them to help develop further interventions, as would many other doctors and academics.

Doctors

Medics, in my view, need to think and talk about these issues much more, share what they know, and act. This might mean a number of things, as discussed earlier in the book: individuals could avoid industry marketing; declare what they've had to their patients; decline sandwiches and free flights; and so on. They could also engage with the senior figures in their professional societies and Royal Colleges, to try and encourage them to step back from the current dangerous positions that most hold, especially with the odd claims of the Ethical Standards in Health and Life Sciences Group. All these organisations should have a clear policy on publishing trials, and on the ethics of other encounters with industry, based on the best currently available evidence. Where evidence is missing, it should be gathered.

Medical schools

Medical schools can teach medical students about how to spot bad evidence from the pharmaceutical industry, and in particular, how its marketing techniques work. There is some evidence from the US that students taught about these techniques are better able to spot distortions in promotional material, and this deserves much more concerted work: the current generation of trainee doctors will go on to practise medicine independently for at least three decades, without any further formal teaching. If we do not future-proof them, they will be taught by industry, with the encouragement of the government and – reading the latest collaborative document – with the encouragement of every eminent medical body in the UK. If there is to be any hope of defending the medical profession against the technical distortions used as marketing techniques by industry, young doctors must be trained to identify them.

Ghostwriters

Commercial medical writers – and the International Committee of Medical Journal Editors – need to fix their ridiculous guidelines, because everybody knows that they still permit ghostwriting to happen. Commercial medical writers could organise an amnesty where they expose every paper they've covertly written, and every ghost author they've ever paid, on ethical grounds, for the protection of patients. They won't, but they could.

Lawyers

In the US, individuals and the state are better able to take action against those who have harmed them, often reframing the issue in terms of financial fraud. Drug companies are not the only target here, and many have argued recently that ghostwritten articles also present an opportunity.[6] If a patient is injured, when their doctor is relying on the content of an article that was covertly manipulated, then the commercial medical writers, the 'ghost authors', could be held liable. More than that, the 'guest authors' – the academics who allowed their names to be put onto these papers, despite minimal contribution, often in exchange for money – might also bear responsibility. If an academic paper is used by Medicare or Medicaid, in the US, to justify off-label use of a drug, but that paper subsequently turns out to be ghostwritten and distorted, then, again, the authors may be liable for this act of fraud, perpetrated on the government. There are also anti-kickback laws to consider, and a clear precedent that the First Amendment right to freedom of speech does not shield fraud. This could rapidly become an interesting avenue.

Journal editors
Journal editors are the current gatekeepers for medical evidence, and they have dropped the ball. All journals should declare their industry income, in full, and no journal should permit any trial to deceitfully switch its primary outcomes: this practice misleads doctors and harms patients. All journal articles reporting unregistered trials should state this fact clearly, and the ICMJE should publicly declare that they have failed to police this practice, so that others know to fix it properly.

The pharmaceutical industry
There is a great deal to cover here, and much has already been said in the book, but there is a special call I would like to make, to the many good people who work in industry. It's possible for companies and professions to be structured such that good people participate in projects which overall do great harm,

without necessarily ever knowing. There are countless people working in industry who have never stopped to think about how all these issues fit together in the round: that's no surprise.

Of course, I strongly encourage you to become a whistle-blower, whenever you see things being done that are wrong, using three main vehicles, in ascending order of melodrama. At the simplest level, if you can handle the secrecy properly, write an anonymous blog, explaining what you see on a daily basis: the banal, mild distortions, the days when you're asked to hunt in a data file and cherry-pick any pattern that makes your company's drug look good, the unofficial sales advice you're given as a drug rep, and so on. Next, I'd appreciate specific leaks, on ben@badscience.net (but please don't send me confidential information from your work email address). Last, many of you reading this have access to large quantities of data or documents that would change the lives of patients, and help to prevent ongoing suffering and harm. I would appreciate a data dump, on the scale of the US Iraq and Afghanistan war records, and if I'm honest, I'm surprised and disappointed this hasn't happened yet. If you need any help, just ask: I will do everything I can to assist you.

But more than that, I would also strongly encourage you to talk about these issues at work, and do small things to help foster an environment where they can be more openly addressed. This book is inevitably combative: because it's all the problems, in one place, and right now, these problems aren't discussed enough in public. But we also have a great deal in common.

Professional bodies
The medical associations, the faculties, and the societies have failed us. Almost none have had the clarity of thought to stand up and declare that withholding trial data is unethical, immoral and grounds for expulsion of its members. If these organisations have integrity, they will fix this; alternatively, if the senior

members truly think that withholding trial data is fine and acceptable, then they should come out and say so, clearly, to their members and to patients. I'm not the first to raise this, so I have low expectations. If you are a member of one of these organisations, you could write and ask why it does not have a policy of chastising and expelling people who harm patients by withholding trial data. Please, send me their responses, or better still, simply post them online.

Funders

Money is short, outside of industry-funded research. Organisations like the National Institute for Health Research already do a great job, funding trials on important questions that drug companies won't want to pay for: it assesses the benefits of older drugs, for example, or of treatments that don't involve commercial products (I should say that I sit on one of these funding committees). But I believe that public funders should have two further priorities. First, there are some small outstanding evidence gaps in our knowledge of how industry distorts medical prescribing practice, and this is not a field that the pharmaceutical industry is likely to fund. More than that, as I argued at the beginning of the marketing section: the great horizon for evidence-based medicine is finding new methods and tools for disseminating the evidence we already have, to doctors and other decision-makers. This will require innovative collaboration between coders, pharmacists, librarians, doctors and academics. At the simplest level, I'm always frustrated that I can't press a button, when reading a systematic review, to say 'notify me when this summary is updated with new trial results'. In more elaborate terms, it's clear that we should be moving medical knowledge and trial results into structured databases; and perhaps embedding high-quality contextual advice into a doctor's routine workflow, as they sit at the computer in their office.

Academics and nerds

This book is filled with exciting untapped areas. There are the big jobs, on organising medical knowledge, from the section above, but also an avalanche of smaller studies, many of which could be done as undergraduate dissertations. Do a fact-checking audit of claims from drug reps; gather quantitative evidence on industry sponsorship in your medical school; find out about your university's policies on hiding trial data (or anything else); and then collaborate with other medical schools, to gather comparative national data. Share your ideas, and publish your results: we are all cheering you on.

AFTERWORD: WHAT HAPPENED NEXT?

A lot has happened in the year since *Bad Pharma* was first published. On the issue of missing trials, there is the beginning of a popular movement that could deliver real change, with your help, as we will see. But first, it may be useful to look at how the book was received. While colourful idiocy makes for fun stories, most reactions have been very sensible (with notable exceptions). *Bad Pharma* is not an easy book to read if you work in the pharmaceutical industry, or even in medicine; this combative thoughtfulness has been truly encouraging.

From medics and academics – in letters, emails, conversations and at public events – the response has been positive and pragmatic, with a gung-ho approach to fixing the flaws. Those with concerns, for the most part, had few serious objections to the factual content. It would be graceless to address individuals in a book, where they can't answer back, but I can give you the flavour of their concerns.

Some felt it was wrong to frighten patients by telling them about the problems in medicine. This, I think, is a matter of culture. There are two strands around 'public engagement' activity for science. One strand involves telling everyone how fantastic science is, with colourful images of molecules and the galaxy, exciting explosions, and so on. The other is more collaborative: discussing the problems, the shortcomings and challenges, so that an informed public can help us overcome barriers to change. I put myself firmly in the latter category, mainly because – after meeting a lot of them as a doctor – you know that the public aren't stupid. This book is a clear description, for a lay audience, of shortcomings in medicine that have been widely discussed and documented in academic journals for years. I believe we have a duty to share the problems in science, especially when we have failed to fix these problems ourselves; and I believe we need the public's help to deliver change.

Some objected to the book's tone, in part. One extremely senior British doctor, whom I shall not name, asked if it was really necessary to use 'inflammatory' terms like 'avoidable suffering and death' when the correct words are 'morbidity and mortality'. This one comment has stayed with me, because it speaks volumes about why the fixable problems described in these pages have remained unaddressed for so long: we have allowed the abstract world of effect sizes, diagrams and primary outcomes to be walled off from the real world of suffering, bereavement and death.

This should come as no surprise. Like many doctors, I was frankly traumatized by some of the experiences I had early on in my career. When you lean over a patient in an emergency room, trying to bring a dead body back to life, you are entirely focused on the job at hand. On the other side of a thin curtain, you can hear that person's husband or wife howling and wailing, knowing that the person they loved and lived with for fifty years is dying, begging the staff to do all they can, phoning their children, struggling to speak through tears to form the words and communicate the horror, telling them to *come*, quickly. I have memories from cubicles that I will never be able to deal with, and they upset me even now. But those moments are all about one single death.

We have to learn, in medicine, to block out these feelings so we can get the job done. With research, we have the opposite challenge: we need to force ourselves to give the numbers the emotional content they deserve. This isn't just true for the problems set out in this book, it's also true for teaching evidence-based medicine. Go back to page 15 and look at the diagram used as the logo for the Cochrane Collaboration. That abstraction speaks to a horror: parents whose babies struggled to breathe and then died, with all the suffering and horror that entails: the bereavement, the scarred relationships, and the pain. Nobody who is taught that graph in medical school experiences one hundredth of the emotional impact from it that they do from telling one mother that her child is dead, yet that graph is a representation of exactly that suffering, on a grand scale.

This is why, when I talk about avoidable 'mortality' and 'morbidity' – as a result of shortcomings in our abstract systems for gathering, synthesizing, disseminating and implementing evidence – I will always use the better words 'suffering' and 'death'. Because only they have even a small chance of nudging us towards feeling the feelings we ought to feel. That is how we will overcome complacency and drive improvement.

Another comment came predominantly from doctors working in the pharmaceutical industry. We should be clear, once again, that these are largely good and thoughtful individuals. They felt wounded, arguing that the book gives too little credit to the people who have tried and failed to fix these problems, especially in the UK, over the past thirty years. On this, I can't fight the reflex to point to the text: pages 48 to 51 cover some of these failed initiatives, enough for a lay reader, who will soon move on to something more cheerful. The successful passage of primary legislation in the US requiring all results to be posted on clinicaltrials.gov within a year of completion – and the ensuing failure to implement or audit that law – is far more significant than the admirable but failed efforts of some people in the UK in the 1990s. I hope that one day they publish a history of their struggle: I will read it.

Less-thoughtful people from the industry sometimes said: 'How can you say these problems harm patients when drugs must always be better than placebo if they are to get onto the market?' This betrays a stunning lack of ambition, but explains, I think, how people have been able to justify promoting mediocre drugs. Let's imagine there are two treatments: both cost the same amount, but one saves six lives for every hundred people taking it, and the other saves eight lives. Which treatment do you want? They're both better than nothing. From a bird's-eye view of the problems, I think that treatments which actively do more harm than good are pretty rare. But again, that's a very low ambition. We are commonly bounced into giving people the second-best treatment in a class through a combination of badly designed trials, trials that are never done, missing results, and biased dissemination of evidence through aggressive marketing. Where this happens, patients are harmed. They're not worse off than if they had been born in the Stone Age – of course they're not. They're just worse off than they need to be. If there are eight people tied to a railway track, with a very slow shunter crushing them one by one, and I only untie the first six before stopping and awarding myself a point, you would rightly think that I had harmed two people. Medicine is no different.

The same people sometimes felt I didn't say often enough that the industry has made some good drugs. I'll say it again: the industry has made some good drugs.

One extremely senior doctor said, on the topic of missing trials (and I am quoting verbatim): 'Look here, *old boy*, we were working well to fix these problems behind closed doors. Was it really necessary to make such a fuss and frighten the public?' We were standing, at the time, in an oak-paneled room. Once again, I have to say: we were not fixing this problem behind closed doors. We were not. We needed the help of the public – of lawyers, civil servants, campaigners, politicians, journalists, and more. We are now starting to get that help, and we should be glad of it, because we failed to prioritise this problem, and we failed to fix it.

On a related note, the Ethical Standards in Health and Life Sciences Group, co-chaired by the ABPI and the Royal College of Physicians, produced documents which claim (quite falsely) that trial results are made available, and so on. An impressive campaign was led by medical students at badguidelines.org, and the response has been mixed, to say the least: the BMA and the *Lancet* have removed their names, which is reassuring, but others have not, and the two documents have remained online, now marked as 'under review', without action, for half a year. All the organizations that signed these documents were approached, and almost all have declined to explain how or why they came to sign up to a series of false statements. I have been clear, throughout this book, that doctors collaborating with the pharmaceutical industry is a good thing (at the very least, you have to agree that it is inevitable) but there are risks to be managed: there is a clear need for good, evidence-based guidelines, and it is a shame that the most eminent organisations in medicine and academia are still failing to deliver. On an upbeat note, the BMA passed a formal motion at its annual conference this year, declaring unambiguously that non-publication of trial results is research misconduct. I hope others will follow.

Some from industry took offence at the title, though they largely relaxed when they read the book. The pharma industry does hurt patients. It's not all it does – of course it's not – but avoidable harm is something we all need to talk and think about. The Francis report into poor care in British hospitals has shown the harm that can come when medics get caught up in denial and obfuscation. If I regret one thing, though, it is that the focus of *Bad Pharma* – of the title, at any rate – was too much on the industry. I don't say this because the title worried people who work in drug companies; I say it because it helped doctors, policymakers, universities and regulators to continue letting themselves off the hook. If I were publishing this book tomorrow, its subtitle would be 'How Medicine Fails to Do the Best for Patients'. My favourite working title before publication, greeted with derision by my publisher, was 'The Information Architecture of Medicine Has Several Interesting Flaws, Many of Which Inflict Avoidable Harm on Patients, but All of Which Are Amenable to Cost-Effective Change, Were There to Be Adequate Public and Political Will'. That remains my preferred option.

From industry PR people I expected a slick sop ('These are important problems that we are working to fix, together with stakeholders') and a competent kick into the long grass. I predicted in the Conclusion to the first edition of the book that the worst elements would respond in three ways: they would issue personal comments; they would point to mountains of legislation that has been shown to be ineffective, as if that were an answer; and they would say that the problems I have described are all in the past. The response from ABPI itself – the official representative body of the pharmaceutical industry in the UK – did not fall short.

I should cover the silliness briefly, but there are more interesting battle grounds than this. I've been shown written comments from Stephen Whitehead, the Chief Executive of the ABPI, that I choose not to reproduce here. They're unpleasant, and essentially try to frame me as some kind of fringe figure, who is not to be taken seriously: it's for you to decide whether he's correct, and whether his behaviour here is stylish or surprising.

In public, both Stephen Whitehead and Deepak Khanna – president of the ABPI – said things like 'during the development of Mr Goldacre's book, the ABPI did try on numerous occasions to make contact with Mr Goldacre, and were disappointed that he did not respond. You can call me 'mister' all you like, but I am open to discussion with anyone: my email address is on my website, in my Twitter profile, at the foot of five hundred newspaper articles, and in all my books, including this one. There can't be many more accessible doctors or writers in the world. My email address is ben@badscience.net. I've searched back ten years, and I can't find these numerous approaches anywhere (for total clarity: one generic email from one ABPI press officer, two years ago, to my *Guardian* email address, saying 'I just started here a couple of weeks ago and thought I'd introduce myself. I know you cover the subject sometimes, so if you're ever looking for stats/information/etc relating to the industry let me know.').

Equally baffling – but much more important – was the ABPI's claim, in a press release on the day of the book's publication, that the cases were all 'long-documented', 'historical' and 'long-addressed'. As a PR strategy, this was unwise, to say the least. Far from being historic, many of the issues discussed are core elements on the medical student curriculum: it would be impossible to pass an undergraduate epidemiology course without knowing about design flaws in clinical trials, or the shortcomings of observational data. The problem of withheld trial results is vast, and plainly ongoing. Suggesting that these cases are historic was also optimistic, because every month there are new examples of these ongoing problems. Medicine isn't perfect, so we document and discuss its imperfections constantly, in an effort to fix them. Follow this asterisk* and you will find a 1,200-word-long footnote: it is a brief walk through the academic literature from just

---

*In the *BMJ*, to take just three examples from the past few months, you could read an editorial on corporate crime in pharma;[2] or a long analysis piece by an eminent diabetes professor on whether several insulin studies were 'marketing' rather than research;[3] or a paper arguing that we perversely incentivize companies to develop new drugs with few new benefits;[4] and so on.

Alongside the analysis of these problems that one would routinely expect – all very much alive – there was also a steady drip of new findings, setting out just how and why these are all such huge problems.

March 2013 saw an excellent round-up of the harms inflicted on patients by our over-reliance on surrogate outcomes – rather than real-world outcomes like death – in clinical research (this you will remember from pp. 132–5).[5] The problem of trials comparing new drugs against hopeless existing treatments, in

the past few months, a small selection of the greatest hits from the newest research about these ongoing problems, which really are the bread and butter of academic discourse. I don't think the many ethical professionals in the pharmaceutical industry are well served by a representative body that is willing to imply that these cases are all in the past. I also don't believe that policy-makers –

---

order to get an impressive results, continues to be widely discussed, both by academics online[6] and in the academic literature.[7] This practice is important: it creates uninformative evidence, but it also exposes trial participants to unnecessary harm. One editorial, accompanying further research on the topic, was unambiguous: 'In fifty-six out of sixty-three trials involving patients with highly active rheumatoid arthritis, potentially helpful treatments were withheld from 9,224 out of 13,095 patients randomized to the control arms. Why? Because placebos or treatments known to be ineffective were used as controls.' The title of this editorial was unusually direct about the impact on patient care: 'Blood on Our Hands: Seeing the Evil in Inappropriate Comparators'. The ABPI's public response to this paper was to say that the industry is very well regulated.[8] Everyone else, mercifully, has continued to have an informed conversation about fixable flaws in the design and analysis of clinical trials.

The problem of drugs being tested in small, brief trials continues as before. A paper in *PloS Medicine* from March 2013 puts hard figures on this. From a representative sample of two hundred newly approved medicines, each was studied on average in 1,708 people before approval; and one in five drugs for long-term use failed to meet the guidelines' recommendations.[9] These guidelines, to be clear, are weak in themselves, requiring only that three hundred people should be followed up on the drug for more than six months: for effectiveness on long-term outcomes, this is plainly inadequate.

As well as tiny studies, patients also continue to suffer because of research that is simply never done: a paper published in August 2013, as this chapter went to press, explains that for half of all the recommendations in the current major cardiology guidelines, the level of evidence is only 'grade C'.[10] This is astoundingly poor for one of the biggest specialties in medicine.

In July 2013 the team running ClinicalTrials.gov wrote an influential editorial in *Annals of Internal Medicine* about the harm that arises from the ongoing problem of trials switching their primary outcomes.[11] A paper one month earlier looked at trials in surgery and found that half weren't registered before they began, while half of all primary outcomes were either switched, omitted or changed.[12]

Notably, none of these papers received significant mainstream media attention, and that illustrates one clear reason why such issues remain who need to talk to industry – will feel enthusiastic about an industry body that is willing to gloss over live problems. But finally, as explained throughout this book, even if all this shoddy practice had miraculously ceased in August 2012, then we would still need to know about it. This is not for some kind of moral reckoning: it's because the medicine of today is practised on the

---

unresolved. Huge fuss and effort are deployed on writing newspaper stories about any single 'rogue surgeon', but there can be no doubt that our tolerance of poor data – which quietly undermines the quality of clinical decision-making throughout the world – permits far more avoidable suffering and death than any one individual could ever match. Discussion of this problem gets little air-time, because it requires background knowledge about how evidence is created and used; eliciting any outrage around that subject takes more time and imagination than 'bad surgeon hurts patient'.

Research into biased dissemination of evidence through marketing has also continued, as one would expect it to, regardless of implausible industry denials. One paper from May 2013 – it happens to be on my desk – found that 34 per cent of all industry advertisements in Sweden breached that country's voluntary code of practice.[13]

In April 2013 a paper was published in the *Journal of General Internal Medicine* reporting on a survey of 255 doctors, in four different countries, describing 1,692 visits from drug reps.[14] Less than 2 per cent of visits gave 'minimally adequate safety information', and at only 6 per cent of visits were serious adverse events mentioned, even though 45 per cent of promotions were for drugs with 'black box' warnings. In the same month, President Barack Obama was attacked for his 'Academic Detailing' initiative in the USA. This is a new project, designed to run in parallel to the system of 'drug reps' (called 'detailing' in the US): clinician consultants visit doctors, nurses and pharmacists, disseminating unbiased summaries of evidence about which treatments work best. Pfizer CEO Ian Read said – with no shred of irony – that the Academic Detailing project is a waste of public money, because government has a 'conflict of interest' when it disseminates evidence to clinicians.[15]

Stories and court cases from whistleblowers about kickbacks and aggressive marketing continue, similarly, at the usual pace.[16] A paper in May 2013 found that fewer than 6 per cent of biomedical journals have detection and response procedures in relation to ghostwriting.[17] If you read the accompanying editorial – although it's behind the usual paywall, which obstructs popular discussion of all science – you'll find all the arguments from this book.[18] That's not because anyone has copied anyone: it's because these issues are entirely mainstream, and routinely discussed behind the paywall.

evidence of yesterday, of the past: the past five years and the past five decades.

Once again, this is not simply about the pharma industry: it is about a culture of complacency in medicine, an inappropriate lack of urgency about solving the everyday problems we face, and an 'excuses culture' that perpetuates them. In the deluge of mail from academics that followed *Bad Pharma*'s publication, I received endless everyday tales that spoke to this issue. One academic forwarded the response he'd had from a journal

---

The *Journal of Medical Ethics* carried a fascinating description, in June 2013, of an industry-supported medical school lecture series on opioid-prescribing for pain management, with questionable content, undeclared conflicts of interest and more. The course was only stopped after individuals in the medical school protested.[19] Senior staff in medical institutions should remember that this kind of action, by junior staff, is risky, brave, and beneficial to patients.

June 2013 brought a remarkable article in *JAMA Internal Medicine* by a whistleblower who had himself engaged in 'disease-mongering', not just in the academic literature, but also directly to the public, writing in everyday supermarket consumer magazines:

'Low T' (low testosterone level, aka hypogonadism) is high-profile these days. Sales of testosterone replacement therapies (TRTs) for Low T have more than doubled since 2006 and are expected to triple to $5 billion by 2017, according to forecasts by Global Industry Analysts. Driving these sales is a sophisticated marketing effort to define low testosterone level as a disease for which the treatment is TRT. I know this because, as a professional medical writer, I have helped craft that message for transmission in a range of media to both physicians and consumers.

The challenges of regulating trials in the developing world continue to be discussed widely, though not so much in the UK. In July 2013, India's *Economic Times* (you've probably never heard of it, but it has a readership of 800,000) wrote: 'Many desperate and poor people in India are unwittingly taking part in clinical trials for drugs by Indian and multinational pharmaceutical companies that outsource the work to unregulated research organisations.'[20]

I could go on. For the ABPI to even suggest that this stuff is historical, and all long addressed, is absurd, but is also dangerous. We need to recognise that these problems exist so we can work – together or apart – to address them.

editor after submitting a paper documenting clear bad practice. The editor's comments betray a world-view that is holding medicine back from doing the best we can for patients. I reproduce one sentence with permission:

> There is nothing unusual about drug manufacturers producing one-sided promotional material, or about there being uncertainty about a medication's place in therapy because of a scarcity of adequate data, so the indignity expressed in the commentary seems out of proportion to the commonness of the problems.

Similarly, one reader forwarded a letter from his member of the European Parliament, who had previously been a clinician, in response to a letter about shortcomings in the evidence for prescribing decisions: 'The objections you raise do not endanger patient safety in any way, [they] only question clinical efficacy compared to existing similar medicines.' This reveals, once more, a dangerous lack of ambition in medicine: it is not enough that we prescribe medicines that are better than nothing; we need to ensure that we are prescribing the best treatment. Where we fail to do that, we expose patients to avoidable harm, as surely as if we had inflicted that harm recklessly and directly, through incompetence or malice.

The AllTrials campaign

There has, however, been some small movement on the key issue of missing trials. Some of this has been triggered by *Bad Pharma*, but it has acted only as a catalyst (if I'm honest, the outlandish denials from industry have also helped, in some quarters: people love a scrap, and a bad opponent can do a lot to galvanise busy people into action, or remind them that an issue is still alive). But most important, while the formal policy units of medical and academic professional bodies had largely neglected the issue of transparency, there was huge pent-up demand for more action in the wider community, going back for at least three decades.

So: last September, an MP and GP called Sarah Wollaston asked me in for a meeting, and then tabled a parliamentary question on missing trial results. These questions are given to the Minister in advance, so as is common, the Health Minister responded in advance, at 7 a.m., in a slot on the BBC Radio 4 *Today* programme. He explained that the problem simply doesn't exist: 'By law, companies must report any adverse reactions experienced in trials, as well as all results both positive and negative, and any abandoned or incomplete studies.' Five hours later, the formal response in Parliament more closely reflected reality:

> The government support transparency in publishing results of clinical trials, and they recognise that more can, and should, be done . . . Greater transparency can only serve to further public confidence in the safety of medicines. My Hon. Friend raises absolutely legitimate concerns, which have been raised by others, including Ben Goldacre.

This turnaround is a microcosm of the battle of missing trial results: outright denials and reassurance before a slow erosion to more serious engagement. The Minister agreed to meet a group of us, so we went along: Sarah Wollaston MP, Fiona Godlee (editor of the *BMJ*), Iain Chalmers (co-founder of the Cochrane Collaboration), Carl Heneghan (director of the Centre for Evidence Based Medicine in Oxford), and myself. The meeting was positive, and we organised a letter to *The Times*, signed by the great and good of British medicine, from the president of the Royal College of GPs to the editor of the *Lancet*. You might think these letters are trivial or pompous, but this one was helpful for two reasons: first, it was a line in the sand against widespread denialism, and called for clear action over an ongoing problem; second, it showed a broader front for the issue, which makes it harder to smear the discussion away.

Coverage picked up in all UK newspapers, from the *Mail* and the *Telegraph* to the *Independent* and others. The House of Commons Science and Technology Select Committee announced an inquiry into the issue (citing this book in its announcement, and also, I'm pleased to say, covering the bureaucratic barriers to getting trials done, which is equally important, as you will know from Chapter 5). This is yet to report its findings, and it is always possible that it might end up holding progress back. However, one important outcome has already been delivered: both MRC and NIHR, major UK academic research funders, have conducted internal audits on their rate of publication – and found that they publish 89 per cent and just under 100 per cent of all trials, respectively. Regardless of the specific numbers, since academic trials, just like industry trials, have been racked by publication bias – and clear public audit is the only way forward – the very existence of these figures is a small victory.

Next, a group of MPs (including David Davis, Sarah Wollaston and Julian Huppert) wrote to the hugely influential Public Accounts Committee, and as a result the National Audit Office is now conducting a formal inquiry into Tamiflu. The Health Select Committee had clear words for NICE and the pharmaceutical industry on ensuring that all results are made available. There were further questions in Parliament, including a Prime Minister's Question: this was answered with a blasé promise that everything was about to be fixed in Europe. At that point, it became clear that we needed a proper campaign.

The author Simon Singh is shy, and doesn't want people to know that he had £5,000 to spare, so this is our secret: he very generously gave us the money we needed to cover the costs of an intern at Sense About Science for a few months. Simon and I had both worked with them before, on quackery, bad science reporting, and the incredibly successful Libel Reform campaign. The experience at Sense About Science of lobbying, drawing together interested parties and running a public campaign was exactly what our small group of academics and doctors needed. The AllTrials campaign launched in January 2013, calling for three things: all trials should be registered, all results should be reported, and clinical study reports should be made publicly available. The campaign's central group includes the *BMJ*, Cochrane, the Centre for Evidence Based Medicine, patient advocates, and the open-access publisher PLoS. Within a few months, from nothing, the AllTrials campaign has grown to represent the mainstream, normative position in the UK. We now have over 50,000 individual supporters, and more than two hundred organisations have also signed up, including the Medical Research Council, NICE, the Wellcome Trust, and effectively all the major medical and academic faculties, societies, Royal Colleges and so on. There is also extensive international support, from organisations as widespread as the National Physicians' Alliance in the US to the European Public Health Association. Among the earliest signatories were patient groups, which now number well over a hundred, with eighty signing up on one day alone. This is key, because patients are showing leadership on this issue, helping professionals to find their way. The National Rheumatoid Arthritis Society – which came in for some criticism on page 270 – was one of our earliest supporters, and I am hugely impressed by what I have seen in this whole community (NRAS in particular have now shown real leadership on increasing transparency about their finances, as well as for clinical trials).

Then, in March, GlaxoSmithKline signed up to the AllTrials campaign. That is hugely significant: like most other pharmaceutical companies, GSK has received enormous regulatory fines over the past few years, and you will remember some of the stories involving it from earlier in this book. In October 2012 it had already promised to share some individual patient data with individual research groups, behind closed doors, after review by a panel, for trials going back to 2007. By signing up to AllTrials in 2013, GSK has committed to something different: it will make all clinical study reports publicly available within a year of completion of the trial; and it will apply this retrospectively, going back to the beginning of the company. There are several things that make me believe GSK is sincere, rather than merely making a quick PR move. First, in a meeting with chief executive Andrew Witty the week before GSK signed up, he was combative, with specific questions about what information we wanted in the public domain, and why. From the moment the policy was announced, it was clear that GSK had thought through the specifics, such as how to prioritise trial reports for release and how to redact genuinely confidential patient information. Last, the GSK team delivering this commitment has come to AllTrials meetings, presented on progress, and given timelines that are in advance of the suggested minimums. We can judge them by results, next year, but that is my risk assessment so far.

GSK has done something far more than simply promising to share specific pieces of information on its own drugs, though: it has shown leadership and formally recognised that transparency matters for patient care at a time when others in the industry are hoping the issue can be dodged. It has shown that delivering on transparency is possible, and has undermined other companies' claims to the contrary. In fact, GSK committed to share all clinical study reports in the very same week that PhRMA, the US industry body, put out a colourful press release saying that transparency was expensive, impractical, dangerous and impossible.

If AllTrials has had an impact, it is because it has encouraged organisations and companies to make a clear commitment on transparency, and declare where they stand. It has also provided a focus for lobbying activity and a venue for discussions.

Crucially, it has also allowed individuals to feel safe about speaking out. There is one comment we heard from policy staff in certain signatory organisations that I found chilling: we knew that withholding trial data was common, they said, and we knew that it harms patients, but we felt embarrassed to talk about it, because even raising the issue seemed somehow subversive. This is a terrifying state of affairs, and a testament to the power of aggressive lobbying by the industry. It is also, perhaps, a testament to the capture of key opinion leaders, and to the dangers of longstanding inaction at senior levels in the medical establishment. In the UK, we have seen the same phenomena during prominent enquiries into failing hospitals: many senior staff, in numerous organisations, saw a problem, but most were too busy – or too anxious about workplace conflict – to put patients first.

Europe, and beyond

It was crucial that the campaign grew when it did, because there were several events happening at the same time which needed to be harnessed and influenced. Here the story becomes so detailed that I can only give an outline: I strongly encourage you, if you are interested, to visit www.alltrials.net and follow it through.

First, the European Union Clinical Trials Regulation was already up for revision, with a Labour member of the European Parliament, Glenis Willmott, as the 'lead rapporteur' (who guides the legislation through the European Parliament). We were able to work with her at a time when she was under phenomenal pressure from hundreds of industry lobbyists in Brussels. She has been excellent, making a clear public commitment to transparency and introducing several crucial amendments to a deeply flawed piece of legislation. There is the prospect of ensuring that all trials conducted within a European country after 2014 will report results within a year, with a clear audit trail of changes to the protocol, and so on. This is far less than patients need, since it leaves all trials from before 2014 (and all those conducted outside Europe) untouched, but it is at least some small progress. The battle, however, is not yet won, and the key votes will come over the next year, with the second reading in October.

Through AllTrials there have been various meetings in the UK on how to practically deliver transparency. There has also been a series of other meetings, hosted by people like Wellcome, and even the ABPI, although these have often acted more as a notice-board for opponents of transparency to publicise their objections. Even here there has been progress, though. Roche has announced a policy to share more clinical study reports, and sent some new information to Cochrane that is currently being assessed (though we must remember its previous broken promises – see pp. 82–91). At the recent All Party Parliamentary Group meeting, when the Roche representative agreed that 'transparency is a must', everybody nodded enthusiastically. Julian Huppert MP (a rare scientist in Parliament) said: 'Myself and Sarah Wollaston MP are smiling, because there is a stark difference between that accord and the ABPI meeting we went to, where the discussion was all about how impossible this is.' Is it now fashionable to say you support more transparency? The answer may be yes: this is why we must judge people on what they deliver.

Some regulators have moved firmly in the right direction. In May 2013 the UK Health Research Authority sent out a ground-breaking set of new proposals to improve transparency around trials conducted in the UK, with a range of impressive and solid practical suggestions (as I said, rather cheesily, in the HRA press release). In July its action plan was announced.

Meanwhile, in November 2012 the European Medicines Agency made a remarkable announcement: it is planning to share individual patient data, for all the trials it is given, starting from 2014. There is no doubt about whether the EMA will do this, it explained; the only thing to be discussed is how. The outstanding questions are complex. Sharing individual patient data is fairly common, especially within academia, but unlike clinical study reports, where personal data can be easily redacted, individual patient data poses a genuine risk of reidentification. It is also not clear whether this rule will be imposed retrospectively, or only for trials completing (or maybe even starting) after 2014.

This has created an outpouring of activity from the industry: the changes have even been raised by PhRMA lobbyists with the US team working on EU/US trade treaty negotiations. At the launch of the new EMA policy, I sat on a discussion panel alongside Peter Gøtzsche of Cochrane and Ginny Barbour from PLoS, up against two representatives from EFPIA, the industry body. EFPIA argued that individual patient data should — at most — only be shared with people who can prove that they have a serious academic question. Crucially, they said that everyone wanting access to data sets should be compelled to disclose their full protocol and statistical analysis plan beforehand. This is a sensible demand to make, in some respects, because where people don't post their protocol and analysis plan, there is room for them to go on long statistical 'fishing trips', changing their analysis until it provides an answer that suits their wishes or preconceptions (as you now know, from reading this book). That is why many of us have asked that everybody running a trial should themselves post their full protocols and statistical analysis plans (and a log of all changes) in public. We need the people running trials to post their analysis plans for their own data, publicly, for all the same reasons that we want outsiders working on other people's data to do so.

This suggestion was not met with enthusiasm. I was struck most by an industry lawyer, who spoke without a shred of self-doubt or subsequent regret. Look, he said: we pay for the trial to be done; it should be our choice how that data is used. There was no sharp intake of breath, no sense of astonishment, and I felt very strongly that all of us in that room, in our suits, had become acculturated to accepting some truly remarkable positions. This man was charming and friendly, with no hint of bad intentions. I don't believe for one moment that he could sit down with a patient at breakfast and tell them that they had no right to access the most complete possible information about the tablet they were about to swallow. He is, in my view, both the product and a beneficiary of the complexity around this whole area: he was making his case in a room full of people who were no longer shocked by his position, with everyday patients several hours' reading away from being able to understand its implications. In any case, since then there have been a series of advisory groups on how this should work (with many of us from AllTrials – and fellow travellers – involved, as well as many from industry); in September 2013 the EMA's draft proposal will go out for consultation, and by November it will be finalised.

All has not been roses in Europe, however. You will remember (pp. 72–9) that the EMA was forced to share clinical study reports after the EU Ombudsman made a ruling of maladministration against the agency. This policy has, in fact, been a quiet success, and over 1.9 million pages of documents were shared up until May 2013, when the programme was forced to a dramatic halt. Many of the applicants for these documents were from industry, and many in industry took objection to this. In my view, companies seeing one another's trials is good for transparency: if anyone is likely to spot a hole in your claims, it is a competitor, and this is a common phenomenon in complex regulatory areas. Many complaints to the medicines marketing regulator (or rather, self-regulator) come from rival drug companies' sales reps; and even with the Advertising Standards Authority it's common to find that the complaint about a technically inaccurate claim from one paint shop was made, in triumph, by the paint shop down the road.

In any case, two companies, InterMune and AbbVie, sued the EMA, demanding that it stop releasing clinical study reports on their drugs. An interim ruling has now been issued by the General Court of the European Union forcing the EMA to stop. Although the agency initially promised to be combative and encouraged people to continue applying, the legal advice we received at AllTrials was clear: it would find it impossible to release any further information with this court order hanging over them. Within a month, the first Cochrane request for a clinical study was rejected. It is now clear that this is one of the most significant backward steps that has ever been seen in the battle for transparency: where briefly CSRs were available, they are now once again kept under lock and key. This is extremely serious, since we now know – from the work done on Tamiflu, but also other work published just this year on a series of gabapentin trials[21] — that brief trial reports in academic journals can be incomplete or misleading about the methodological shortcomings in a study's design. Wealthy industries are fond of saying, with huge, expensive court cases like these, that they simply want clarity on the law. It will take several years for the full hearing to pass through the court system, and during that time, medicine will be gravely held back.

Despite its pivotal role in testament to the complexity of this area, the AbbVie and InterMune case has received almost no mainstream media coverage. Both PhRMA and EFPIA supported the two companies in their court action, while at the same time proclaiming that they are in favour of transparency. Similarly, while professing to support transparency in principle, the ABPI has campaigned hard against AllTrials, calling it, formally and on the record, a 'publicity stunt' (a remarkable response to a group supported by over a hundred patient groups, representing over a million people – the very people who participate in trials).

Then, in July 2013, an extraordinary document was leaked: it was an internal PhRMA and EFPIA memo, sent to leaders of major pharma companies around the world, setting out their strategy on transparency.[22] Under the heading 'Advocacy', it discussed a plan for 'mobilizing patient groups to express concern about the risk to public health by non-scientific re-use of data'. The notion of using patient groups as an advocacy tool against transparency is one that many people will find abhorrent. Many patient groups – including the European Patients' Forum, the European Cancer Patient Coalition and a wide coalition of AIDS charities – were appalled by this notion, and sent out clear statements confirming their commitment to information being made available.[23] It will be interesting to see which patient groups – if any – raise objections to greater transparency over the coming year.

During all this time, evidence on the scale of the problem has continued to mount. The most recent paper, from July 2013, tracked 8,907 trials registered on clinicaltrials.gov over three years: for over half of them they could find no evidence of any results.[24] This is a huge piece of work, and one of the biggest studies yet. Another group followed hundreds of cancer trials registered at clinicaltrials.gov and found that only one in five had results available at twelve months, and only half at three years. A *BMJ* paper surveyed what happens when Cochrane authors try to get unpublished data from trialists: half had their requests rejected.[25] Manufacturers were the least helpful, with requests resulting in longer delays, greater difficulties, more frequent follow-up requests, and ultimately, less data. In October 2012 IQWiG – the German equivalent of NICE – published a paper describing how the manufacturer of Exubera (an inhaled insulin) refused to share clinical study reports.[26] In the very same week, a *New England Journal of Medicine* editorial explained: 'Substantial instances of missing data are a serious problem that undermines the scientific credibility of causal conclusions from clinical trials.'[27] The evidence on issues upstream has also continued to flow out. In April *PLoS One* carried a paper showing that only a third of academic journal reviewers bother to check whether a trial is properly registered when considering it for publication.[28]

At the same time, the industry has tried desperately to undermine these findings. The data on FDA Amendment Act compliance for posting results at clinicaltrials.gov (only 22 per cent managed to do so) has been challenged, and in bizarre terms that make the need for simple, routine public audit by the FDA even clearer. The sticking point is that companies can ask for an extension, or argue that their specific trial should be exempt from reporting results at all. This discussion is conducted in secret, and the industry has lobbied against having adjudications in public on clinicaltrials.gov, so nobody can know for sure why a trial hasn't reported. This broken system has inevitably created chaos and contradiction where there should be simple clarity. In one interview, the FDA told *Nature* that the 22 per cent figure is wrong, and that it is only pursuing fifteen cases of overdue results: a tiny proportion. *Nature* then asked the US government research agency NIH for an informal audit, which found 50 per cent compliance for the industry (and even worse for NIH: it found that it reported only one in five of its own trials). Meanwhile, some internal FDA slides being circulated among industry people give the compliance for that period as 65 per cent. The industry brandishes this triumphantly, but a figure of 65 per cent is still abysmal, and furthermore, it comes from a secret audit that has never been published or publicly discussed, available only as some slides, with unknown methods and no data sharing, from the same organisation that has also claimed that a mere fifteen trials in total are breaching. Alongside this mess, the Prayle audit, published in the *BMJ*, is fully explicit about the methods it used, and more than that, its entire data set – huge files listing every single trial it looked at, and every individual adjudication it made on compliance – is freely downloadable for anyone who wishes to review it.[29] While all studies may have flaws, this is the only audit whose flaws we can see for ourselves and discuss.

But it has not all been denial and gloom. Some companies have chosen a much more impressive path, alongside GSK, and this is where there is hope. Roche has said that it will look at sharing clinical study reports, though the mechanism is not yet clear and it has felt unable to join GSK in actively campaigning for transparency throughout the industry by signing up to AllTrials. I hope it will change its mind, and I hope you will encourage it to do so.

Others have gone much further. The US company MedTronic has shared individual patient data – on its treatment to improve outcomes in spinal surgery – with a group of researchers from Yale, and very nobly, because the results show that the product has only very limited benefits.[30] Was this an act of corporate suicide? I'm not so sure. I've always been in favour of one simple, clear regulation, across the board, requiring all companies to share all results and to compete on a level playing field of high ethical standards. I still believe this is the optimal outcome, but now I can also see that diversity in the market, with respect to transparency, could actively create a financial incentive for companies to do the right thing. It's easy to see why, if you follow through one example.

Let's imagine there are two treatments for early dementia, made by two different companies, and both have identical benefits. This is the last bit of maths in the book. Let's say, for example, that your chances of staying well, without deterioration, are exactly 35 per cent higher if you're on either of these two drugs than if you get a placebo pill. But next, let's say that one of these drugs is made by a company with a proven track record of transparency: it doesn't just say the right things, it has actually delivered full clinical study reports on all the trials it has done (we're imagining this is in 2016, or later); and it has also made good on promises to share data with outside researchers. The other drug, meanwhile, is made by a company whose staff actively mock and deride the very notion that you should even *dare* to ask it to share all trial results; it participates in industry dirty tricks to rubbish such calls; it sues and lobbies regulators and governments to prevent it from sharing trial information; and so on.

In that situation, which drug do you want to take? Remember, both have numerically identical benefits: both increase your chances of staying disease-free by exactly 35 per cent. But are those figures really identical? We know there is a background rate of missing trials, and we might reasonably assume that this applies to all trials, across the board, regardless of where they come from. We know that overall, when results go missing, that tends to exaggerate the benefits of treatments. We cannot possibly know for sure that the specific companies lobbying, denigrating, suing and campaigning have themselves hidden any unflattering information. But overall, if we were playing safe and all other things were equal, would you really want to take a drug manufactured by such a company in preference to a drug for which there were no such concerns?

I know what I would want.

But really, this can only be fixed by you. After years of appallingly slow progress on missing trials, wider public attention has begun to move things forward. The peculiar smears unloaded onto people daring to share this problem with the public are clear proof, to me, that public engagement is the most powerful approach. The phenomenal progress with patient groups has been truly heartening. But only a united front and a large public campaign can put this issue on the agenda for politicians and fix it for good. So please sign up to the AllTrials campaign. If you are a member of a professional body, or a patient group, or a university, or a company, ask them to do the same. If they have a shred of doubt, ask them to get in touch, through alltrials.net, and we will talk with them.

Afterword: What Happened Next?

There is a huge amount happening right now: enquiries in multiple countries, European legislation, a groundswell of popular opinion, a barrage of new research findings, cultural shifts in regulators, isolated individuals in industry showing leadership, and more. But this has all happened before, without delivering success. The window will only open briefly, and there are many, many people trying to wrestle it shut. For the good of patients, for the reputation of medicine, and for the reputation of the pharmaceutical industry, we have one new chance, right now, to deliver real change. I hope you can help us to take it.

GLOSSARY

ABPI – The Association of the British Pharmaceutical Industry.

ACCME – The Accreditation Council for Continuing Medical Education.

BNF – British National Formulary.

CME – Continuing medical education.

CSR – Clinical study report.

EMA – The European Medicines Agency, the European drugs regulator.

FDA – The Food and Drug Administration, the US drugs regulator.

GCP – Good Clinical Practice guidelines.

Generic drug – When a drug is first invented, it is owned by the company, and no one else is allowed to manufacture and sell it. A drug usually comes off patent eighteen years after it was first registered, and about ten years after it comes onto the market. After that, any company can make a copy of it, as with very old drugs such as paracetamol or aspirin. When this happens, you can't make much profit out of selling it.

GMC – General Medical Council.

ICH – International Conference on Harmonisation.

ICMJE – International Committee of Medical Journal Editors.

JAMA – Journal of the American Medical Association.

KOL – 'Key opinion leader'.

MHRA – Medicines and Healthcare products Regulatory Agency.

NEJM – New England Journal of Medicine.

Off-label – A drug is allowed onto the market by a regulator on the condition that it can only be marketed for use in a specific medical condition. Doctors can use the drug for other medical problems if they wish, but this use is 'off-label'.

Off patent – A drug that is out of patent (see generic drug).

PMCPA – Prescriptions Medicines Code of Practice Authority.

Publication bias – The phenomenon whereby trials with results regarded as unflattering or uninteresting are left unpublished.

SSRI – Selective serotonin reuptake inhibitor.

Surrogate outcome – Real-world problems are what we really want to change with medicines: things like death, heart attack or stroke. A surrogate outcome is something that is easier to measure, like blood pressure or cholesterol levels, that we hope is a proxy for those outcomes. Often they are not as good at predicting real-world outcomes as we think.

ACKNOWLEDGEMENTS, FURTHER READING AND A NOTE ON ERRORS

I have been taught, corrected, calibrated, cajoled, entertained, encouraged and informed by a very large number of people, including John King, Liz Parratt, Steve Rolles, Mark Pilkington, Shalinee Singh, Alex Lomas, Liam Smeeth, Josie Long, Ian Roberts, Tim Minchin, Ian Sample, Carl Heneghan, Richard Lehman, Dara Ó Briain, Paul Glasziou, Hilda Bastian, Simon Wessely, Cicely Marston, Archie Cochrane, William Lee, Brian Cox, Sreeram Ramagopalan, Hind Khalifeh, Martin McKee, Cory Doctorow, Evan Harris, Muir Gray, Amanda Burls, Rob Manuel, Tobias Sargent, Anna Powell-Smith, Tjeerd van Staa, Robin Ince, Roddy Mansfield, Rami Tzabar, Phil Baker, George Davey-Smith, David Pescovitz, Charlotte Wattebot-O'Brien, Patrick Matthews, Giles Wakely, Claire Gerada, Andy Lewis, Suzie Whitwell, Harry Metcalfe, Gimpy, David Colquhoun, Louise Burton, Simon Singh, Vaughan Bell, Richard Peto, Louise Crow, Julian Peto, Nick Mailer, Rob Aldridge, Milly Marston, Tom Steinberg, Mike Jay, Amber Marks, Reg, Mum, Dad, Josh, Raph, Allie, and Lou. I'm hugely indebted to the late Pat Kavanagh, Rosemary Scoular, Lara Hughes-Young, Zoe Ross, and especially Sarah Ballard, who is amazing. Robert Lacey has copy-edited my last two books, he is gentle. Louise Haines has been mighty.

There are tools that made this book better, especially Zotero, Scrivener, Evernote, ReaditLater, IntervalTimer and Repligo. AntiSocial is a piece of software that irreversibly disables Twitter and Gmail on your computer when you're working: I highly recommend it. In recent years I've had day jobs supported by the National Institute for Health Research, the Scott Trust, the Wellcome Trust, Nuffield College, Oxford, and the NHS, and a bursary from the Oxford University Business Economics Programme. Readers are hugely important, and various people got in touch with thoughts on the text after publication, helping me to understand where things weren't clear, spot typos, and shape small changes, for which I'm hugely grateful.

It's a common joke in evidence-based medicine that whenever you think you've had an idea, Iain Chalmers has probably written it up in an essay fifteen years earlier. He helped to formulate many of the key ideas in evidence-based medicine, and to spot the problems, and I hope I've attributed enough to him. There are also many other academics whose work comes up repeatedly: some of them I've met, most of them I've not, but where you see recurring names in the references, it's fair to say that we all stand on their shoulders. There are huge rewards in medicine, but they're not reliably in the right places. Many of the people whose work is cited in this book have taken a personal hit, on income and eminence, to work on the serious systemic problems in medicine. They are quiet heroes. It's an honour to spread their work more widely.

There are many excellent review papers on the issues raised in this book, and I have highlighted them in the references wherever possible. I have specifically tried to seek out papers that are freely accessible (look for the references to a journal called *PLoS*, in particular), though some sadly are still behind academic journal paywalls.

There are also many excellent books that have covered some

of the issues around bad behaviour by the pharmaceutical industry, though all are US-focused, many are now almost a decade old, and none have focused on missing data. If you are keen to read more on any specific area, there are several books that have influenced my thinking over the years.

Jerome Kassirer was editor of the *NEJM*, and his *On the Take* (2004) is great on the marketing issues, and the way that continuing medical education in the US has been captured by industry. Marcia Angell was also *NEJM* editor, and her book *The Truth About Drug Companies* (2005) first brought the story of marketing, institutional corruption and bad evidence to a larger audience. Richard Smith is the previous editor of the *BMJ*, and his book *The Trouble With Medical Journals* (2006) explains itself. Ray Moynihan's various books on medicalisation are all excellent. Donald Light recently edited *The Risks of Prescription Drugs* (2010), which is a good collection on contemporary problems, especially the lack of innovation. Melody Petersen, previously of the *New York Times*, wrote *Our Daily Meds* (2008), which is excellent on marketing in the US. Daniel Carlat is a bioethicist, and his *White Coat Black Hat* (2007) is great on the ethical issues around drug testing. Tom Nesi's book on Vioxx is magnificent.

While criticising bad behaviour from industry is important, I'm also struck that the public have little opportunity to read about the basic techniques used to appraise new treatments, find out what works, and spot what harms. *Testing Treatments* (2006, second edition 2011) by Imogen Evans, Hazel Thornton, Iain Chalmers and Paul Glasziou remains the single go-to book on this topic in my view, published in several languages and also available free online at testingtreatments.org (I should mention that I wrote the foreword). *Powerful Medicines* (2005) by Jerry Avorn is the first attempt I've seen by a pharmacoepidemiologist to explain the science of side effects monitoring to the public. *How to Read a Paper* by Trisha Greenhalgh remains the medical-student bible on critically appraising academic papers, and it can be understood by anyone.

Finally, I have no doubt that there will be some mistakes in this book, whether they are small slips or misinterpretations. I should say that I've written it to illustrate underlying themes, not to rubbish any particular drug or company; and so I hope the criticism has been roughly evenly distributed, perhaps according to market share. I certainly don't think any one company is any better than any other. If you do find an error of fact, do let me know, and if it's a genuine error, I will happily correct the text in future editions. In the unlikely event that any example I've used is simply wrong, there will be another to take its place. If you like – if it's in your nature, and if that is how you'd like others to see you – you can point out mistakes with self-righteous fury. Or you can just point them out. I'll be cheerful in either case, but more than anything: I'm certain there is no error that will change the argument of the book, so your feedback will help to make the argument stronger.

On a related note, in the UK (especially) there is a vogue for large companies to sue writers over critical concerns they have raised, in the public interest, on matters of science and health. I've experienced success in libel cases, and have helped to drive a partially successful campaign to change the libel laws in Britain. Even where libel cases have been technically successful – though for clarity, none against me has ever been – these have often backfired on the litigant's reputation. There is a strong sense among the public that libel is used to dissuade people from raising legitimate concerns or to create anxiety, encouraging writers to censor themselves and steer away from anything critical. I mention this because, as I said, I've tried very hard to be accurate in this book.

If you genuinely feel that something in here is plainly untrue, I encourage you to drop me a note so we can look at your concerns and change the text, if appropriate, or clarify in future editions. I offer this freely, with no sense of either fear or threat: I just think it's how things should work. As I have repeatedly reiterated throughout this book, the problems it describes are systemic and widespread. The specific stories I have included are intended to illustrate points of methodology, and those points would only make sense if they were attached to real studies. I hope you'll view any story that relates to you in the spirit in which it is intended, and recognise the genuine concern and public interest in the issues raised, as well as the room for improvement in your industry.

NOTES

Chapter 1: Missing Data

1. Bourgeois FT, Murthy S, Mandl KD. Outcome Reporting Among Drug Trials Registered in ClinicalTrials.gov. *Annals of Internal Medicine* 2010;153(3):158–66.
2. Bero L, Oostvogel F, Bacchetti P, Lee K. Factors Associated with Findings of Published Trials of Drug–Drug Comparisons: Why Some Statins Appear More Efficacious than Others. *PLoS Med* 2007 Jun 5;4(6):e184.
3. Kelly RE Jr, Cohen LJ, Semple RJ, Bialer P, Lau A, Bodenheimer A, et al. Relationship between drug company funding and outcomes of clinical psychiatric research. *Psychol Med* 2006 Nov;36(11):1647–56.
4. Bekelman JE, Li Y, Gross CP. Scope and impact of financial conflicts of interest in biomedical research: a systematic review. *JAMA* 2003;289:454–65. Lexchin J, Bero LA, Djulbegovic B, Clark O. Pharmaceutical industry sponsorship and research outcome and quality: systematic review. *BMJ* 2003;326:1167–70.
5. Sismondo S. Pharmaceutical company funding and its consequences: A qualitative systematic review. *Contemporary Clinical Trials* 2008 Mar;29(2):109–13.
6. Eyding D, Lelgemann M, Grouven U, Harter M, Kromp M, Kaiser T, et al. Reboxetine for acute treatment of major depression: systematic review and meta-analysis of published and unpublished placebo and selective serotonin reuptake inhibitor controlled trials. *BMJ* 2010 Oct 12;341:c4737–c4737.

7. Suntharalingam G, Perry MR, Ward S, Brett SJ, Castello-Cortes A, Brunner MD, et al. Cytokine storm in a phase 1 trial of the anti-CD28 monoclonal antibody TGN1412. *NEJM* 2006 Sep 7;355(10):1018–28.

8. Expert Group on Phase One Clinical Trials: Final report [Internet]. 2006 [cited 2012 Apr 5]. Available from: http://www.dh.gov.uk/en/Publicationsandstatistics/Publications /PublicationsPolicyAndGuidance/DH_063117

9. Decullier E, Chan A-W, Chapuis F. Inadequate Dissemination of Phase I Trials: A Retrospective Cohort Study. *PLoS Med* 2009 Feb 17;6(2):e1000034.

10. Cowley AJ, Skene A, Stainer K, Hampton JR. The effect of lorcainide on arrhythmias and survival in patients with acute myocardial infarction: an example of publication bias. International journal of cardiology. 1993;40(2):161–6. Iain Chalmers was the first to raise TGN1412 and anti-arrhythmics as examples of the harm done when individual early trials are left unpublished. They are the best illustrations of this problem, but you should not imagine that they are unusual: the quantitative data shows that they are just two among many, many similar cases.

11. Antman EM, Lau J, Kupelnick B, Mosteller F, Chalmers TC. A comparison of results of meta-analyses of randomized control trials and recommendations of clinical experts. Treatments for myocardial infarction. *JAMA* 1992 Jul 8;268(2):240–8.

12. Turner EH, Matthews AM, Linardatos E, Tell RA, Rosenthal R. Selective Publication of Antidepressant Trials and Its Influence on Apparent Efficacy. NEJM 2008 Jan 17;358(3):252–60.

13. Here is the classic early paper arguing this point: Chalmers Iain. Underreporting Research Is Scientific Misconduct. *JAMA* 1990 Mar 9;263(10):1405–1408.

14. Sterling T. Publication decisions and their possible effects on inferences drawn from tests of significance – or vice versa. *Am Stat Assoc J* 1959;54:30–4.

15. Sterling TD, Rosenbaum WL, Weinkam JJ. Publication decisions revisited – the effect of the outcome of statistical tests on the decision to publish and vice-versa. *Am Stat* 1995;49:108–12.

16. Bacon F (1645). Franc Baconis de Verulamio/Summi Angliae Cancellarii/Novum organum scientiarum. [Francis Bacon of St. Albans Lord Chancellor of England. A 'New Instrument' for the sciences] Lugd. Bat: apud Adrianum Wiingaerde, et Franciscum Moiardum. Aphorism XLVI (p.45–46).

17. Fowler T (1786). Medical reports of the effects of arsenic in the cure of agues, remitting feveres and periodic headachs. London: J Johnson, pp. 105–107.

18. Hemminki E. Study of information submitted by drug companies to licensing authorities. *BMJ* 1980 Mar 22;280(6217):833–6.

19. Lee K, Bacchetti P, Sim I. Publication of clinical trials supporting successful new drug applications: a literature analysis. *PLoS Med* 2008;5(9):e191.

20. Melander H, Ahlqvist-Rastad J, Meijer G, Beermann B. Evidence b(i)ased medicine – selective reporting from studies sponsored by pharmaceutical industry: review of studies in new drug applications. *BMJ* 2003;326:1171–3.

21. Rising K, Bacchetti P, Bero L. Reporting Bias in Drug Trials Submitted to the Food and Drug Administration: Review of Publication and Presentation. *PLoS Med* 2008 Nov 25;5(11):e217.

22. Scherer RW, Langenberg P, von Elm E. Full publication of results initially presented in abstracts. *Cochrane Database Syst Rev* 2007; 2: MR000005.

23. Song F, Parekh S, Hooper L, Loke YK, Ryder J, Sutton AJ, et al. Dissemination and publication of research findings: an updated review of related biases. *Health Technol Assess* 2010 Feb;14(8):iii, ix–xi, 1–193.

24. Dickersin K. How important is publication bias? A synthesis of available data. *Aids Educ Prev* 1997;9(1 SA):15–21.

25. Ioannidis J. Effect of the statistical significance of results on the time to completion and publication of randomized efficacy trials. *JAMA* 1998;279:281–6.

26. Bardy AH. Bias in reporting clinical trials. *Brit J Clin Pharmaco* 1998;46:147–50.

27. Dwan K, Altman DG, Arnaiz JA, Bloom J, Chan AW, Cronin E, et al. Systematic review of the empirical evidence of study publication bias and outcome reporting bias. *PLoS ONE* 2008;3(8):e3081.

28. Decullier E, Lhéritier V, Chapuis F. Fate of biomedical research protocols and publication bias in France: retrospective cohort study. *BMJ* 2005;331:19. Decullier E, Chapuis F. Impact of funding on biomedical research: a retrospective cohort study. *BMC Public Health* 2006;6:165.

29. Cronin E, Sheldon T. Factors influencing the publication of health research. *Int J Technol Assess* 2004;20:351–5.

30. Song F, Parekh S, Hooper L, Loke YK, Ryder J, Sutton AJ, et al. Dissemination and publication of research findings: an updated review of related biases. *Health Technol Assess* 2010 Feb;14(8):iii, ix–xi, 1–193.

31. This was first pointed out to me by Jamie Heywood from PatientsLikeMe, who spent huge resources trying and failing to replicate research findings in another area of medicine. The last time I saw him we talked about writing up his idea that the likelihood of a claim being true is proportional to the cost of making it, and inversely proportional to the cost of refuting it. We've not done so, and until then, a description of our conversation is the only reference for this neat idea.

32. Begley CG, Ellis LM. Drug development: Raise standards for preclinical cancer research. *Nature* 2012 Mar 28;483(7391): 531–3.

33. Harrabin R et al (2003). *Health in the News*, The King's Fund, London, UK.

34. Forsyth, Alasdair J. M. 2001. Distorted? a quantitative exploration of drug fatality reports in the popular press. *International Journal of Drug Policy* 12, no. 5–6 (November 1): 435–453.

35. Dickersin K, Min YI, Meinert CL: Factors influencing publication of research results: follow-up of applications submitted to two institutional review boards. *JAMA* 1992, 267:374–378.

36. Olson CM, Rennie D, Cook D, Dickersin K, Flanagin A, Hogan JW, Zhu Q, Reiling J, Pace B: Publication bias in editorial decision making. *JAMA* 2002, 287:2825–2828.

37. Lee KP, Boyd EA, Holroyd-Leduc JM, Bacchetti P, Bero LA. Predictors of publication: characteristics of submitted manuscripts associated with acceptance at major biomedical journals. *Med J Aust* 2006;184:621–6. Lynch JR, Cunningham MRA, Warme WJ, Schaad DC, Wolf FM, Leopold SS.

Commercially funded and United States-based research is more likely to be published; good-quality studies with negative outcomes are not. *J Bone Joint Surg Am* 2007;89:1010–8. Okike K, Kocher MS, Mehlman CT, Heckman JD, Bhandari M. Publication bias in orthopaedic research: an analysis of scientific factors associated with publication in the *Journal of Bone and Joint Surgery. J Bone Joint Surg Am* 2008;90:595–601.

38. Epstein WM. Confirmation response bias among social work journals. *Sci Technol Hum Values* 1990;15:9–38.

39. Mahoney MJ. Publication prejudices: an experimental study of confirmatory bias in the peer review system. *Cognitive Ther Res* 1977;1:161–75.

40. Ernst E, Resch KL. Reviewer bias – a blinded experimental study. *J Lab Clin Med* 1994;124:178–82.

41. Abbot NE, Ernst E. Publication bias: direction of outcome less important than scientific quality. *Perfusion* 1998;11:182–4.

42. Emerson GB, Warme WJ, Wolf FM, Heckman JD, Brand RA, Leopold SS. Testing for the Presence of Positive-Outcome Bias in Peer Review: A Randomized Controlled Trial. *Arch Intern Med* 2010 Nov 22;170(21):1934–9.

43. Smith, R. *The Trouble with Medical Journals*. RSM Books, UK; 2006.

44. Weber EJ, Callaham ML, Wears RL, Barton C, Young G. Unpublished research from a medical specialty meeting: why investigators fail to publish. *JAMA* 1998;280:257–9.

45. Kupfersmid J, Fiala M. A survey of attitudes and behaviors of authors who publish in psychology and education journals. *Am Psychol* 1991;46:249–50.

46. Song F, Parekh S, Hooper L, Loke YK, Ryder J, Sutton AJ, et al. Dissemination and publication of research findings: an updated review of related biases. *Health Technol Assess* 2010 Feb;14(8):iii, ix–xi, 1–193.

47. Gøtzsche PC, Hróbjartsson A, Johansen HK, Haahr MT, Altman DG, Chan A-W: Constraints on publication rights in industry-initiated clinical trials. *JAMA* 2006, 295:1645–1646.

48. Gornall, J. 'Industry attack on academics.' *BMJ* 338, no. mar09 1 (March 9, 2009): b736–b736.

49. Ibid.

50. Steinbrook R. Gag clauses in clinical-trial agreements. *NEJM* 2005 May 26;352(21):2160–2.
51. Mello MM, Clarridge BR, Studdert DM. Academic medical centers' standards for clinical-trial agreements with industry. *NEJM* 2005;352(21):2202.
52. This is one of many stories for which I recommend delving into the horrible details, if you're interested. A good place to start here is Prof David Colquhoun's blog on the topic, with many links http://www.dcscience.net/?p=193 and this *BMJ* piece written by a lawyer, to keep the lawyers reading this book happy: Dyer C. Aubrey Blumsohn: Academic who took on industry. *BMJ* 2009 Dec 15;339(dec15 1):b5293–b5293.
53. Wendler D, Krohmal B, Emanuel EJ, Grady C, for the ESPRIT Group. Why Patients Continue to Participate in Clinical Research. *Arch Intern Med* 2008 Jun 23;168(12):1294–9.
54. McDonald AM, Knight RC, Campbell MK, Entwistle VA, Grant AM, Cook JA, et al. What influences recruitment to randomised controlled trials? A review of trials funded by two UK funding agencies. *Trials* 2006;7:9.
55. Simes RJ. Publication bias: the case for an international registry of clinical trials. *Journal of Clinical Oncology* 1986 Oct 1;4(10):1529–1541.
56. Clarke M, Clarke L, Clarke T. Yes Sir, No Sir, Not Much Difference Sir. *JRSM* 2007 Dec 1;100(12):571–572.
57. Chalmers Iain. Underreporting Research Is Scientific Misconduct. *JAMA* 1990 Mar 9;263(10):1405–1408.
58. Chalmers I. From optimism to disillusion about commitment to transparency in the medico-industrial complex. *JRSM* 2006 Jul 1;99(7):337–341.
59. Their delegation was led by Frank Wells: his textbook on fraud is fantastic. I tell you this because you should understand that these are not all bad people with inherently secretive natures.
60. Sykes R. Being a modern pharmaceutical company. *BMJ* 1998 Oct 31;317(7167):1172–80.
61. http://www.bmj.com/content/339/bmj.b4330
62. De Angelis C, Drazen JM, Frizelle FA, Haug C, Hoey J, Horton R, et al. Clinical trial registration: a statement from the

International Committee of Medical Journal Editors. *The Lancet* 2004 Sep 11;364(9438):911–2.

63. Mathieu S, Boutron I, Moher D, Altman DG, Ravaud P. Comparison of Registered and Published Primary Outcomes in Randomized Controlled Trials. *JAMA* 2009 Sep 2;302(9):977–84.

64. Wieseler B, McGauran N, Kaiser T. Still waiting for functional EU Clinical Trials Register. *BMJ* 2011 Jun 20;342(jun20 2):d3834–d3834.

65. Prayle AP, Hurley MN, Smyth AR. Compliance with mandatory reporting of clinical trial results on ClinicalTrials.gov: cross sectional study. *BMJ* 2012;344:d7373.

66. A good (but brief) overview of how to try and get info from non-academic sources is here: Chan A-W. Out of sight but not out of mind: how to search for unpublished clinical trial evidence. *BMJ* 2012 Jan 3;344(jan03 2):d8013–d8013.

67. You can read the letters and the report online. It's a gripping read, with many interesting and nefarious details, so I highly recommend doing so: Medicines and Healthcare products Regulatory Agency (MHRA) www. mhra. gov. u. GSK investigation concludes [Internet]. [cited 2012 Apr 29]. Available from: http://www.mhra.gov.uk/Howweregulate/Medicines/Medicines regulatorynews/CON014153

68. This was SmithKline Beecham, before they merged with GlaxoWellcome and became GSK.

69. Strech D, Littmann J. Lack of proportionality. Seven specifications of public interest that override post-approval commercial interests on limited access to clinical data. Trials. 2012 Jul 2;13(1):100.

70. Lenzer J, Brownlee S. Antidepressants: an untold story? *BMJ* 2008;336:532–4.

71. Wood AJ. Progress and deficiencies in the registration of clinical trials. *NEJM* 2009;360(8):824–830

72. O'Connor AB. The need for improved access to FDA reviews. *JAMA* 2009;302(2):191.

73. http://www.prescrire.org/editoriaux/EDI33693.pdf

74. Decision of the European Ombudsman closing his inquiry into complaint 2560/2007/BEH against the European Medicines Agency. November 2010. http://www.ombudsman.europa.eu /cases/decision.faces/en/5459/html.bookmark

75. UK drug regulator destroys all the evidence after 15 years/BMI[Internet]. Available from http://www.bmj.com /rapid-response/2011/11/03/uk-drug-regulator-destroys-all -evidence-after-15-years

76. You might be unsurprised to hear that no large drug company has ever been prosecuted under the safety monitoring regulations in the UK.

77. This story is spread over various publications by the Cochrane team, and the account here is taken from their work, published responses from Roche, and discussions with the Cochrane team. The best place to get the early half of this story is this paper: Doshi P. Neuraminidase inhibitors – the story behind the Cochrane review. *BMJ* 2009;339. And for the second half, I recommend this open-access paper: Doshi P, Jefferson T, Del Mar C (2012) The Imperative to Share Clinical Study Reports: Recommendations from the Tamiflu Experience. *PLoS Med* 9(4): e1001201. doi:10.1371/journal.pmed.1001201 http://bit.ly/HIbwqO

78. This is a fascinating and messy new area. The paper below gives a good summary of the importance of analysing full trial programmes, and the discrepancies found on Tamiflu between papers and Clinical Study Reports: Jefferson T, Doshi P, Thompson M, Heneghan C, Group CARI. Ensuring safe and effective drugs: who can do what it takes? *BMJ* 2011 Jan 11;342(jan11 1):c7258–c7258.

79. This is all from: Jefferson T, Doshi P, Thompson M, Heneghan C, Group CARI. Ensuring safe and effective drugs: who can do what it takes? *BMJ* 2011 Jan 11;342(jan11 1):c7258–c7258.

80. Tom Jefferson, Lecture on Tamiflu, *BMJ* Evidence 2011, London.

81. Tramèr MR, Reynolds DJ, Moore RA, McQuay HJ. Impact of covert duplicate publication on meta-analysis: a case study. *BMJ* 1997 Sep 13;315(7109):635–40.

82. Doshi P, Jefferson T, Del Mar C (2012) The Imperative to Share Clinical Study Reports: Recommendations from the Tamiflu

Experience. *PLoS Med* 9(4): e1001201. doi:10.1371/journal
.pmed.1001201 http://bit.ly/HIbwqO

83. Cohen D (2009) Complications: tracking down the data on
oseltamivir. *BMJ* 339: b5387.

84. If you're interested in this story, the links to primary documents
are all here: Diabetes drug 'victory' is really an ugly story about
incompetence. Ben Goldacre, *The Guardian* 2010 Jul 17 [cited
2012 May 2]. Available from: http://www.badscience.net
/2010/07/pharmaco-epidemiology-would-be-fascinating
-enough-even-if-society-didnt-manage-it-really-really-badly/

85. Nissen SE. Setting the record straight. *JAMA* 2010 Mar 24;
303(12):1194–5

86. Eichler H-G, Abadie E, Breckenridge A, Leufkens H, Rasi G.
Open Clinical Trial Data for All? A View from Regulators. *PLoS
Med* 2012 Apr 10;9(4):e1001202.

87. This is a vast story, told well elsewhere. Start with Curfman GD,
Morrissey S, Drazen JM. Expression of concern reaffirmed.
NEJM 2006 Mar 16;354(11):1193.

88. Opinion: Misleading Drug Trials. *The Scientist* [Internet].
[cited 2012 May 15]. Available from: http://the
-scientist.com/2012/05/14/opinion-misleading-drug-trials/

89. The Yale Open Data Archive project, or YODA, is one good
example of how this might look in the future.

Chapter 2: Where Do New Drugs Come From?

1. I recommend the classic medical student textbook 'Rang and
Dale': *Rang & Dale's Pharmacology*. 6th ed. Churchill
Livingstone; 2007. But also this, on the regulatory process
around early drug development: Friedhoff LT. *New Drugs: An
Insider's Guide to the FDA's New Drug Approval Process for
Scientists, Investors and Patients.* 1st ed. PSPG Publishing,
2009.

2. Elliott C, Abadie R. Exploiting a research underclass in phase 1
clinical trials. *NEJM* 2008;358(22):2316.

3. Cohen LP. To screen new drugs for safety, Lilly pays homeless
alcoholics: it's 'quick cash' to habitues of Indianapolis shelters;
it vanishes quickly, too. *Wall St J* (East Ed). 1996 Nov 14;A1,
A10.

4. Abadie R. *The Professional Guinea Pig: Big Pharma and the Risky World of Human Subjects.* 1st ed. Duke University Press, 2010.

5. Helms R, editor. *Guinea Pig Zero: An Anthology of the Journal for Human Research Subjects.* 1st ed. Garrett County Press; 2006.

6. Tucker T. *Great Starvation Experiment: Ancel Keys and the Men Who Starved for Science.* 1st University of Minnesota Press Ed. University of Minnesota Press; 2008.

7. Gorkin L, Schron EB, Handshaw K, Shea S, Kinney MR, Branyon M, et al. Clinical trial enrollers vs. nonenrollers: The Cardiac Arrhythmia Suppression Trial (CAST) Recruitment and Enrollment Assessment in Clinical Trials (REACT) project. *Controlled Clinical Trials* 1996 Feb;17(1):46–59.

8. Sheppard VB, Cox LS, Kanamori MJ, Cañar J, Rodríguez Y, Goodman M, et al. Brief Report: If You Build It, They Will Come. *J Gen Intern Med* 2005 May;20(5):444–7.

9. ACRO – CRO Market [Internet]. [cited 2012 Feb 11]. Available from: http://www.acrohealth.org/cro-market1.html

10. MacDonald T, Hawkey C, Ford I. Time to treat as independent. *BMJ* 2010 Nov 30;341(nov30 2):c6837–c6837.

11. Kassirer J. *On the Take: How Medicine's Complicity with Big Business Can Endanger Your Health.* Ch 8. 1st ed. Oxford University Press, USA; 2004.

12. Pharmaceutical CSO – Pharmaceutical Commercialization – Quintiles [Internet.] Available from: http://www.quintiles.com/commercial-services/

13. Drug Testing Goes Offshore – August 8, 2005 [Internet]. [cited 2012 Feb 11]. Available from: http://money.cnn.com/magazines/fortune/fortune_archive /2005/08/08/8267653/index.htm

14. Thiers FA, Sinskey AJ, Berndt ER. Trends in the globalization of clinical trials. *Nature Reviews Drug Discovery* 2008 Jan; 7(1):13–4.

15. All of the issues around trials in developing countries are well covered in two books, Shah S. *The Body Hunters, Testing New Drugs on the World's Poorest Patients.* The New Press; 2007. And Petryna A. *When Experiments Travel: Clinical Trials and the Global Search for Human Subjects.* 1st ed. Princeton University Press; 2009.

16. Ethical and Scientific Implications of the Globalization of Clinical Research. Seth W. Glickman, M.D., M.B.A., John G. McHutchison, M.D., Eric D. Peterson, M.D., M.P.H., Charles B. Cairns, M.D., Robert A. Harrington, M.D., Robert M. Califf, M.D., and Kevin A. Schulman, M.D. *NEJM* 2009; 360:816–823. February 19, 2009.

17. Bansal N. The opportunities and challenges in conducting clinical trials globally. *Clinical Research and Regulatory Affairs* 2012 Feb 9;1–6.

18. Ethical and Scientific Implications of the Globalization of Clinical Research. Seth W. Glickman, M.D., M.B.A., John G. McHutchison, M.D., Eric D. Peterson, M.D., M.P.H., Charles B. Cairns, M.D., Robert A. Harrington, M.D., Robert M. Califf, M.D., and Kevin A. Schulman, M.D. *NEJM* 2009; 360:816–823. February 19, 2009.

19. Hyder AA, Wali SA, Khan AN, Teoh NB, Kass NE, Dawson L. Ethical review of health research: a perspective from developing country researchers. *J Med Ethics* 2004 Feb;30(1):68–72.

20. Zhang D, Yin P, Freemantle N, Jordan R, Zhong N, Cheng KK. An assessment of the quality of randomised controlled trials conducted in China. *Trials* 2008;9:22.

21. FDA Requires Foreign Clinical Studies be in Accordance with Good Clinical Practice to Better Protect Human Subjects by W. Thomas Smith – American Bar Association Health eSource October 2008 Volume 5 Number 2 [Internet]. [cited 2012 Feb 11]. Available from: http://www.americanbar.org/newsletter /publications/aba_health_esource_home/Volume5_02_smith.html

22. WikiLeaks cables: Pfizer 'used dirty tricks to avoid clinical trial payout' – Business – *The Guardian* [Internet]. [cited 2012 Feb 11]. Available from: http://www.guardian.co.uk/business/2010/dec/09/wikileaks -cables-pfizer-nigeria

23. US embassy cable Monday 20 April 2009, 16:00, Abuja 000671 'Pfizer reaches preliminary agreement for $75m settlement' [cited 2012 Feb 11]. Available from: http://www.guardian.co.uk /world/us-embassy-cables-documents/203205

24. WikiLeaks cables: Pfizer 'used dirty tricks to avoid clinical trial payout' – Business – *The Guardian* [Internet]. [cited 2012 Feb

11]. Available from: http://www.guardian.co.uk/business /2010/dec/09/wikileaks-cables-pfizer-nigeria

25. Jonathan Kimmelman, Charles Weijer, and Eric M Meslin, 'Helsinki discords: FDA, ethics, and international drug trials', *The Lancet* 373, no. 9657 (January 3, 2009): 13–14.

26. Goodyear MDE, Lemmens T, Sprumont D, Tangwa G. Does the FDA have the authority to trump the Declaration of Helsinki? *BMJ* 2009 Apr 21;338(apr21 1):b1559–b1559.

Chapter 3: Bad Regulators

1. Royal College of Physicians, London UK. Innovating for health. Patients, physicians, the pharmaceutical industry and the NHS. February 2009. Report of a Working Party.

2. If you're very confused about the European Medicines Agency, and the UK MHRA, and how they relate to each other, it's fairly simple. The MHRA used to approve drugs, now the EMA do, but they farm out some of the local work to the old national regulators, especially monitoring and communication, as well as some of the approval stuff.

3. I recommend the work of John Abraham, collected here: http://www.sussex.ac.uk/profiles/6

4. Owen BM, Braeutigam R. *The Regulation Game: Strategic Use of the Administrative Process.* Ballinger Pub Co; 1978. Via Abraham J. On the prohibition of conflicts of interest in pharmaceutical regulation: Precautionary limits and permissive challenges. A commentary on Sismondo (66:9, 2008, 1909–14) and O'Donovan and Lexchin. *Social Science & Medicine* 2010 Mar;70(5):648–51.

5. http://www.alter-eu.org/sites/default/files/documents/lonngren -doc.pdf

6. http://www.alter-eu.org/sites/default/files/documents/lonngren -doc.pdf

7. http://www.alter-eu.org/fr/press-releases/2011/02/25/conflict -of-interest-case-involving-ex-ema-director

8. http://www.corporateeurope.org/publications/block-revolving -door

9. Lurie, P., Almeida, C., Stine, N., Stine, A. R., & Wolfe, S. M. (2006). Financial conflict of interest disclosure and voting

patterns at food and drug administration drug advisory committee meetings. *JAMA* 295, 1921e1928.

10. If you're interested in starting on this, the following are a good place to start: http://www.nytimes.com/2009/09/25/health /policy/25knee.html?_r=1; http://www.nytimes.com /2005/02/25/politics/25fda.html. And the work of the Project On Government Oversight is excellent, led by the researcher who worked on Senator Grassley's reports into the pharmaceutical industry: http://www.pogo.org/investigations/public -health/fda.html; http://pogoblog.typepad.com/pogo/2011/08 /fdas-janet-woodcock-the-substance-behind-her -nonsubstantive-substantive-ties-to-industry.html

11. Light D, editor. *The Risks of Prescription Drugs* (A Columbia/SSRC Book. Columbia University Press; 2010).

12. Survey of FDA Scientists Shows They Feel Pressure to Exclude . . . : *Oncology Times* [Internet]. [cited 2012 May 6]. Available from: http://journals.lww.com/oncology-times/Fulltext/2006/08250/Survey_of_FDA_Scientists_Shows _They_Feel_Pressure.8.aspx

13. USATODAY.com – Survey: FDA scientists question safety [Internet]. [cited 2012 May 6]. Available from: http://www.usatoday.com/news/health/2004-12-16-fda-survey -usat_x.htm

14. *European Journal of Clinical Pharmacology* 2011 10.1007/s00228-011-1052-1 Anything new in EU pharmacovigilance? Silvio Garattini and Vittorio Bertele'.

15. Goldberg NH, Schneeweiss S, Kowal MK, Gagne JJ. Availability of Comparative Efficacy Data at the Time of Drug Approval in the United States. *JAMA* 2011 May 4;305(17):1786–9.

16. Bertele' V, Banzi R, Gluud C, Garattini S. EMA's reflection on placebo does not reflect patients' interests. *European Journal of Clinical Pharmacology* [Internet]. 2011 Dec 2 [cited 2012 Feb 13]; Epub ahead of print. Available from: http://www.springerlink.com/content/4j733734v35381jk/

17. Garattini S, Chalmers I. Patients and the public deserve big changes in evaluation of drugs. *BMJ* 2009 Mar 31;338(mar31 3):b1025–b1025.

18. Van Luijn JCF, Gribnau FWJ, Leufkens HGM. Availability of comparative trials for the assessment of new medicines in the European Union at the moment of market authorization. *Br J Clin Pharmacol.* 2007;63(2):159–162.

19. Stafford RS, Wagner TH, Lavori PW. New, but Not Improved? Incorporating Comparative-Effectiveness Information into FDA Labeling. *NEJM* 2009 Aug 12;NEJMp0906490.

20. Echt DS, Liebson PR, Mitchell LB, Peters RW, Obias-Manno D, Barker AH, et al. Mortality and morbidity in patients receiving encainide, flecainide, or placebo. The Cardiac Arrhythmia Suppression Trial. *NEJM* 1991 Mar;324(12):781–788.

21. ALLHAT Collaborative Research Group. Major cardiovascular events in hypertensive patients randomized to doxazosin vs chlorthalidone: the antihypertensive and lipid-lowering treatment to prevent heart attack trial (ALLHAT). *JAMA* 2000 Apr;283(15):1967–1975.

22. Lenzer J. Spin doctors soft pedal data on antihypertensives. *BMJ* 2003 Jan 18;326(7381):170.

23. Vilsboll T, Christensen M, Junker AE, Knop FK, Gluud LL. Effects of glucagon-like peptide-1 receptor agonists on weight loss: systematic review and meta-analyses of randomised controlled trials. *BMJ* 2012 Jan 11;344(jan10 2):d7771–d7771.

24. Grimes DA, Schulz KF. Surrogate end points in clinical research: hazardous to your health. *Obstet Gynecol* 2005;105:1114–8.

25. This graph is from Chapter 7 of an excellent (though long and serious) history of the FDA: Carpenter D. *Reputation and Power: Organizational Image and Pharmaceutical Regulation at the FDA.* Princeton University Press; 2010.

26. Mitka M. FDA Takes Slow Road Toward Withdrawal of Drug Approved With Fast-Track Process. *JAMA* 2011 Mar 9;305(10):982–4.

27. Press Announcements > FDA Proposes Withdrawal of Low Blood Pressure Drug [Internet]. [cited 2012 May 7]. Available from: http://www.fda.gov/NewsEvents/Newsroom/PressAnnouncements/ucm222580.htm

28. United States Government Accountability Office. September 2009. New Drug Approval. FDA Needs to Enhance Its

Oversight of Drugs Approved on the Basis of Surrogate Endpoints. http://www.gao.gov/new.items/d09866.pdf

29. Davis C, Abraham J. Desperately seeking cancer drugs: explaining the emergence and outcomes of accelerated pharmaceutical regulation. *Sociology of Health & Illness* 2011 Jul 1;33(5):731–47.

30. Barbui C, Cipriani A, Lintas C, Bertel V, Garattini S. CNS drugs approved by the centralised European procedure: true innovation or dangerous stagnation? *Psychopharmacology* 2006 Nov 22;190(2):265–8.

31. There is a good, free summary of the issues around this area in this PDF from the WHO: Aidan Hollis. Me Too Drugs: is there a problem? http://www.who.int/intellectualproperty/topics/ip/Me-tooDrugs_Hollis1.pdf

32. NICE, 'CG17 Dyspepsia: full guideline,' Guidance/Clinical Guidelines, http://guidance.nice.org.uk/CG17/Guidance/pdf/English. But also, if the NICE guideline and its references aren't enough for you (it's from 2004) please do waste an hour of your time browsing other more recent trials. There's nothing magical happening here for esomeprazole.

33. http://www.nytimes.com/2004/10/12/business/media/12drug.html

34. http://www.mediapost.com/publications/?fa=Articles.showArticle&art_aid=92473

35. http://www.forbes.com/forbes/2010/0412/opinions-healthcare-nexium-hmo-prescriptions-heads-up.html

36. Here I should declare an interest: I sit on the funding panel to address this exact question, every quarter, for the NHS 'Health Technology Assessment' programme. This funding stream exists specifically to identify trials that need to be done, but which no company would fund, comparing one drug against another, and if you are aware of any important areas where we don't know which of two important treatments is best, you should submit a request (or if you're lazy, email it to me).

37. http://www.isdbweb.org/publications/view/pharmacovigilance-data ('Broadening access to signal is a positive step, but access to VigiBase is also needed', ISDB, February 15th 2012).

38. Hazell L, Shakir SAW. Under-reporting of adverse drug reactions : a systematic review. *Drug Saf* 2006;29(5): 385–96.

39. L. Härmark and A. C. Grootheest, 'Pharmacovigilance: methods, recent developments and future perspectives', *European Journal of Clinical Pharmacology* 64, no. 8 (June 4, 2008): 743–752.

40. FDA warns Pfizer for not reporting side effects – Reuters [Internet]. [cited 2012 May 8]. Available from: http://www.reuters.com/article/2010/06/10/us-pfizer-fda -idUSTRE6586PE20100610

41. Healy D: *Let Them Eat Prozac.* New York: New York University Press; 2004. Furukawa TA: All clinical trials must be reported in detail and made publicly available. *BMJ* 2004, 329:626. Via Gøtzsche PC. Why we need easy access to all data from all clinical trials and how to accomplish it. *Trials* 2011 Nov 23;12(1):249.

42. Serena Frau et al., 'Risk Management Plans: are they a tool for improving drug safety?', *European Journal of Clinical Pharmacology* 66, no. 8 (June 25, 2010): 785–790.

43. Giezen TJ, Mantel-Teeuwisse AK, Straus SMJM, Egberts TCG, Blackburn S, Persson I, et al. Evaluation of post-authorization safety studies in the first cohort of EU Risk Management Plans at time of regulatory approval. *Drug Saf* 2009;32(12):1175–87.

44. Andrew Herxheimer, 'Looking at EU pharmacovigilance', *European Journal of Clinical Pharmacology* 67, no. 11 (November 2011): 1201–1202.

45. Schwartz LM, Woloshin S. Communicating Uncertainties About Prescription Drugs to the Public: A National Randomized Trial. *Arch Intern Med* 2011 Sep 12;171(16):1463–8.

46. EMA Press Office, 2 February 2012, EMA/30803/2012

47. Garattini S, Bertele' V. Anything new in EU pharmacovigilance? *European Journal of Clinical Pharmacology* 67, no. 11 (November 2011): 1199–1200.

48. Garattini S, Bertele' V (2010) Rosiglitazone and the need for a new drug safety agency. *BMJ* 341:c5506. doi:10.1136/bmj.c5506

49. http://www.nap.edu/catalog.php?record_id=11750#orgs

50. http://www.iom.edu/Reports/2006/The-Future-of-Drug-Safety
    -Promoting-and-Protecting-the-Health-of-the-Public.aspx

51. Carpenter D. *Reputation and Power: Organizational Image and
    Pharmaceutical Regulation at the FDA*. Princeton University
    Press; 2010.

52. *European Journal of Clinical Pharmacology* 2011
    10.1007/s00228-011-1052-1 Anything new in EU
    pharmacovigilance? Silvio Garattini and Vittorio Bertele'.

53. Uncommon knowledge. *Drug and Therapeutics Bulletin* 2009
    Nov 1; 47(11):121

54. Schwartz LM, Woloshin S, Welch HG. Using a Drug Facts Box
    to Communicate Drug Benefits and Harms Two Randomized
    Trials. *Ann Intern Med* 2009 Apr 21;150(8):516–27.

55. Germany's tough reimbursement rules cause drug companies to
    consider alternative drug trial solutions – FT.com [Internet].
    [cited 2012 Feb 15]. Available from: http://www.ft.com/cms/s/2
    /d458d470-4696-11e1-89a8-00144feabdc0.html#axzz1mTzZ2jdb.

Chapter 4: Bad Trials

1. Anesthesiology News – Fraud Case Rocks Anesthesiology
   Community [Internet]. [cited 2012 Feb 12]. Available from:
   http://www.anesthesiologynews.com/ViewArticle.aspx?d=
   Policy per cent2B per cent26amp per cent3B per
   cent2BManagement&d_id=3&i=March per
   cent2B2009&i_id=494&a_id=12634&ses=ogst

2. This story is well covered in Wells F. Fraud and Misconduct in
   Biomedical Research. Chapter 5, Fourth ed. RSM Books; 2008. I
   highly recommend this book if you want to get started reading
   about fraud detection and management. Beware that it's an
   academic book, and therefore appallingly expensive.

3. Rothwell PM. External validity of randomised controlled trials:
   'To whom do the results of this trial apply?' *The Lancet* 2005
   Jan 1; 365(9453):82–93.

4. Pratt, C.M. & Moye, L.A., 1995. The Cardiac Arrhythmia
   Suppression Trial : Casting Suppression in a Different Light.
   Circulation, 91(1), pp. 245–247.

5. Travers, J. et al., 2007. External validity of randomised controlled trials in asthma: to whom do the results of the trials apply? *Thorax*, 62(3), pp. 219 –223.

6. Zimmerman, M., Chelminski, I. & Posternak, M.A., 2004. Exclusion criteria used in antidepressant efficacy trials: consistency across studies and representativeness of samples included. *The Journal of Nervous and Mental Disease*, 192(2), pp. 87–94.

7. Keitner, G.I., Posternak, M.A. & Ryan, C.E., 2003. How many subjects with major depressive disorder meet eligibility requirements of an antidepressant efficacy trial? *The Journal of Clinical Psychiatry*, 64(9), pp. 1091–3.

8. Jarvinen TLN, Sievanen H, Kannus P, Jokihaara J, Khan KM. The true cost of pharmacological disease prevention. *BMJ* 2011 Apr 19;342(apr19 1):d2175–d2175.

9. Van Staa T-P, Leufkens HG, Zhang B, Smeeth L. A Comparison of Cost Effectiveness Using Data from Randomized Trials or Actual Clinical Practice: Selective Cox-2 Inhibitors as an Example. *PLoS Med* 2009 Dec 8;6(12):e1000194.

10. Safer DJ. Design and reporting modifications in industry-sponsored comparative psychopharmacology trials. *J. Nerv. Ment. Dis* 2002 Sep;190(9):583–92.

11. Califf RM, DeMets DL. Principles From Clinical Trials Relevant to Clinical Practice: Part I. *Circulation* 2002 Aug 20;106(8):1015–21.

12. Mueller PS, Montori VM, Bassler D, Koenig BA, Guyatt GH. Ethical Issues in Stopping Randomized Trials Early Because of Apparent Benefit. *Annals of Internal Medicine* 2007 Jun 19;146(12):878–881.

13. Bassler D, Briel M, Montori VM, Lane M, Glasziou P, Zhou Q, et al. Stopping Randomized Trials Early for Benefit and Estimation of Treatment Effects: Systematic Review and Meta-regression Analysis. *JAMA* 2010 Mar 24;303(12):1180–7.

14. Montori VM, Devereaux PJ, Adhikari NKJ, Burns KEA, Eggert CH, Briel M, et al. Randomized Trials Stopped Early for Benefit: A Systematic Review. *JAMA* 2005 Nov 2;294(17):2203–9.

15. Trotta F, Apolone G, Garattini S, Tafuri G. Stopping a trial early in oncology: for patients or for industry? *Annals of Oncology*

[Internet]. 2008 Jan 1 [cited 2012 Feb 14]; Available from: http://annonc.oxfordjournals.org/content/early/2008/04/09/ann onc.mdn042.full

16. Lurie P, Wolfe SM. Misleading data analyses in salmeterol (SMART) study. *The Lancet* 2005 Oct;366(9493):1261–2.

17. Rickard KA. Misleading data analyses in salmeterol (SMART) study – GlaxoSmithKline's reply. *The Lancet* 2005 Oct;366(9493):1262.

18. Garcialopez F, Dealvaro F. INSIGHT and NORDIL. *The Lancet* 2000 Dec 2;356(9245):1926–1926.

19. Safer DJ. Design and reporting modifications in industry-sponsored comparative psychopharmacology trials. *J. Nerv. Ment. Dis* 2002 Sep;190(9):583–92.

20. Gilbody S, Wahlbeck K, Adams C. Randomized controlled trials in schizophrenia: a critical perspective on the literature. *Acta Psychiatr Scand* 2002;105:243–251.

21. Montori VM, Jaeschke R, Schünemann HJ, Bhandari M, Brozek JL, Devereaux PJ, et al. Users' guide to detecting misleading claims in clinical research reports. *BMJ* 2004 Nov 6;329(7474):1093–6.

22. Shaughnessy AF, Slawson DC. What happened to the valid POEMs? A survey of review articles on the treatment of type 2 diabetes. *BMJ* 2003 Aug 2;327(7409):266.

23. Melander H, Ahlqvist-Rastad J, Meijer G, Beermann B. Evidence b(i)ased medicine – selective reporting from studies sponsored by pharmaceutical industry: review of studies in new drug applications. *BMJ* 2003;326:1171–3.

24. Vedula, S Swaroop, Lisa Bero, Roberta W Scherer, and Kay Dickersin. 'Outcome reporting in industry-sponsored trials of gabapentin for off-label use.' *NEJM* 361, no. 20 (November 12, 2009): 1963–1971.

25. Chan A-W, Hróbjartsson A, Haahr MT, Gøtzsche PC, Altman DG: Empirical evidence for selective reporting of outcomes in randomized trials: comparison of protocols to published articles. *JAMA* 2004, 291:2457–2465.

26. Jon N. Jureidini, Leemon B. McHenry, Peter R. Mansfield. Clinical trials and drug promotion: Selective reporting of study 329. *International Journal of Risk & Safety in Medicine* 20 (2008) 73–81 73 DOI 10.3233/JRS-2008-0426

27. K L Lee et al., 'Clinical judgment and statistics. Lessons from a simulated randomized trial in coronary artery disease', *Circulation* 61, no. 3 (March 1980): 508–15.

28. On subgroup analysis, I recommend this excellent 2005 review article by Peter Rothwell: Rothwell PM. Subgroup analysis in randomised controlled trials: importance, indications, and interpretation. *The Lancet* 2005;365(9454):176–86. Currently available free online at http://apps.who.int/rhl/Lancet_365 -9454.pdf

29. EuropeanCarotidSurgeryTrialists'CollaborativeGroup. Randomised trial of endarterectomy for recently symptomatic carotid stenosis: final results of the MRC European Carotid Surgery Trial (ECST). *The Lancet* 1998; 351: 1379–87.

30. The Canadian Cooperative Study Group. A randomised trial of aspirin and sulfinpyrazone in threatened stroke. *NEJM* 1978; 299: 53–59.

31. Ioannidis JPA, Karassa FB. The need to consider the wider agenda in systematic reviews and meta-analyses: breadth, timing, and depth of the evidence. *BMJ* 2010 Sep;341(sep341): c4875.

32. Hill KP, Ross JS, Egilman DS, Krumholz HM. The ADVANTAGE Seeding Trial: A Review of Internal Documents. *Annals of Internal Medicine* 2008;149(4):251–8.

33. Sox HC, Rennie D. Seeding Trials: Just Say "No." *Annals of Internal Medicine* 2008;149(4):279–80.

34. Krumholz SD, Egilman DS, Ross JS. Study of Neurontin: Titrate to Effect, Profile of Safety (STEPS) Trial: A Narrative Account of a Gabapentin Seeding Trial. *Arch Intern Med* 2011 Jun 27;171(12):1100–7.

35. I recommend this book as an introduction to 'shared decision making' (I helped on one chapter): Gigerenzer G, Muir G. *Better Doctors, Better Patients, Better Decisions: Envisioning Health Care 2020*. 1st ed. MIT Press; 2011.

36. Malenka DJ, Baron JA, Johansen S, Wahrenberger JW, Ross JM. The framing effect of relative and absolute risk. *J Gen Intern Med* 1993 Oct;8(10):543–8.

37. Bucher HC, Weinbacher M, Gyr K. Influence of method of reporting study results on decision of physicians to prescribe drugs to lower cholesterol concentration. *BMJ* 1994 Sep 24;309(6957):761–4.

38. Fahey T, Griffiths S, Peters TJ. Evidence based purchasing: understanding results of clinical trials and systematic reviews. *BMJ* 1995 Oct 21;311(7012):1056–9.

39. Express.co.uk New wonder heart pill that may save millions [Internet]. [cited 2012 Feb 12]. Available from: http://www.express.co.uk/posts/view/70343

40. Drug could save thousands from heart attacks – Science – *The Guardian* [Internet]. [cited 2012 Feb 12]. Available from: http://www.guardian.co.uk/science/2008/nov/10/drugs -medical-research

41. Boutron I, Dutton S, Ravaud P, Altman DG. Reporting and Interpretation of Randomized Controlled Trials With Statistically Nonsignificant Results for Primary Outcomes. *JAMA* 2010 May 26;303(20):2058–64.

42. Alasbali, T. et al., 2009. Discrepancy between results and abstract conclusions in industry – vs nonindustry-funded studies comparing topical prostaglandins. *American Journal of Ophthalmology*, 147(1), pp. 33–38.e2.

43. Jørgensen AW, Hilden J, Gøtzsche PC. Cochrane reviews compared with industry-supported meta-analyses and other meta-analyses of the same drugs: systematic review. *BMJ* 2006 Oct 14;333(7572):782.

Chapter 5: Bigger, Simpler Trials

1. Staa T-P v., Goldacre B, Gulliford M, Cassell J, Pirmohamed M, Taweel A, et al. Pragmatic randomised trials using routine electronic health records: putting them to the test. *BMJ* 2012 Feb 7;344(feb07 1):e55–e55.

2. Edwards P, Arango M, Balica L, Cottingham R, El-Sayed H, Farrell B, et al. Final results of MRC CRASH, a randomised placebo-controlled trial of intravenous corticosteroid in adults with head injury-outcomes at 6 months. *The Lancet* 2005 Jun 4;365(9475):1957–9.

3. Dresden GM, Levitt MA. Modifying a Standard Industry Clinical Trial Consent Form Improves Patient Information Retention as Part of the Informed Consent Process. *Academic Emergency Medicine* 2001;8(3):246–52.

Chapter 6: Marketing

1. Alper BS, Hand JA, Elliott SG, Kinkade S, Hauan MJ, Onion DK, et al. How much effort is needed to keep up with the literature relevant for primary care? *J Med Libr Assoc* 2004;92:429–37

2. Moon JC, Flett AS, Godman BB, Grosso AM, Wierzbicki AS. Getting better value from the NHS drug budget. *BMJ* 2010 Dec 17;341(dec17 1):c6449–c6449.

3. Marketing spend is a contested area, as the industry is keen to play it down. I recommend the following paper as it's open access and offers a summary figure, methods from which it was derived, and a critical discussion of other estimates: Gagnon M-A, Lexchin J. The Cost of Pushing Pills: A New Estimate of Pharmaceutical Promotion Expenditures in the United States. *PLoS Med* 2008 Jan 3;5(1):e1.

4. Gilbody S, Wilson P, Watt I. Benefits and harms of direct to consumer advertising: a systematic review. *Quality and Safety in Health Care* 2005;14(4):246–50.

5. Kravitz RL, Epstein RM, Feldman MD, Franz CE, Azari R, Wilkes MS, et al. Influence of patients' requests for direct-to-consumer advertised antidepressants: a randomized controlled trial. *JAMA* 2005 Apr 27;293(16):1995–2002.

6. Iizuka T. What Explains the Use of Direct-to-Consumer Advertising of Prescription Drugs? *The Journal of Industrial Economics* 2004;52(3):349–79.

7. NICE. CG17 Dyspepsia: full guideline [Internet]. [cited 2011 Jan 4]. Available from: http://guidance.nice.org.uk/CG17/Guidance/pdf/English

8. Law MR, Soumerai SB, Adams AS, Majumdar SR. Costs and Consequences of Direct-to-Consumer Advertising for Clopidogrel in Medicaid. *Arch Intern Med* 2009 Nov 23;169(21):1969–74.

9. I first saw the Reynolds, Bacall, Lowe and *Serial Mom* examples in: Petersen M, p. 32: *Our Daily Meds: How the Pharmaceutical Companies Transformed Themselves into Slick Marketing Machines and Hooked the Nation on Prescription Drugs.* Picador; 2009.

10. Eisenberg D. It's an Ad, Ad, Ad World. *Time* [Internet]. 2002 Aug 26 [cited 2012 Mar 25]. Available from: http://www.time.com /time/magazine/article/0,9171,344045,00.html

11. Stars Profit From Covert Drug Pitches – CBS News [Internet]. [cited 2012 Mar 25]. Available from: http://www.cbsnews.com /2100-207_162-520196.html

12. Ibid.

13. Alzheimer's Campaign Piques Public and Media Interest: PR News May 21, 2001. Available from: http://www.prnewsonline .com/news/4782.html

14. Keidan J. Sucked into the Herceptin maelstrom. *BMJ* 2007 Jan 6;334(7583):18.

15. Wilson PM, Booth AM, Eastwood A, Watt IS. Deconstructing Media Coverage of Trastuzumab (Herceptin): An Analysis of National Newspaper Coverage. *J R Soc Med* 2008 Mar 1;101(3):125–32.

16. The selling of a wonder drug – Science – *The Guardian* [Internet]. [cited 2012 Mar 26]. Available from: http://www .guardian.co.uk/science/2006/mar/29/medicineandhealth. health

17. Ibid.

18. To be absolutely clear, there is no evidence that a company was involved in promoting Barbara Moss to the media. This case simply illustrates the melodramatic ineptitude of coverage for new cancer drugs.

19. Castrén E. Is mood chemistry? *Nature Reviews Neuroscience* 2005 Mar 1;6(3):241–6.

20. *The Pittsburgh Tribune Review* (4/2/07).

21. Lacasse JR, Leo J. Serotonin and Depression: A Disconnect between the Advertisements and the Scientific Literature. *PLoS Med* 2005 Nov 8;2(12):e392.

22. Petersen M, p. 102: *Our Daily Meds: How the Pharmaceutical Companies Transformed Themselves into Slick Marketing*

Machines and Hooked the Nation on Prescription Drugs.
Picador; 2009.

23. Ibid.

24. Leo J, Lacasse J. The Media and the Chemical Imbalance Theory of Depression. *Society* 2008 Feb 19;45(1):35–45.

25. The test has now been altered; the original description is preserved online at: WebMD's Depression Test Has Only One (Sponsored) Answer: You're 'At Risk' – CBS News [Internet]. [cited 2012 Mar 26]. Available from: http://www.cbsnews .com/8301-505123_162-42844266/webmds-depression-test-has -only-one-sponsored-answer-youre-at-risk/?tag=bnetdomain

26. Ebeling M. 'Get with the Program!': Pharmaceutical marketing, symptom checklists and self-diagnosis. *Social Science & Medicine* 2011 Sep;73(6):825–32.

27. Laumann EO, Paik A, Rosen RC. Sexual Dysfunction in the United States Prevalence and Predictors. *JAMA* 1999 Feb 10;281(6):537–44.

28. The Nation: Better Loving Through Chemistry; Sure, We've Got a Pill for That – *New York Times* [Internet]. [cited 2012 Mar 27]. Available from: http://www.nytimes.com/1999/02/14/weekinreview/the-nation -better-loving-through-chemistry-sure-we-ve-got-a-pill-for -that.html?pagewanted=all&src=pm

29. Moynihan R. The making of a disease: female sexual dysfunction. *BMJ* 2003;326(7379):45–47.

30. Moynihan R. Company launches campaign to 'counter' *BMJ* claims. *BMJ* 2003 Jan 18;326(7381):120.

31. Tiefer L. Female Sexual Dysfunction: A Case Study of Disease Mongering and Activist Resistance. *PLoS Med* 2006 Apr 11;3(4):e178.

32. Ibid.

33. Ibid.

34. Testosterone Patches for Female Sexual Dysfunction. *DTB* 2009 Mar 1;47(3):30–4.

35. Durand M. Pharma's Advocacy Dance [Internet]. Successful Product Manager's Handbook. 2006 [cited 2012 Mar 26]. Available from: http://www.pharmexec.com/pharmexec

/Articles/Pharmas-Advocacy-Dance/ArticleStandard/Article
/detail/377999

36. 11 August 2010 – HAI Europe Research Article – Patient &
    Consumer Organisations at the EMA: Financial Disclosure &
    Transparency. Written by Katrina Perehudoff and Teresa
    Leonardo Alves. 11 August 2010 – HAI Europe Factsheet –
    Patient & Consumer Organisations at the EMA: Financial
    Disclosure & Transparency.

37. HAI. The Patient & Consumer Voice and Pharmaceutical
    Industry Sponsorship [Internet]. [cited 2012 Mar 26]. Available
    from: http://apps.who.int/medicinedocs/en/m/abstract
    /Js17767en/

38. 'Drug firms bankroll attacks on NHS'. *Independent*, 1 October
    2008.

39. 'Analysis: Are patient protests being manipulated?' *Independent*,
    1 October 2008.

40. Health chief attacks drug giants over huge profits – UK
    news – *The Observer* [Internet]. [cited 2012 Mar 26]. Available
    from: http://www.guardian.co.uk/uk/2008/aug/17
    /pharmaceuticals.nhs

41. Spurling GK, Mansfield PR, Montgomery BD, Lexchin J, Doust
    J, Othman N, et al. Information from Pharmaceutical
    Companies and the Quality, Quantity, and Cost of Physicians'
    Prescribing: A Systematic Review. *PLoS Med* 2010 Oct
    19;7(10):e1000352.

42. Azoulay P. Do pharmaceutical sales respond to scientific
    evidence? *Journal of Economics & Management Strategy*
    2002;11(4):551–94.

43. Heimans L, van Hylckama Vlieg A, Dekker FW. Are claims of
    advertisements in medical journals supported by RCTs? *Neth J
    Med* 2010 Jan;68(1):46–9.

44. Villanueva P, Peiro S, Librero J, Pereiro I. Accuracy of
    pharmaceutical advertisements in medical journals. *The Lancet*
    2003 Jan;361(9351):27–32.

45. Spielmans GI, Thielges SA, Dent AL, Greenberg RP. The
    accuracy of psychiatric medication advertisements in medical
    journals. *J. Nerv. Ment. Dis* 2008 Apr;196(4):267–73.

46. Van Winkelen P, van Denderen JS, Vossen CY, Huizinga TWJ, Dekker FW, for the SEDUCE study group. How evidence-based are advertisements in journals regarding the subspecialty of rheumatology? *Rheumatology* 2006 Sep 1;45(9):1154–7.

47. Othman N, Vitry A, Roughead EE. Quality of Pharmaceutical Advertisements in Medical Journals: A Systematic Review. *PLoS ONE* 2009 Jul 22;4(7):e6350.

48. Gibson L. UK government fails to tackle weaknesses in drug industry. *BMJ* 2005 Sep 10;331(7516):534–40.

49. Wilkes MS, Kravitz RL. Policies, practices, and attitudes of North American medical journal editors. *J Gen Intern Med* 1995 Aug;10(8):443–50.

50. Via: Cooper RJ, Schriger DL, Wallace RC, Mikulich VJ, Wilkes MS. The Quantity and Quality of Scientific Graphs in Pharmaceutical Advertisements. *J Gen Intern Med* 2003 Apr;18(4):294–7. 'Polling of the audience occurred as part of the discussion of the oral presentation of this abstract'. Fourth International Congress on Peer Review [Internet]. Available from: http://www.ama-assn.org/public/peer/prc_program2001.htm#ABSTRACTS

51. You might also enjoy some of the books written by retired drug reps, such as: Reidy J. *Hard Sell: The Evolution of a Viagra Salesman.* 1st ed. Andrews McMeel Publishing; 2005.

52. Rockoff JD. Drug Reps Soften Their Sales Pitches. *Wall Street Journal* [Internet]. 2012 Jan 10 [cited 2012 Mar 22]; Available from: http://online.wsj.com/article/SB10001424052970204331304577142763014776148.html

53. Fugh-Berman A, Ahari S. Following the Script: How Drug Reps Make Friends and Influence Doctors. *PLoS Med* 2007 Apr;4(4).

54. Soyk, C., B. Pfefferkorn, P. McBride, and R. Rieselbach. 2010. Medical student exposure to and attitudes about pharmaceutical companies. *World Medical Journal* 109: 142–148.

55. Fischer MA, Keough ME, Baril JL, Saccoccio L, Mazor KM, Ladd E, et al. Prescribers and Pharmaceutical Representatives: Why Are We Still Meeting? *J Gen Intern Med* 2009 Jul;24(7):795–801.

56. Morgan MA, Dana J, Loewenstein G, Zinberg S, Schulkin J. Interactions of doctors with the pharmaceutical industry. *J Med Ethics* 2006 Oct;32(10):559–63.

57. B Hodges, 'Interactions with the pharmaceutical industry: experiences and attitudes of psychiatry residents, interns and clerks', *CMAJ: Canadian Medical Association Journal = Journal De l'Association Medicale Canadienne* 153, no. 5 (September 1, 1995): 553–559.

58. Spurling GK, Mansfield PR, Montgomery BD, Lexchin J, Doust J, Othman N, et al. Information from Pharmaceutical Companies and the Quality, Quantity, and Cost of Physicians' Prescribing: A Systematic Review. *PLoS Med* 2010 Oct 19;7(10):e1000352.

59. MM Chren and CS Landefeld, 'Physicians' behavior and their interactions with drug companies. A controlled study of physicians who requested additions to a hospital drug formulary,' *JAMA* 271, no. 9 (March 2, 1994): 684–689.

60. Ladd EC, Mahoney DF, Emani S. 'Under the radar': nurse practitioner prescribers and pharmaceutical industry promotions. *Am J Manag Care* 2010;16(12):e358–362.

61. Zipkin DA, Steinman MA. Interactions Between Pharmaceutical Representatives and Doctors in Training. *J Gen Intern Med* 2005 Aug;20(8):777–86.

62. Spingarn RW, Berlin JA, Strom BL. When pharmaceutical manufacturers' employees present grand rounds, what do residents remember? *Acad Med* 1996 Jan;71(1):86–8.

63. Wazana A. Physicians and the Pharmaceutical Industry: Is a Gift Ever Just a Gift? *JAMA* 2000 Jan 19;283(3):373–80.

64. Lurie N, Rich EC, Simpson DE, Meyer J, Schiedermayer DL, Goodman JL, et al. Pharmaceutical representatives in academic medical centers: interaction with faculty and housestaff. *J Gen Intern Med* 1990 Jun;5(3):240–3.

65. Fugh-Berman A, Ahari S. Following the Script: How Drug Reps Make Friends and Influence Doctors. *PLoS Med* 2007 Apr;4(4).

66. Ibid.

67. Sismondo S. How pharmaceutical industry funding affects trial outcomes: Causal structures and responses. *Social Science & Medicine* 2008;66(9):1909–14.

68. Completed Cases – PMCPA Website [Internet]. [cited 2012 Mar 26]. Available from: http://www.pmcpa.org.uk /?q=node/868

69. Completed Cases – PMCPA Website [Internet]. [cited 2012 Mar 26]. Available from: http://www.pmcpa.org.uk/?q=node/883

70. Orlowski JP, Wateska L. The effects of pharmaceutical firm enticements on physician prescribing patterns. There's no such thing as a free lunch. *Chest* 1992 Jul;102(1):270–3.

71. Steinbrook R. For sale: physicians' prescribing data. *NEJM* 2006 Jun 29;354(26):2745–7.

72. Physician Data Restriction Program (PDRP) [Internet]. [cited 2012 Mar 22]. Available from: http://www.ama-assn.org/ama /pub/about-ama/physician-data-resources/ama-database -licensing/amas-physician-data-restriction-program.page

73. Outterson K. Higher First Amendment Hurdles for Public Health Regulation. *NEJM* 2011 Aug 18;365(7):e13.

74. Zipkin DA, Steinman MA. Interactions Between Pharmaceutical Representatives and Doctors in Training. *J Gen Intern Med* 2005 Aug;20(8):777–86.

75. Wislar JS, Flanagin A, Fontanarosa PB, DeAngelis CD. Honorary and ghost authorship in high impact biomedical journals: a cross sectional survey. *BMJ* 2011 Oct 25;343(oct25 1):d6128–d6128.

76. Gøtzsche PC, Hróbjartsson A, Johansen HK, Haahr MT, Altman DG, Chan A-W. Ghost Authorship in Industry-Initiated Randomised Trials. *PLoS Med* 2007 Jan 16;4(1):e19.

77. 'Ghost writing in the medical literature' 111th Congress, United States Senate Committee on Finance Sen. Charles E. Grassley, 2010. [cited 2012 Mar 24]. Available from: http://www.grassley.senate.gov/about/upload/Senator-Grassley -Report.pdf

78. Richard Horton PI 108, House of Commons – Health – Minutes of Evidence [Internet]. [cited 2012 Mar 24]. Available from: http://www.publications.parliament.uk/pa/cm200405 /cmselect/cmhealth/42/4121604.htm

79. Galanter M, Galanter M, Felstiner WLF, Friedman LM, Girth M, Goldstein P, et al. Why the haves come out ahead: Speculations on the limits of legal change. *Law Society Review* 1974;9:95–169.

80. Lilly 'Ghostwrote' Articles to Market Drug, Files Say (Update2) – Bloomberg [Internet]. [cited 2012 Mar 24]. Available from:

http://www.bloomberg.com/apps/news?pid=newsarchive&sid=
a6yFu_t9NyTY

81. http://www.psychiatrynorthwest.co.uk/general_adult
_psychiatry/spr_posts/salford-haddad/index.html

82. Medical Press Pre-Launch Feature Outline, Zyprexa MDL 1596,
confidential subject to protection order ZY200187608.
http://zyprexalitigationdocuments.com/per cent5Cdocuments
per cent5CConfidentiality-Challenge per cent5CDocs
-challenged-in-10-3-list per cent5C145-ZY200187608-7614.pdf

83. Drug Industry Document Archive [Internet]. [cited 2012 Mar
24]. Available from: http://dida.library.ucsf.edu/

84. Drug Industry Document Archive – Search Results [Internet].
[cited 2012 Mar 24]. Available from: http://dida.library.ucsf
.edu/tid/anu38h10

85. Ibid.

86. Ross, J.S., K.P. Hill, D.S. Egilman, and H.M. Krumholz. 2008.
Guest authorship and ghostwriting in publications related to
rofecoxib: A case study of industry documents from rofecoxib
litigation. *JAMA* 299: 1800–1812.

87. POGO Letter to NIH on Ghostwriting Academics [Internet].
Project On Government Oversight. [cited 2012 Mar 24].
Available from: http://www.pogo.org/pogo-files/letters/public
-health/ph-iis-20101129.html

88. http://www.nytimes.com/2010/11/30/business/30drug.html

89. http://pogoblog.typepad.com/pogo/gw-attachment-e.html

90. Lacasse JR, Leo J. Ghostwriting at Elite Academic Medical
Centers in the United States. *PLoS Med* 2010 Feb 2;7(2):e1000230.

91. Matheson A. How Industry Uses the ICMJE Guidelines to
Manipulate Authorship – And How They Should Be Revised.
PLoS Med 2011;8(8):e1001072.

92. Dyer O. Journal rejects article after objections from marketing
department. *BMJ* 2004 Jan 31;328(7434):244-b–244.

93. Fugh-Berman A, Alladin K, Chow J. Advertising in Medical
Journals: Should Current Practices Change? *PLoS Med* 2006
May 2;3(6):e130.

94. Becker A, Dörter F, Eckhardt K, Viniol A, Baum E, Kochen MM,
et al. The association between a journal's source of revenue and
the drug recommendations made in the articles it publishes.

CMAJ 2011 Feb 28 Available from:
http://www.cmaj.ca/content/early/2011/02/28/cmaj.100951

95. Smith R. Medical Journals Are an Extension of the Marketing Arm of Pharmaceutical Companies. *PLoS Med* 2005 May 17;2(5):e138.

96. AUTH/2424/8/11 and AUTH/2425/8/11 – General Practitioner v Boehringer Ingelheim and Lilly. Available from: http://www.pmcpa.org.uk/?q=node/998.

97. Handel AE, Patel SV, Pakpoor J, Ebers GC, Goldacre B, Ramagopalan SV. High reprint orders in medical journals and pharmaceutical industry funding: case-control study. *BMJ* 2012 Jun 28;344(jun28 1):e4212–e4212.

98. Jefferson T, Di Pietrantonj C, Debalini MG, Rivetti A, Demicheli V. Relation of study quality, concordance, take home message, funding, and impact in studies of influenza vaccines: systematic review. *BMJ* 2009 Feb 12;338(feb12_2):b354.

99. http://classic.the-scientist.com/blog/display/55679/

100. http://elsevier.com/wps/find/authored _newsitem.cws_home/companynews05_01203

101. Bowman MA. The impact of drug company funding on the content of continuing medical education. *Möbius: A Journal for Continuing Education Professionals in Health Sciences* 1986 Jan 1;6(1):66–9.

102. Bowman MA, Pearle DL. Changes in drug prescribing patterns related to commercial company funding of continuing medical education. *Journal of Continuing Education in the Health Professions* 1988 Jan 1;8(1):13–20.

103. The Carlat Psychiatry Blog: PRMS [Internet]. [cited 2012 Mar 31]. Available from: http://carlatpsychiatry.blogspot.co.uk /search/label/PRMS

104. Stephan Sahm, 'Of mugs, meals and more: the intricate relations between physicians and the medical industry,' *Medicine, health care, and philosophy* (2011).

105. Avorn J, Choudhry NK. Funding for Medical Education: Maintaining a Healthy Separation From Industry. *Circulation* 2010 May 25;121(20):2228–34.

106. L. Garattini et al., 'Continuing Medical Education in six

European countries: A comparative analysis', *Health policy* 94, no. 3 (2010): 246–254.

107. Eckardt VF. Complimentary journeys to the World Congress of Gastroenterology – an inquiry of potential sponsors and beneficiaries. *Z Gastroenterol* 2000 Jan;38(1):7–11.

108. http://www.pmlive.com/find_an_article/allarticles/categories /General/2011/november_2011/features/cme_continuing_medi cal_education_change

109. US Senate Committee on Finance. Committee Staff Report to the Chairman and Ranking Member: Use of Educational Grants by Pharmaceutical Manufacturers. Washington, DC: Government Printing Office; 2007.

110. Hensley S, Martinez B. To sell their drugs, companies increasingly rely on doctors. *Wall St J* (East Ed). 2005 Jul 15;A1,A2.

111. Tabas JA, Boscardin C, Jacobsen DM, Steinman MA, Volberding PA, Baron RB. Clinician Attitudes About Commercial Support of Continuing Medical Education: Results of a Detailed Survey. *Arch Intern Med* 2011 May 9;171(9):840–6.

112. Amy T Wang et al., 'Association between industry affiliation and position on cardiovascular risk with rosiglitazone: cross sectional systematic review', *BMJ* 340, no. 18 (March 18, 2010): c1344.

113. Rothman KJ, Evans S (2005) Extra scrutiny for industry funded trials. *BMJ* 331: 1350–1351.

114. Wager E, Mhaskar R, Warburton S, Djulbegovic B (2010) *JAMA* Published Fewer Industry-Funded Studies after Introducing a Requirement for Independent Statistical Analysis. *PLoS ONE* 5(10): e13591. doi:10.1371/journal.pone.0013591

115. Chalmers TC, Frank CS, Reitman D. Minimizing the Three Stages of Publication Bias. *JAMA* 1990 Mar 9;263(10):1392–5.

116. Samena Chaudhry et al., 'Does declaration of competing interests affect readers' perceptions? A randomised trial', *BMJ* 325, no. 7377 (December 14, 2002): 1391–1392. (below).

117. Reporting of Conflicts of Interest in Meta-analyses of Trials of Pharmacological Treatments. *JAMA* 2011;305(10):1008–1017. doi: 10.1001/jama.2011.257

118. Loewenstein G, Sah S, Cain DM. The Unintended Consequences of Conflict of Interest Disclosure. *JAMA* 2012 Feb 15;307(7):669–70.

119. Cain, D. M., Loewenstein, G., & Moore, D. A. (2005). The dirt on coming clean: perverse effects of disclosing conflicts of interest. *Journal of Legal Issues*, 34, 1e25.

120. Campbell EG, Weissman JS, Ehringhaus S et al. Institutional academic industry relationships. *JAMA* 2007;298:1779–86.

121. http://www.propublica.org/series/dollars-for-docs

122. http://www.propublica.org/article/doctors-dine-on-drug -companies-dime

123. http://www.propublica.org/article/dollars-for-docs-sparks -policy-rewrite-at-colorado-teaching-hospitals

124. http://www.propublica.org/article/medical-schools-plug-holes -in-conflict-of-interest-policies

125. http://www.propublica.org/article/dollars-to-doctors -physician-disciplinary-records/single

126. http://www.propublica.org/article/drug-companies-reduce -payments-to-doctors-as-scrutiny-mounts

127. http://www.propublica.org/article/piercing-the-veil-more -drug-companies-reveal-payments-to-doctors

128. Carlowe J. Drug companies to declare all payments made to doctors from 2012. *BMJ* 2010 Nov 5;341(nov05 1):c6290–c6290.

129. Tuffs A. Two doctors in Germany are convicted of taking bribes from drug company. *BMJ* 2010 Nov 9;341(nov09 2):c6359–c6359.

130. http://www.fcaalert.com/2011/02/articles/dojhhs-releases-new -statistics-about-sealed-qui-tam-cases/

131. Sweet M. Experts criticise industry sponsorship of articles on health policy in Australian newspaper. *BMJ* 2011 Oct 25;343(oct25 2):d6903–d6903.

132. http://www.pmcpa.org.uk/?q=node/499

133. http://www.propublica.org/documents/item/87376-heart -rhythm-society

134. http://www.propublica.org/article/medical-groups-shy-about -detailing-industry-financial-support

135. JP Kassirer. *On the Take: How Medicine's Complicity with Big Business Can Endanger Your Health.* 1st ed. Oxford University Press, USA; 2004.
136. http://www.eatright.org/corporatesponsors/
137. JP Kassirer. *On the Take: How Medicine's Complicity with Big Business Can Endanger Your Health.* 1st ed. Oxford University Press, USA; 2004, p105.
138. Choudhry NK, Stelfox HT, Detsky AS. Relationships between authors of clinical practice guidelines and the pharmaceutical industry. *JAMA* 2002 Feb 6;287(5):612–7.

Conclusion: Better Data

1. Department of Justice, Office of Public Affairs. GlaxoSmithKline to Plead Guilty and Pay $3 Billion to Resolve Fraud Allegations and Failure to Report Safety Data. Monday, July 2, 2012. http://www.justice.gov/opa/pr/2012/July/12-civ-842.html
2. Glaxo executives cited in case now lead Sanofi, Actelion. Bloomberg News, 3/7/12. http://www.businessweek.com/news/2012-07-03/glaxo-executives-cited-in-case-now-lead-sanofi-actelion
3. Inpharm 4/7/12. GSK ruling: another failing, but will the industry learn? http://www.inpharm.com/news/173307/gsk-ruling-another-failing-will-industry-learn
4. Glaxo Agrees to Pay $3 Billion in Fraud Settlement. *New York Times*, July 2 2012. http://www.nytimes.com/2012/07/03/business/glaxosmithkline-agrees-to-pay-3-billion-in-fraud-settlement.html
5. Level playing field push to continue despite setback – 8 December 2011. Medicines Australia. http://medicinesaustralia.com.au/2011/12/08/level-playing-field-push-to-continue-despite-setback/
6. Bosch X, Esfandiari B, McHenry L. Challenging Medical Ghostwriting in US Courts. *PLoS Med* 2012 Jan 24;9(1):e1001163.

Afterword: What Happened Next?

1. The citation here is, for obvious reasons, 'personal communication'. The email from the academic continued: 'From my perspective, I don't think we should be anything but indignant!'
2. Davis C, Abraham J. Is there a cure for corporate crime in the drug industry? *BMJ*. 2013;346:f755.
3. Gale EAM. Post-marketing studies of new insulins: sales or science? *BMJ* 2012;344:e3974.
4. Light DW, Lexchin JR. Pharmaceutical research and development: what do we get for all that money? *BMJ*. 2012;345:e4348.
5. Svensson S, Menkes DB, Lexchin J. Surrogate Outcomes in Clinical Trials: A Cautionary Tale. *JAMA Intern Med*. 2013;173(8):611–612.
6. "*JAMA*, Integrity, Accessibility, and Social vs. Scientific Peer Review". *Emergency Medicine Literature of Note*, Feb 26, 2013.
7. Abbasi, K. Blood on our hands: seeing the evil in inappropriate comparators. *J R Soc Med*. 2013 January;106(1): 1.
8. Inside Health, BBC Radio 4, January 2013.
9. Duijnhoven RG, Straus SMJM, Raine JM, de Boer A, Hoes AW, et al. (2013) Number of Patients Studied Prior to Approval of New Medicines: A Database Analysis. *PLoS Med* 10(3):e1001407. doi:10.1371/journal.pmed.1001407
10. Ioannidis JPA. How Many Contemporary Medical Practices Are Worse Than Doing Nothing or Doing Less? *Mayo Clinic Proceedings*. 2013 Aug;88(8):779–81.
11. Zarin DA, Tse T. Trust but Verify: Trial Registration and Determining Fidelity to the Protocol. *Ann Intern Med*. 2013;159(1):65–67.
12. Rosenthal R, Dwan K. Comparison of randomized controlled trial registry entries and content of reports in surgery journals. Ann Surg. 2013 Jun;257(6):1007–15.
13. Zetterqvist AV, Mulinari S (2013) Misleading Advertising for Antidepressants in Sweden: A Failure of Pharmaceutical Industry Self-Regulation. *PLoS ONE* 8(5): e62609.doi:10.1371/journal.pone.0062609
14. Mintzes B, Lexchin J, Sutherland JM, Beaulieu M-D, Wilkes MS, Durrieu G, et al. Pharmaceutical Sales Representatives and

Patient Safety: A Comparative Prospective Study of Information Quality in Canada, France and the United States. *J Gen Intern Med*. 2013 Apr 5.

15. http://dailycaller.com/2013/04/27/critics-see-conflict-of interestas-obama-admin-advises-doctors-on-prescriptions/

16. http://www.propublica.org/article/pay-to-prescribe -twodozendoctors-named-in-novartis-kickback-case

17. Bosch X, Hernández C; Pericas JM, Doti P. Ghostwriting Policies in High-Impact Biomedical Journals: A Cross-Sectional Study. *JAMA Intern Med*. 2013;173(10):920–921.

18. Nancarrow, CM. Editorial Policies to Ensure Honesty and Transparency: Comment on "Ghostwriting Policies in High-Impact Biomedical Journals: A Cross-Sectional Study". *JAMA Intern Med*. 2013;173(10):921–922.

19. Persaud N. Questionable content of an industry-supported medical school lecture series: a case study. *J Med Ethics*. doi:10.1136/medethics-2013-101343

20. India's poor duped into clinical drug trials. *Economic Times*, 7th July 2013.

21. Vedula SS, Li T, Dickersin K. Differences in Reporting of Analyses in Internal Company Documents Versus Published Trial Reports: Comparisons in Industry-Sponsored Trials in Off-Label Uses of Gabapentin. *PLoS Med*. 2013 Jan 29;10(1):e1001378.

22. Sample, Ian. Big pharma mobilizing patients in battle over drugs trials data. *Guardian*, 21st July 2013.

23. http://www.alltrials.net/2013/responses-to-leaked-memo/

24. Huser V, Cimino JJ (2013) Linking ClinicalTrials.gov and PubMed to Track Results of Interventional Human Clinical Trials. *PLoS ONE* 8(7): e68409. doi:10.1371/journal.pone.0068409

25. Schroll JB, Bero L, Gøtzsche PC. Searching for unpublished data for Cochrane reviews: cross sectional study. *BMJ* 2013;346:f2231.

26. Wieseler B, McGauran N, Kerekes MF, Kaiser T. Access to regulatory data from the European Medicines Agency: the times they are a-changing. *Syst Rev*. 2012 Oct 30;1:50.

27. Little RJ, D'Agostino R, Cohen ML, Dickersin K, Emerson SS,

Farrar JT, et al. The Prevention and Treatment of Missing Data in Clinical Trials. *N Engl J Med*. 2012;367:1355–1360.

28. Mathieu S, Chan A-W, Ravaud P (2013) Use of Trial Register Information during the Peer Review Process. *PLoS ONE* 8(4): e59910. doi:10.1371/journal.pone.0059910

29. Prayle AP, Hurley MN, Smyth AR (2012) Data from: Compliance with mandatory reporting of clinical trial results on ClinicalTrials.gov: cross sectional study. Dryad Digital Repository.doi:10.5061/dryad.j512f21p

30. Simmonds MC, Brown JVE, Heirs MK, Higgins JPT, Mannion RJ, Rodgers MA, et al. Safety and Effectiveness of Recombinant Human Bone Morphogenetic Protein-2 for Spinal Fusion: A Meta-analysis of Individual-Participant Data. *Ann Intern Med*. 2013;158(12):877–889.

Illustrations

Fig. 1 (p. 15): http://www.cochrane.org/about-us/history/our-logo%23files

Fig. 2 (p. 17): Mulrow CD. Rationale for systematic reviews. *BMJ* 1994 Sep 3;309(6954):597–9. Available at: http://www.ncbi.nlm.nih.gov/pmc/articles/PMC2541393/?page=1;

Fig. 3 (p. 64): Ranibizumab and pegaptanib for the treatment of age-related macular degeneration: a systematic review and economic evaluation. *NICE* 2006. Available at: http://www.nice.org.uk/nicemedia/live/11700/34991/34991.pdf

Fig. 4 (p. 77): http://www.prescrire.org/editoriaux/EDI33693.pdf

Fig. 5 (p. 136): Carpenter D. *Reputation and Power: Organizational Image and Pharmaceutical Regulation at the FDA*. Princeton University Press, 2010.

Fig. 6 (p. 166): Schwartz LM, Woloshin S, Welch HG. Using a Drug Facts Box to Communicate Drug Benefits and Harms Two Randomized Trials. *Ann Intern Med* 2009 Apr 21;150(8):516–27. Available at: http://dartmed.dartmouth.edu/spring08/pdf/disc_drugs_we/lunesta_box.pdf

Fig. 7 (p. 167): http://www.lunesta.com/PostedApproved LabelingText.pdf

Fig. 8 (p. 190): Lurie P, Wolfe SM. Misleading data analyses in salmeterol (SMART) study. *The Lancet* 2005 Oct;366(9493): 1261–2.

Fig. 9 (p. 209): Rothwell PM. Subgroup analysis in randomised controlled trials: importance, indications, and interpretation. *The Lancet* 2005;365(9454):176–86. Available at: http://apps.who.int /rhl/Lancet_365-9454.pdf

Fig. 10 (p. 263): Moynihan R. The making of a disease: female sexual dysfunction. *BMJ* 2003;326(7379):45–47. Available at: http:// www.ncbi.nlm.nih.gov/pmc/articles/PMC1124933/table/TN0x95f 06b0.0x98eca30/

Fig. 11 (p. 295), fig. 12 (p. 296 top), fig. 13 (p. 296 middle), fig. 14 (p. 296 bottom): http://zyprexalitigationdocuments.com/%5 Cdocuments%5CConfidentiality-Challenge%5CDocs -challenged-in-10-3-list%5C145-ZY200187608-7614.pdf

Fig. 15 (p. 297): Drug Industry Document Archive [Internet]. Available at: http://dida.library.ucsf.edu/pdf/vou38h10

Fig. 16 (p. 299): http://pogoblog.typepad.com/pogo/gw-attachment -e.html

Fig. 17 (p. 349): 15 August 2012, http://uk.finance.yahoo.com /echarts?s=GSK.L#symbol=GSK.L;range=1y

INDEX

Abadie, Robert, 106
Abbott, 119–20, 347
AbbVie, 384
ABPI, *see* Association of the British Pharmaceutical Industry
Abraham, John, 140, 142
'Academic Detailing' initiative, 373
academic journals, 16, 29–37, 51, 179, 242, 304–12, 338, 366; advertising in, 244, 245, 271–74, 305–307, 312; bias favoring positive results in, *see* publication bias; bundled data published in, 88*n*; conflict of interest of, xi, 305–309, 312, 330; documents withheld from, 76, 90; fake fixes of, 51–55, 80; fictitious or spoof papers in, 34–35; fraudulent papers in, 174–75; ghostwriting in, xv, 245, 287–304, 310, 360, 373*n*; guidelines for treatment in, 63; imitation, industry-sponsored, 310–11; impact factor of,

309–310; methodological rigour lacking in, 199, 203–204, 384; multi-centre trial coverage in, 114; open-access, xvii, 36–37, 394; peer review for, 34–36, 274, 304–305, 328, 386; rare adverse events reported in, 153; reprint income of, 307–309, 312; researcher reasons for withholding papers from publication in, 29–32, 34; seeding trial denounced by, 214; 'spin' in, 220–22; suggestions for corrective measures for, 98, 312, 361; *see also specific publications*
accelerated approval, 134–43
Accreditation Council for Continuing Medical Education (ACCME), 318, 319
ACE inhibitors, 210
Actelion, 346
ACT UP, 137
Advair, 346
ADVANTAGE trial, 213–14

anti-platelet drugs, 250
antipsychotic drugs, 180–82;
    see also specific drugs
Antony, Veena, 334
anxiety, 334, 357; medico-legal, 12;
    separation, in dogs, 260; sexual,
    262; social, 259, 260
Aondoakaa, Michael, 117–18
Apotex, 138
Appeals Court, US, 284
ARB drugs, 243–44
Argentina, 113
Aricept, 252
arrhythmia, 10–11, 133, 176, 400n10
arthritis, 58, 252; rheumatoid, 270,
    372, 378
Arthritis and Musculoskeletal
    Alliance (ArMA), 270
aspirin, 101, 208, 220, 391
Association of the British
    Pharmaceutical Industry
    (ABPI), 49, 346, 369–71,
    371–74n, 391; AllTrials and,
    381, 384; Cochrane
    Collaboration and, 49; code
    of, 272, 280, 334–35
asthma, 177, 187–88, 349, 346
Association of British Science
    Writers, 323
AstraZeneca, 50, 145; accelerated
    approval for drug
    manufactured by, 140–42;
    advertising strategy for me-
    again drug released by, 147–48,
    250; fined for fraudulent
    practices, 347; payments to
    doctors by, 332, 333
atorvastatin, 225, 226, 233, 243
atypical drugs, 181
Australia, 81–82, 310–11, 350

*Australian Journal of Bone and
    Joint Medicine*, 311
Australian Medicines Industry, 337
Australian (newspaper), 337
Avastin, 255–56
Aventis, 50

Bacall, Lauren, 252
Bacon, Francis, 21
Banderas, Antonio, 252
Barber, Elaine, 254
Barbour, Ginny, 382
Baseline Observation Carried
    Forward method, 69
Bayh-Dole Act (1980), 326
Bem, Daryl, 30–31
bevacizumab, 211–12
Bextra, 247
Bible, 13, 75
bird flu, 91–93
bisoprolol, 185
bisphosphonates, 177–78
bladder: overactive, 252; unstable,
    260
bleeding, 197, 329
blobbogram, 14–18
blood clotting, 8, 232, 250
blood pressure: drop in, 8–9, 138;
    high, drugs for controlling, 109,
    115, 132–34, 149, 193, 225, 228,
    243–44, 317, 392
blood sugar, control of,
    see diabetes drugs
blood vessel surgery, 185
Bloomberg News, 108
Blumsohn, Aubrey, 42
Boehringer Ingelheim, 308
bone density, loss of,
    see osteoporosis

Index

Index

Serono, 320

Serostim, 320

serotonin hypothesis, 256–58, 260

Sheffield University, 42

shingles, 330

Shipman, Harold, 313

Shire, 138

Shirky, Clay, 241

side effects, ix, xiv, 62, 73, 96, 101, 121, 299; in animal versus human trials, 103; of antidepressants, 56, 59–60, 92, 147, 180, 260, 283; of antipsychotics, 180–82, 317; of cancer drugs, 22, 211, 252–53; comparative trials on, 170; of diabetes drugs, 324; in drug facts boxes, 164–65; idiosyncratic, 92, 145, 151, 226; monitoring, 150–65, 227, 395; of painkillers, 174, 216, 248; patients withdrawing from trials due to, 7, 68, 69, 106, 144, 198; in phase 2 and phase 3 trials, 109; in randomised trials, 237; in real versus ideal patients, 86; in shared decision-making, 218; of statins, 163–64, 219–20, 226; surrogate outcomes and, 134; in systematic reviews, 70; of testosterone patches, 265; in trials in poorer countries, 113, 115; in truncated trials, 185; of weight-loss drugs, 72, 183; worst-ever, for determining outcomes, 194–95

Signal (WHO publication), 154n

Simes, Robert, 22–23

Similac infant formula, 349

simvastatin, 225, 226, 233, 243

Singh, Simon, 377–78

Singin' in the Rain (film), 251

'single enantiomer preparations', 146–47

Smith, Richard, 36, 308, 395

SmithKlineBeecham, 50, 427n68

Smithsonian Institution, 257

smoking, 33, 251

social anxiety disorder, 259, 260

South Korea, 247

Spain, 273

spin, 111, 216, 220–22, 343, 347

SSRIs, 55, 79, 144, 256–58, 392

Stanford University, 333

statins, xiii, 1–2, 243; adverse reactions to, 163; comparison among, 193–94, 224–26, 233–36; and reduced risk of heart attack and death, 216–20; surrogate outcomes for, 132; trials of, 193, 227, 255

Sterling, Theodore, 21

Stern (magazine), 114

steroids, 9; for head injuries, 231–32; for premature births, 13–16

STI medical writing company, 298

streptokinase, 16–20

streptomycin, 237–38

stroke, 197, 224, 226, 227, 392; diabetes and, 196–97; prevention of, 210, 219

subgroup analysis, 4, 96, 205–10, 212, 221

Subject Access Requests, 278

suffering, avoidable death and, xvi, 27, 47, 97, 366–67, 373n

suicide, 59–60, 66, 75, 158–59, 258

Sunshine Act (2013), 334, 339, 340

Index